Mega Mammals in Ancient India

Mega Mammals in Ancient India

Rhinos, Tigers, and Elephants

SHIBANI BOSE

OXFORD
UNIVERSITY PRESS

OXFORD
UNIVERSITY PRESS

Oxford University Press is a department of the University of Oxford.
It furthers the University's objective of excellence in research, scholarship,
and education by publishing worldwide. Oxford is a registered trademark of
Oxford University Press in the UK and in certain other countries.

Published in India by
Oxford University Press
22 Workspace, 2nd Floor, 1/22 Asaf Ali Road, New Delhi 110002, India

First Edition published in 2020

ISBN-13 (print edition): 978-0-19-012041-2
ISBN-10 (print edition): 0-19-012041-X

ISBN-13 (eBook): 978-0-19-909987-0
ISBN-10 (eBook): 0-19-909987-1

Typeset in Adobe Garamond Pro 11/13
by Tranistics Data Technologies, New Delhi 110044
Printed in India by Replika Press Pvt. Ltd

For

Ma and Baba

CONTENTS

FIGURES

FOREWORD

HISTORY AS A CHRONICLE OF human activity is as old as the hills, the saying goes. Yet, history of the environment and other life forms we share our planet with is a recent phenomenon. The toll taken by the industrial revolution, imperial formations, and the two world wars on humans and the environment was no doubt the impulse that led to the rise of the new genre of historical writing in the second half of the twentieth century. Since other life forms cannot communicate their experience of anthropomorphic depredations, we have only our perceptions of what happened and what is happening to them and their habitats.

Coming to the history of faunal experience and ecology in India, first off the block was Jean Philippe Vogel, who, in 1962, wrote of the ubiquitous *haṃsa*, the goose, in art and culture in ancient times. In 1977, P. Thampakkan Nair chronicled the place of the peacock, India's national bird, in art, culture, religion, and history. The histories of these birds were soon followed by those of the cheetah and Asia's lions. Their passage through history, their suffering due to human depredations, and their place in Indian art and culture were chronicled in 1995 and 2005 respectively. It is not out of place to mention that Asia's lions have had many votaries as well, including Mattias and Monika Klum, who wrote about them in Swedish in 2000! Also, Raman Sukumar gave us his seminal work on Asia's elephants in 2011 in the same genre. Tomes have been written about the tiger, India's national animal, and the subject of India's most ambitious effort in faunal conservation. Yet, surprisingly, it awaits a serious biographer who traces its travails through ancient times to the present.

Bose's research is a crucial addition to this corpus. It has a somewhat different approach though, insofar as she has chosen three megafauna: the greater one-horned rhinoceros, which is endemic to India, the tiger, and Asia's elephants. The biographies of the first two are being chronicled for the first time. Her treatment of the elephant's travails adds a different perspective to Sukumar's work. Unlike other works mentioned here, Bose has confined herself to the period from the Pleistocene to c. 300 CE, and geographically to north India. These parameters of time and space have enabled her to dig deeper into varied sources where others have not ventured, particularly in the case of the tiger and rhinoceros.

The author traces the past distribution of the three animals through faunal remains, proto-historic artefacts, and historical literature up to c. 300 CE. From this it becomes evident that the rhinoceros's extreme range included Gujarat and Rajasthan, from where it has long since disappeared though it was recorded in Punjab until the first half of the sixteenth century. Evidence for the tiger, on the other hand, is somewhat scarce. It is prominent in the proto-historic period though, as witnessed on the seals of the Harappan civilization. It lasted in this area till 1886, when the last tiger was reportedly shot on the banks of the Indus in Sindh. It continued to thrive in Iran as the Caspian tiger, and survived there till 1953, when the last one there too was shot.

Asia's elephants were found in the Indian subcontinent as far west as Mehrgarh in Pakistan in the proto-historic period. It may be noted, however, that Pharaoh Tuthmosis III (1479–1425 BCE) hunted 120 Syrian elephants, a subspecies of Asia's elephants: *Elephas maximus asurus.* But by the time we come to the *Arthaśāstra*, Gujarat is its extreme western range.

A detailed analysis of the proto-historic finds and historical literature illustrates the intimate knowledge the ancient world had of these three species. For example, the solitary character of the rhinoceros, references to *musth* in elephants, and the touching interrelationship between the tiger and forests, where one could not last without the other, are but a few illustrations. Even today, this is startlingly so. The impulse to save the tiger in India by initiating Project Tiger in 1972–3 is the evidence. Can one imagine the protection of the tiger without its habitat?

These animals have had a profound impact on our cultural and religious beliefs throughout the period. The rhinoceros's solitary nature is extolled in the *Khaggavisāṇa Sutta* of the *Sutta Nipāta*. A king is compared to an elephant, the premier inhabitant of the natural world, and the most extolled animal of all in India's literature. The tiger, on the other hand, is important in royal ceremonies. Its skin is required to adorn a royal throne and because of it a king's royal power prospers.

By integrating a range of sources, Bose has been able to draw a kaleidoscopic picture of the forests and grasslands of early north India, which by itself is of crucial importance. Additionally, by writing the species history of the rhinoceros, tiger, and the Asian elephant, she has laid a firm foundation (particularly in the context of the first two animals) for scholars to pursue their travails from c. 300 CE up to the present. She has also given us a new perspective in the case of the elephant.

Bose's survey of ancient Indian literature highlights numerous interactions between lions and elephants. Equally thought-provoking are descriptions and comparisons between the lion and tiger, where the former usually emerges the better off. Lions and elephants must have shared common landscapes to evoke such frequent attention in Indian literature. To this day, they share the same savannahs of Eastern Africa and elsewhere. Lions and tigers did share the same landscape with distinct niches as Kota and Bundi paintings of the eighteenth century testify. How did the changes in landscapes cause the separation of the two cats, causing the disappearance of the lion leading to its near extinction? How did the lion and elephant become so separated and when? Bose has laid the foundation for such enquiries which would give us a holistic view of the changes brought about by anthropomorphic interference in our environment. I hope scholars shall soon pick up the baton and run with it.

This study is a welcome and important addition to the nascent field of faunal history and environmental changes.

Divyabhanusinh Chavda
Former President of the Worldwide Fund
for Nature, India (WWF India), and author of widely
acclaimed titles including *The End of a Trail: The Cheetah in India* (1995) and *The Story of Asia's Lions* (2005)

ACKNOWLEDGEMENTS

THIS BOOK WOULD BE INCOMPLETE without a mention of the individuals and institutions which have been an integral part of its transformation from a doctoral thesis to a monograph.

It is with immense gratitude that I look back at the time when I embarked on my research career with Nayanjot Lahiri as my mentor. From underlining the scruples of rigorous research to a strong emphasis on an accessible style of writing, her meticulous guidance has been an enormous learning experience for me.

The consistent encouragement and valuable insights that I received from Divyabhanusinh Chavda, Mahesh Rangarajan, Upinder Singh, and Bhairabi Prasad Sahu at different stages not only boosted my morale but also enhanced my understanding of the nuances this work encompassed.

My research career has been a journey fraught with the challenge of familiarizing myself with scientific data. With a background in humanities, this has been a great learning experience, which was aided by the generous help I received from specialists who deal with methodological issues in the discipline. While M.D. Kajale (Deccan College, Pune) helped me with his insights on the vegetational backdrop—a prerequisite for understanding faunal records—the opportunity to have an audience with faunal experts such as Umesh C. Chattopadhyaya (Allahabad University), Pramod Joglekar,

Arati Deshpande-Mukherjee, and Vijay Sathe (Deccan College, Pune) exposed me to the practical nuances involved in engaging with faunal remains. G.L. Badam helped me understand the importance of palaeontology as a starting point for the development of archaeo-zoology, and made me aware of morphological issues involved in the identification of faunal remains.

I am grateful to V.H. Sonawane, Rakesh Tewari, Erwin Neumayer, and Omar Khan for patiently responding to requests for permissions to use images from their works and collections. Asko Parpola very kindly provided me with a high resolution version of the cover image for this book.

Sincere thanks are due to the officials and staff of the Deccan College Library, and the Department of History, University of Delhi, whose cooperation and all-out efforts enabled me to make optimum use of my time when it was limited. I will always be deeply indebted to Mr Rizvi and Mr Kushawaha at the American Institute of Indian Studies (AIIS), Gurgaon. I also cannot fail to express my gratitude to the Indian Council of Historical Research (ICHR), New Delhi, for the Junior Research Fellowship, which provided the much-needed financial support for research. The library staff at ICHR, National Museum, Indira Gandhi National Centre for the Arts (IGNCA), and the Central Archaeological Library, New Delhi, was always willing to help.

A note of gratitude also goes out to Amanda Seligman at the University of Wisconsin-Milwaukee (UWM) for her wholehearted effort in making sure that I had every facility required to support my academic pursuits during my stint there as a visiting scholar. At a time when I had just transitioned to a new land and culture, the acceptance and encouragement that I received at UWM is fondly remembered. I am also grateful for the library facilities which I received at the Ames Library of South Asia, University of Minnesota.

The team at Oxford University Press has been extremely supportive, and I wholeheartedly thank them for that.

On my mind are also the friendships forged over the years. Miranda House gave me concerned and loving colleagues such as Snigdha Singh, Bharati Jagannathan, Radhika Chadha, and Srimanjari. Also warmly thought of are the friends who have been an integral part of this journey—Prachi Sharma, Niharika K. Sankrityayan, Nisha P.R., Sanjukta Datta, Vibha Tayal, Kanika Kishore Saxena, Deeksha Bhardwaj, Devika Rangachari, Aditi Mann Dahiya, Meera Visvanathan, Tishyarakshita A. Nagarkar, Shirley Khoirom, and Rajeshree Dutta Kumar.

Most crucially, the fact that I have been able to see this through is because of the unconditional encouragement, love, and warmth of family, which has been the single most source of strength for me. This book was my late mother's dream, and I earnestly hope that it is at least close to being a fitting tribute to her continuing presence in my life, and to all the values she strove hard to instill in me. My father has battled health frailties with exemplary fortitude, and remarkably donned the role of both parents. My brother has consistently gone way beyond being just an elder sibling. My sister-in-law stood by steadfastly in so many ways, while my precious little nephew has been the ultimate stress buster in the darkest of hours. I feel extremely fortunate that in the same breath as family, I can mention Shraddha Sharma Sen, Shikha Durlabhji, and Natasha Raina-Kanwar, who make all the difference by just being around. Finally, I seize this opportunity to let my husband Rajib know that he has been a remarkably patient and caring companion, and that for this as well as his presence in my life, I feel truly blessed.

1
INTRODUCTION

SINCE ANTIQUITY, BIG MAMMALS HAVE inspired fear as well as fantasy, forcibly impressing themselves upon the imaginations of the people coming in contact with them, and firing the inventive genius of storytellers who described fantastic and terrifying beasts of the primordial past.

Notwithstanding the varied emotions evoked by them, it would serve us well to remember that the importance of these mega mammals is not merely cultural. Apart from pervading the domains of religion, art, literature, and folklore, the utility of megafauna for portrayals of the environment is now widely acknowledged. This is particularly in view of the growing academic interest in and engagement with applying multidisciplinary approaches to the understanding of India's ancient ecological past.

Research amply demonstrates that consciousness regarding transient ecologies, including changing landscapes and shifting river courses, is not new. Yet, engagements with India's ancient ecological past have at best comprised broad surveys of the ancient period. These have seldom woven in the large body of published archaeozoological and archaeobotanical data, thereby inhibiting the drawing of inferences beyond the level of loose generalizations. This, for instance, is clearly exemplified in Madhav Gadgil and Ramachandra Guha's *This*

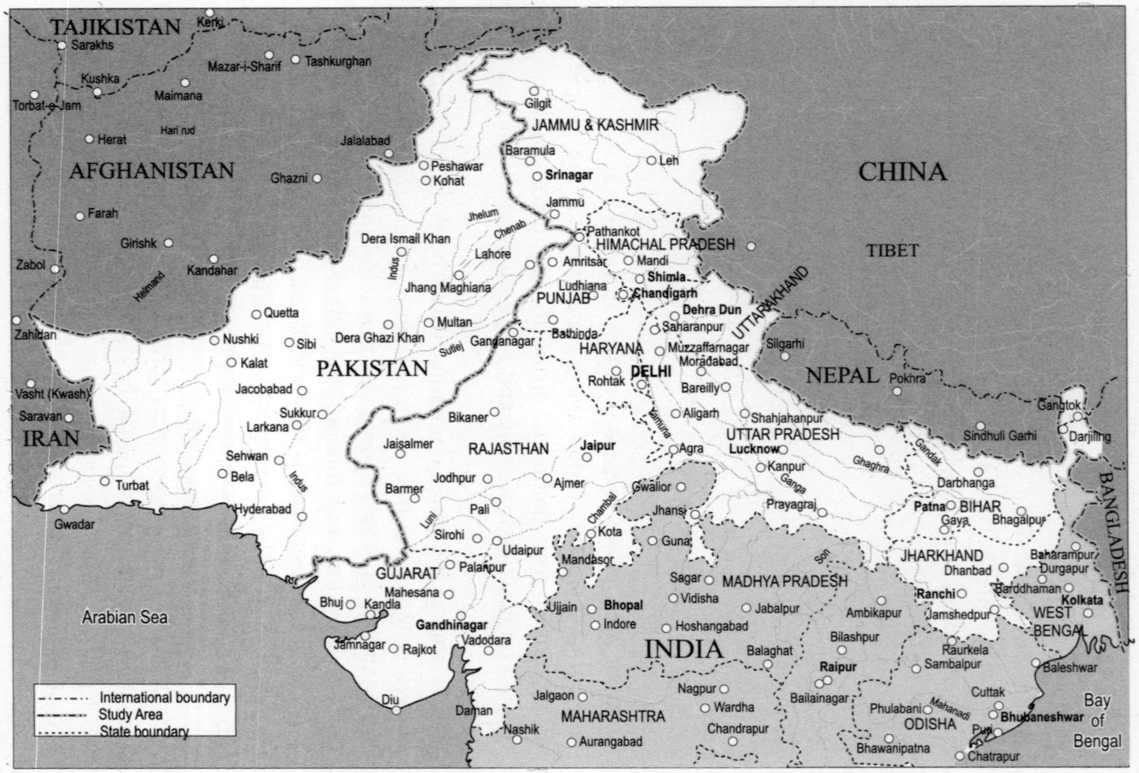

Figure 1.1 Study area with administrative divisions (for representative purposes only)

Source: Courtesy of the author.

Fissured Land: An Ecological History of India (1992). The same gap is also evident in the case of faunal histories, with few exceptions, (Divyabhanusinh 1995, 2005; Sukumar 2011) that have moved beyond the general focus on the Mughal and the colonial periods.

Against this backdrop, I venture to look into eras bygone in order to chart the passage of three mega species—rhinoceros (*Rhinoceros unicornis*), elephant (*Elephas maximus*), and tiger (*Panthera tigris tigris*)—across millennia in early north India (Figure 1.1), and to situate them within the realms of both culture and ecology.

While the term 'megafauna' may sound technical and intimidating, the endeavour will be to show how when explored it opens up a fascinating world that can be accessed through an archive comprising archaeology and a gamut of literary texts as well as visual depictions.

Coming to what determines the need to study megafauna, it is crucial to underline that their charisma apart, studies indicate that megafauna species play decisive roles in the ecosystems they inhabit, and are often critical to ecological processes including seed dispersal, nutrient cycling, and predator–prey cycles (Estes et al. 2011; Ripple et al. 2014, 2015 cited in Lindsey et al. 2017, 244). Not just this, because of their sizeable spatial requirements, they can act as umbrella species whose conservation indirectly benefits a range of other species (Macdonald et al. 2012 cited in Lindsey et al. 2017, 244). However, despite their importance, many large mammals are rendered vulnerable due to their low reproductive rates, large spatial requirements, and threats such as poaching and habitat destruction by human agencies. Their conservation, therefore, has been a major concern for biologists, ecologists, and conservationists.

As for the choice of the animals focused on in the study, it may be worthwhile to point out that when it comes to wildlife histories, particularly in the context of ancient India, the rhinoceros as well as the tiger have received scant attention as against the iconic status enjoyed by elephants, cheetahs, and lions. Therefore, while that partly determines my choice of the first two animals, the elephant seemed inescapable in view of its all-pervasive presence in Indian history and culture.

Having said this, it is necessary to reiterate that while all three species have enthralled as well as challenged humans since time immemorial, we need to expand the lens to look beyond their magnitude and magnificence. The approach, therefore, will be one which goes beyond treating them as mere cultural icons to one that is also sensitive to their value as markers of ecology.

It, therefore, becomes imperative to acknowledge the debt owed to works which have served as indispensable springboards for the study of mammals by informing us about their taxonomy, distribution, behaviour, and ecology. Thomas Caverhill Jerdon, a medical surgeon with the East India Company, gave us the first comprehensive volume on Indian mammals titled *The Mammals of India—A Natural History of All the Animals Known to Inhabit Continental India* (1867). The book is a compilation of the observations of naturalists, and served as a blueprint for later mammalian studies. Subsequently, Robert Armitage Sterndale's *Natural History of the Mammalia of India and Ceylon* (1884) was followed by the volumes titled *Mammalia* by William Thomas Blanford (1888–91) and Reginald Innes Pocock (1939, 1941) in the series *The Fauna of British India, Including Ceylon and Burma* (later *Fauna of India*). Drawing upon the first-hand observations of naturalists, Stanley Henry Prater's *The Book of Indian Animals* (1948) continues to serve as a seminal reference for anyone setting out to understand the habitats, behaviour, and geographical distribution of the more common mammals of India. Carrying brief yet careful descriptions of more than a hundred species found in the subcontinent, the work is a comprehensive handbook of information regarding mammalian species.

Much has come along since then in the form of studies that have focused on individual species. The extensive literature that now exists related to the evolutionary ecology, behaviour, distribution, and conservation of these three mega mammals has greatly aided our understanding of the history of India's wildlife, sensitizing us at the same time to the dangers that confront our faunal wealth, and suggesting long-term strategies for their effective management and conservation. A survey of this body of literature will, however, have to be selective since the purpose is only to have a sense of the behaviour, distribution, and ecology of these megafauna before we historically trail them.

Beginning with our one-horned protagonist, it is well known that the rhinoceros has a highly specialized niche, and typically inhabits alluvial grasslands and riverine floodplains that provide ample wallows and swampy feeding grounds. In fact, the Indian rhino is known to be the most moisture loving of the five rhino species. As an animal of the grasslands, its fortunes are a good index for mapping landscape changes in early India.

In 1966, Charles Albert Walter Guggisberg bemoaned the plight of the animal in *S.O.S. Rhino* after charting its journey, widespread distribution in the past, behaviour in captivity, and physiology. William Andrew Laurie's (1978) doctoral thesis 'The Ecology and Behaviour of the Greater One-Horned Rhinoceros' came at a time when not much was known about the biology of the Indian rhino. Its significance lay in its being the first detailed scientific study of the animal in the wild. Set in the Chitwan Valley of southern Nepal, Laurie sought to map the past and present distribution of the species, and to study its ecology and social behaviour as a means of facilitating management and conservation plans.

While Esmond and Chrysee Bradley Martin (1982) recounted the tragedy of rhino species including the *unicornis* in their lavishly illustrated monograph *Run Rhino Run,* L.C. Rookmaaker's (1983) *Bibliography of the Rhinoceros: An Analysis of the Literature on the Recent Rhinoceroses in Culture, History and Biology* is practically an encyclopedia for rhinoceros studies. Painstakingly listing references to all the five extant species in books and essays, the bibliography is an indispensable guide for anyone attempting to tread the rhino path.

Similarly, Rookmaaker's contributions (1984, 1999, 2000, 2002) on the past distribution of the rhinoceros in northern India, Pakistan, and Afghanistan have been instrumental in helping us map the geographical retreat of the animal to its present-day havens. The details of this retreat are not only integral to understanding aspects of the environmental history of ancient India, but also for recovering echoes of the animal's presence in areas where it is now extinct.

Unicornis: The Great Indian One-Horned Rhinoceros is the sole hero in Arup Kumar Dutta's (1991) narrative. His is a cogent account of the species written with the aim of creating awareness regarding the irrationality behind beliefs which have proved to be the undoing of the animal. Dutta writes accessibly about the myths that surround the animal, its history, behaviour, and habitat, and in doing so, effectively traces the journey of the animal to where it stands today. More significantly, he drives home its importance as an indicator species of the wetland ecosystem.

Eric Dinerstein's (1992, 701–4) field studies at the Royal Chitwan National Park in Nepal showed how browsing and trampling by rhinos inhibited the vertical growth of *Litsea* and *Mallotus* saplings, which occur in high densities in riverine forests. He also pointed out that the trampling of the saplings of *Dalbergia Sissoo* Roxb. was common on the floodplains of the park. More significantly, he emphasized that even at reduced population levels, the interactions between the rhinoceros and woody plants showed the impact of selective browsing by mega herbivores on forest structure and canopy composition.

Subsequently, this conservation biologist came up with a larger project dealing with the evolutionary history, biology, ecology, and recovery of the rhinoceros and other Asian mammals. Dinerstein (2003) reiterated the centrality of these giant plant eaters to the ecological architecture of the tree-dotted tall grassland, and elaborated on their role as 'landscape architects'. Rhinos, he noted, also played a prominent role in the dispersal of the seeds of *Trewia nudiflora*, a common riverine forest tree of southern Nepal.

Similarly, naturalist and ornithologist Maan Barua (2006, 49) argued that the size and feeding habits of the rhino (weighing nearly two tonnes) influence the physical habitat and spatial distribution of other species in the ecological community. He cautioned that the vital 'landscape architecture' phenomenon disappears with the vanishing of rhinos, resulting in swift and profound ecological changes.

Moving on to the striped predator, it is crucial to remember that the tiger is not just an enigmatic carnivore living in a distant forest. This mega carnivore is the top predator in the lands it inhabits in India, which is also home to over half the tiger population in the wilderness. It forms the pinnacle of the food chain, and helps keep in check the population of wild ungulates (hoofed mammals). Not only does this maintain the balance between herbivore prey and the vegetation they feed on, it also ensures the diversity of the ecosystem. Hence, it is not without reason that conservationists refer to the tiger as an umbrella species.

The volume of literature on the tiger in India is so overwhelming that one struggles to look for an entry point. Since Reginald George Burton's (1933) *The Book of the Tiger*, Kenneth Anderson's (1967) *Tiger Roars*, Charles McDougal's (1977) *The Face of the Tiger*, Guy Mountfort's (1981) *Saving the Tiger*—all of which brought alive the majestic predator and warned us of its precarious predicament—literature on tigers has come a long way.

Within this corpus, the first truly scientific book was George Beals Schaller's *The Deer and the Tiger: A Study of Wildlife in India* (1967). At a time deficient in writings on large mammals, Schaller's work on the interrelationship between species based on fourteen months of field studies of the Bengal tiger and the major hoofed animals in the Kanha National Park, Madhya Pradesh, was incisive for several reasons. It offered new information about the ecology and behaviour of ungulates including the chital, barasingha, sambar, blackbuck, gaur, and the most important predator in the park—the tiger—and also provided valuable insights regarding predator–prey relations in the wild, their implications for the density of wildlife in the park, and effective management practices within the park. Hoping that it would act 'as a stimulus for other studies', Schaller (1967, 9) crucially underlined the gap in knowledge regarding these species and their relation with human ecology, and emphasized the need for more scientific research in order to ensure their conservation. The monograph continues to be perceived as a classic.

Subsequently, drawing upon the early issues of the *Journal of the Bombay Natural History Society*, Jivanayakam Cyril Daniel's (2001) *The Tiger in India: A Natural History* came as a readable compilation of information familiarizing us with the morphology, behaviour, and distribution of the predator.

Another name that has been instrumental in opening up the world of the predator is K. Ullas Karanth. *The Way of the Tiger: Natural History and Conservation of the Endangered Big Cat* (Karanth 2002) was a popular but scientific introduction to the mega carnivore in order to create awareness regarding the threats it faced, and helped to garner general support for its conservation. However, far more crucial is the collection of his scientific writings titled *The Science of Saving Tigers* (Karanth 2011). Compiled with the aim of making his authored as well as co-authored essays on science-based tiger conservation accessible to

Asian readers, the volume covers 'the entire spectrum of tiger conservation, ranging from ecology to policy' (Karanth 2011, ix–x), and persuasively demonstrates that complex issues ranging from understanding tiger population dynamics and the factors affecting these dynamics to ways of mitigating human–tiger conflicts can only be addressed through scientific research and methods. The range of writings, for instance, underline the efficacy of using radio-telemetry to measure the movements of tigers and other predators and assessing the intensity of competition amongst them, as also the use of photographic capture–recapture sampling to estimate tiger and prey densities.

Similarly, emanating from nearly four decades of watching tigers in their natural habitat, Valmik Thapar's writings on the predator have been prolific and wide-ranging. Employing the lens as well as the written word to capture wild tigers, Thapar has given us diverse glimpses of the species. Titles such as *Tiger: Portrait of a Predator* (1986), *Tigers: The Secret Life* (1989), *The Land of the Tiger: A Natural History of the Indian Subcontinent* (1997), *The Cult of the Tiger* (2002), *Tiger: The Ultimate Guide* (2004) along with numerous others serve to give us a sense of the canvas he has covered from popular writing to ones meant for specialists. *Tiger Fire: 500 Years of the Tiger in India* (Thapar 2013) effectively illustrates the interest evoked by the animal across centuries by bringing together writings of naturalists, writers, photographers, and tiger enthusiasts.

Elephants, similarly, have generated ample scholarly interest. S.K. Eltringham's (1991) *The Illustrated Encyclopedia of Elephants: From Their Origins and Evolution to Their Ceremonial and Working Relationship with Man* is a telling and popular title which assembles contributions from a team of experts dealing with the origins, evolution, anatomy, physiology, social life, reproduction, ecology, conservation, and other aspects of the animal. The volume also deals with working elephants and the pachyderm in Indian myths and history, thereby putting human–elephant interactions in a historical context. Subsequently, Robert Delort's *Life and Lore of the Elephant* (1992) served as a popular and accessible introduction to aspects of the natural history of the animal and its interactions with humans through time. While J.C. Daniel's *The Asian Elephant: A Natural History* (1998) was a scientific enquiry into the ecology of the elephant, Stephen Alter's (2004) *Elephas Maximus: A Portrait of the Indian Elephant* combined art, literature, and legend along with his sojourns to parks and reserves in search of wild elephants to tell a compelling tale of the pachyderm.

More crucially, the animal's role in the ecosystem was clearly spelt out by Raman Sukumar in *The Living Elephants: Evolutionary Ecology, Behavior, and Conservation* (2003, 252–3), where he underlined the impact of the pachyderm on its habitat through feeding and other activities and enumerated the consequences for plant and animal communities in the ecosystem. For instance, he argued that apart from their role in dispersing fruits and seeds, browsing by elephants keeps trees in a stunted, shrubby stage, facilitating feeding by smaller browsing mammals. Similarly, the paths made by them through dense undergrowth also help smaller mammals. In this sense, elephants act as an umbrella species since their conservation protects a large number of other species occupying the same area.

***

Having emphasized the ecological importance of these mega mammals and the significance of looking at them historically, it is time to underline the aims of my enquiry. The focus will be twofold: to comprehend perceptions, attitudes, and sensibilities oscillating between veneration and persecution in order to reconstruct human–wildlife interactions, as also to use these animals to understand the larger ecology of ancient India.

So how do I go about reconstructing the stories I seek to tell? Animals can be looked at through multiple prisms ranging from faunal remains and visual depictions retrieved from archaeological records to the formidable corpus of Sanskrit, Pali, Prakrit, and classical Western accounts. Carefully sifting through sources that have a bearing on the subject, my focus will be on assessing the cultural dimensions of this interface as also on what the faunal spectrum suggests of the ecological and climatic set-up that prevailed in the past.

For a meaningful peep into the ancient past, one may begin with archaeology. Faunal assemblages being the direct result of the interactions of animals with humans within a specific ecological setting, an entry point would be a survey of the archaeozoological record. Within the subcontinent, the faunal remains of these animals, though scattered and often patchy (as in the case of the tiger), have been retrieved from diverse ecological niches and cultural and temporal contexts. Apart from providing possible clues regarding early human interactions with them, these remains also give us a sense of the ecologies which harboured these animals in the ancient past.

Much before issues of wildlife conservation concerned ecologists, biologists, and environmental historians, palaeontologists, archaeologists, and hydrologists were trying to grapple with faunal evidence not only from a cultural but also an ecological perspective.

An early beginning in this direction, for instance, can be seen in John Hubert Marshall's *Mohenjo-Daro and the Indus Civilization* (1931, I: 2) as he wrote about the climate and landscape of the site and contended that rainfall used to be substantially heavier than it is today. Evidence for this inference, according to him, came from the widespread use of kiln-burnt bricks and the animals engraved on seal amulets. Apart from cattle and fabulous creatures, the animals according to him were such as were commonly found in 'damp, jungly country', namely the tiger, rhinoceros, and elephant. The lion, on the other hand, 'which notoriously prefers a dry zone', he pointed out, was conspicuous by its absence. Even though these assertions were subsequently reviewed and revised, Marshall himself was exercising caution when he qualified that taken by themselves none of these pieces of evidence from Mohenjodaro could be regarded as conclusive, but considered in conjunction with each other, they certainly suggested higher precipitation levels than contemporary Sindh enjoyed.

Three decades later, Robert L. Raikes, a hydrologist, and Robert H. Dyson, an archaeologist, stepped in to review the categories of evidence which had been cited by Marshall to argue in favour of a wetter climate during Harappan times. Relevant for us here is their approach to the question of the relationship of the landscape to the fauna known from Harappa and Mohenjodaro in the form of seals, figurines, and animal bones.

As against Marshall's assertion regarding the tiger, rhinoceros, and the elephant indicating a wetter climate, and the rarity of the camel and the absence of the lion arguing against drier conditions, Raikes and Dyson (1961) sought to review the faunal evidence in order to demonstrate that a wet climate was by no means indicated. They garnered evidence from wildlife literature to show the presence of the tiger, rhinoceros, the elephant, and even lion in the region till fairly recent times, and drew upon the presence of other species such as hog deer, wild bear, water buffalo, and mongoose till at least the end of the nineteenth century. Additionally, they underlined the presence of *Zootecus insularis* (Ehrenberg), a native of arid regions as the only land snail species found at Mohenjodaro. Putting together the faunal evidence, they argued in favour of an environmental regime little changed from that of today.

While the issue of environment during the Harappan civilization continues to be deliberated upon, from a historiographical point of view it would suffice to say that these were early attempts at treating mammals (big and small) as ecological indices. In this sense, the approach (which was to gradually become more organized, and result in full-blown cultural and ecological histories of various species) can be viewed as a precursor to subsequent studies that sought to firmly integrate the use of faunal evidence in reconstructing past ecologies.

An important model regarding how this could be done was provided by Frederick Everard Zeuner in *The Pleistocene Period: Its Climate, Chronology and Faunal Successions* (1959). As a palaeontologist and geological archaeologist, Zeuner (1959, 50) emphasized the importance of fauna and flora enclosed in Pleistocene deposits in discussions of climates in this period, and argued that they served as evidence in two respects. In a strictly stratigraphical sense, the presence or absence of some species that are now extinct indicated the relative age of the deposits within the stratigraphical scale. In the ecological sense, these indicated the environment in which they lived and, therefore, the climate. However, caution was advised on the ground that only a fair knowledge of the biological requirements of the species could facilitate sound conclusions.

Shortly thereafter, Zeuner (1963a) put together a series of lectures demonstrating how past environments could be understood through implements, soils, sediments, fauna, and flora. Emphasizing the importance of applying biological methods to archaeology, Zeuner pressed for the study of plant and animal remains retrieved from early sites in view of their potential to reveal past patterns of resource exploitation. Not just this, he reiterated that these remains, and the combination of species could also be indicative of environment—a prospect, which he lamented was less widely recognized. He emphasized that at times it is even possible to apply very precise biological information to archaeology, for instance, in cases where species indicative of certain types of environment were replaced by other species as the environment changed.

To demonstrate the interpretation of faunal remains from an occupation site, Zeuner took the case of the microlithic site of Langhnaj in northern Gujarat. For him, the most interesting find was a rhinoceros shoulder blade, which had been used as an anvil by the microlith makers. The importance of the find attesting the presence of the rhinoceros at the site, argued Zeuner, was not because of its great size, but due to the fact that the species no longer occurred in western India. While the details of Zeuner's enquiry are discussed in Chapter 2, it is important to note his emphasis on the climatic interpretation of fauna, particularly in Pleistocene contexts, as well as his attempt to use the rhinoceros as an indicator species for reconstructing the landscape at Langhnaj.

An integrated ecological approach (acknowledging the importance of bones as an indicator of human–environment interactions) to the past commenced in the last quarter of the nineteenth century with Richard Lydekker's (1886) study of the faunal assemblage of the Billa Surgam caves in the Kurnool district of Andhra Pradesh. This set the stage for archaeozoologists to document and interpret evidence for human–environment interactions as they pertained to the study of the archaeological remains of animal resources. The reports for Mohenjodaro (Sewell and Guha 1931) and Harappa (Prashad 1936) marked this early endeavour, and were embarked upon with the principal objective of the representation of animal taxa at the sites together with a systematic description of the identified species.

In the post-Independence era, Bhola Nath of the Zoological Survey of India contributed to the field with the study of faunal remains from a number of archaeological sites including Hastinapur (Nath 1954–5), Rangpur (Nath 1962 and 1963), Rupar and Bara (Nath 1968), Alamgirpur (Nath and Biswas 1969), and Chirand (Nath and Biswas 1980) within the study area. Nath's (1963) 'Advances in the Study of Prehistoric and Ancient Animal Remains in India: A Review' was the first of its kind to chart the progress made in the field. He (1969) also analysed 'The Role of Animal Remains in the Early Prehistoric Cultures of India' by collating faunal evidence then available from different cultural contexts. More significantly, Nath went on to underline the presence of species indicative of ecological conditions. For instance, in the context of the Indus civilization, on the basis of animal representations amplified by actual bone remains, he inferred 'that the climatic conditions of the Punjab were such as to favour a suitable habitat for the rhinoceros, the wolf and the tiger as well as the water buffalo (*Bos bubalis* Linn.) and the elephant (*Elephas maximus* Linn.), none of which survives as wild species in the region today (except the tiger which is sometimes found in Sind)' (Nath 1969, 105). The approach, thus, is again a reminder that faunal remains have been considered significant not just for their cultural associations but also their ecological implications.

While Nath's method was largely an extension of the line pursued by Sewell and Guha, this era witnessed gradual changes in objectives and methods. Juliet Clutton-Brock's *Excavations at Langhnaj, Part II: The Fauna* (1965) demonstrated the importance of bone measurements in faunal studies as an aid to distinguish closely related species or wild from domesticated ones.

Gyani Lal Badam's (1979) *Pleistocene Fauna of India* came as a strong reminder of the importance of palaeontology in archaeological research. Credited for being the first systematic synthesis of Pleistocene faunal evidence from different parts of the country, the work went much beyond that in terms of addressing morphological and palaeoecological issues. His listing of the diverse fauna during the Pleistocene not only permitted climatic inferences but also provided crucial evidence for morphological changes in animals from the Plio-Pleistocene transition to the Pleistocene.

The observations of K.R. Alur (1980), a specialist in veterinary science, on the faunal assemblages from Chopani-Mando, Mahagara, and Koldihwa in the Belan Valley, and Sarai Nahar Rai and Mahadaha in the Ganga Valley, showed an engagement with methodological issues. Besides delving into the range of fauna in different periods, and on the process of evolution and domestication, he sought to reconstruct the importance of animal resources for the economy by discerning human-induced deformities on bones.

The study of the faunal assemblages from the sites of Sarai Nahar Rai, Mahadaha, Damdama, Koldihwa, and Mahagara formed a part of Umesh Chandra Chattopadhyaya's (1991) unpublished doctoral thesis 'A Study of Subsistence and Settlement Patterns during the Late Prehistory of North-Central India'. Using the mesolithic of the Ganga Valley as a case study, the contributions of Chattopadhyaya have been significant in pushing the frontiers of archaeozoology through the use of faunal data in addressing a range of archaeological problems such as seasonality of site occupation, hunter-gatherer subsistence and settlement patterns, site functions, and industrial activities. His essay, 'Settlement Pattern and the Spatial Organisation of Subsistence and Mortuary Practices in the Mesolithic Ganges Valley, North-Central India', showed how the 'faunal data, including ageable deer teeth, and grave orientation in relation to solar variation suggest that the sites of Mahadaha and Damdama were logistically organized and residentially stable' (Chattopadhyaya 1996, 461). This inference was subsequently built upon in another contribution titled 'Complementary Partitioned Network System: A Regional Model of Post-Pleistocene Human Adaptations in the Vindhya-Ganga Complex' (Chattopadhyaya 2001), to arrive at an alternative regional model of post-Pleistocene human adaptations in warm regions such as the Vindhyas and the Ganga Valley. Emphasis on the use of taphonomic evidence in order to make faunal studies more meaningful was a significant area of focus in Chattopadhyaya's writings.

The study of the entire faunal assemblage from Damdama was completed by P.K. Thomas, P.P. Joglekar, V.D. Mishra, J.N. Pandey, and J.N. Pal (1995; 1996) from Deccan College, Pune, with observations envisioning 'a cyclical trend in resource management where an increase/decrease in mammalian resources was compensated by parallel changes in the avian fauna and the aquatic fauna' (Thomas, Joglekar, Mishra, Pandey, and Pal 1995, 29).

It is equally important to mention contributions that have focused on spelling out methodological issues in faunal studies, and have also from time to time synthesized their progress in the subcontinent. Bhairabi Prasad Sahu's monograph *From Hunters to Breeders* (1988) delineated patterns of human–animal interactions in terms of resource use, dietary habits of people, and problems related to animal domestication in a panoramic sweep of cultures ranging from prehistoric times to the historical period. While G.L. Badam and Vijay Sathe reviewed the 'Subsistence Economy of the Indus Civilization' (1991), Richard Meadow (1993) devoted a section to zooarchaeology in his review of bio-archaeological studies in Pakistan, with special reference to Mohenjodaro and the Indus Civilization.

P.K. Thomas and P.P. Joglekar's 'Holocene Faunal Studies in India' (1994) critically summarized the existing data on animal remains collected from various archaeological sites in the Holocene context. The change and continuity of animal associations with cultures ranging from the upper palaeolithic to the beginning of the early historic period was also engaged with. Subsequently, their contribution 'Faunal Studies in Archaeology' (1995) dealt with the various concerns of archaeozoological studies ranging from species identification, faunal quantification, activity areas, past environments, season of occupation, ageing and sexing of animals, the importance of metrical analysis, cultural contacts, animal pathology, and animal domestication.

Half a decade later, Thomas (2000–1) reviewed the contribution made by specialists at Deccan College to archaeozoological research, while Chattopadhyaya (2002) traced the evolution of faunal studies, and reviewed Holocene archaeofaunal studies and their main results in the Indian subcontinent.

In 'A Fresh Appraisal of the Animal-based Subsistence and Domestic Animals in the Ganga Valley', Joglekar (2007–8) not only underlined the growing evidence for the faunal economy of the Ganga Valley, but also the application of taphonomic processes and intra-site patterns of animal usage in assessing the evidence. Apart from discussing the limitations of the data and methodological issues, he highlighted the evidence in terms of the economically important domestic species found in the region. A similar but more nuanced approach was in the context of animals in the Harappan civilization, where besides dealing with methodological issues involved in animal identification in the Harappan archaeological record, the authors (Joglekar and Goyal 2014) assessed the quality of the Harappan archaeofaunal database, and emphasized how recent reports such as those of Farmana in Haryana and Kanmer in Gujarat have gone beyond confirming the presence/absence of species. What is more significant is that they analysed the faunal diversity at various Harappan sites to give a comprehensive picture of the animal-based subsistence economy during the period.

It may also be worthwhile to mention another synthesis in the context of the Harappan civilization by Vijay Kumar (2014). Though not exclusively centred on faunal remains, his work ponders over their significance, delves into the forms (representations on pottery, seals, figurines, and actual plant and animal remains) in which they are found, summarizes the evidence available, and based on it, attempts to reconstruct the environment and subsistence.

It may be pointed out that other than bones, faunal experts also acknowledge the importance of sources such as coins, frescos, paintings, and sculptures as tools for understanding animals in ancient India. An attempt to integrate and corroborate other sorts of evidence with archaeozoological remains can, for instance, be seen in Joglekar's study on 'Domestic Animals in Ancient India in the Light of Literary and Archaeological Evidence' (1994–5). This was followed by his essay retrieving glimpses of 'Animal Taxonomy from Ancient and Medieval Indian Literature' (2000). A similar approach is evident in an essay by Vijay Sathe (2010), also of Deccan College, Pune. Though titled 'The Archaeology of Great One-horned Indian Rhinoceros (*Rhinoceros Unicornis* Linnaeus 1758)', the essay is a review of the fossil and archaeological record as also the occurrence of the animal in ancient Sanskrit texts.

As already pointed out, while the application of faunal evidence for ecological reconstruction had early beginnings, the approach emerged more prominently in enquiries that based their hypotheses on the presence or absence of animal species. This, for instance, is exemplified by Y.M. Chitalwala's (1990) attempt to seek possible explanations for the disappearance of the rhinoceros from Saurashtra. Keeping in mind the ecology and behaviour of the animal, he sought to assess the impact of the Harappans on the rhino population of the region.

Joseph Manuel (2004–5), on the other hand, was intrigued by the overwhelming presence of the rhinoceros in terms of bones as well as representations in contrast to the occasional presence of the horse during Harappan times. The explanation, according to him, could be sought in the humid Harappan environment which was conducive for the species. This reasoning was later extended when Manuel (2008b) collated evidence testifying to the presence of the horse (*Equus caballus*) since prehistoric times. His focus was again the Harappan civilization, where despite the animal's presence, neither bones nor terracotta depictions are found in great profusion. Drawing upon proxy evidences for 'a humid, swampy and munificent environment', Manuel (2008b, 172) was emphatic about this being responsible for the paucity of evidence regarding the horse which did not find much use in a 'landscape dotted with swamps and ponds and higher ground level'. Ample evidence for the presence of the rhinoceros, the limited evidence of rice cultivation along with Harappan skeletal remains showing adaptive polymorphism in response to endemic malaria were, according to him, suggestive of wetter environs interspersed with occasional spells of aridity which became more frequent by the later part of the mature phase during which the horse and the camel, already known to the Harappans, gained popularity.

***

Having brought out the application of archaeozoological studies in reconstructing diverse aspects of the human past and reviewed the progress of such approaches, I now move on to other ways of looking at animals.

Animals have a prolific presence in Indian art and sculpture. Tapping the visual archive is, therefore, another way of reconstructing animal histories. The earliest faunal imageries come down to us in the form of rock paintings that tell us of hunts and animals hunted.

As early as 1883, John Cockburn, a government officer in the Opium Department of British India, brought to light a painting representing a rhinoceros hunt in the Ghormangar rock shelter in the Mirzapur region. The representation showed several hunters surrounding the animal with microlithic spears. Nearly half a century later, Manoranjan Ghosh (1932) took us to the Likhunia shelter, also in Mirzapur, to a depiction which, according to him, showed the capture of a wild elephant. I will return at length to these early discoveries in the chapters which follow, but for the moment, it is important to point out that rock paintings are perhaps the earliest pictorial pointers we have to human interactions with these animals.

The survey canvas was broadened as Vishnu Shridhar Wakankar (1973) in his epic contribution *Painted Rock Shelters of India*, brought together his findings in over 1,532 rock shelters covering the states of Uttar Pradesh, Madhya Pradesh, Chhattisgarh, Orissa, and Karnataka. Based on stylistic considerations, Wakankar defined twenty different styles, which for decades, served as a model for the dating of rock paintings. More significantly, Wakankar devoted an entire chapter to plants and animals depicted in rock shelters. Illustrating the range of animals which caught the attention of early hunters, he also gave us important ecological clues. For instance, when he observed rhinoceros depictions being confined to Mirzapur, Adamgad, and Bhopal, he also emphatically underlined the fact that no representations of the animal have been found south of the Narmada and in Western Malwa, these being comparatively drier regions.

Subsequently, Mirzapur was explored by Rakesh Tewari, who also revisited Cockburn's site of Ghormangar. Tewari (1987) went on to point out differences between the original painting and its later representations. He (1990) broadened the geographical canvas of his survey to cover all the rock shelters of the region, resulting in a monograph titled *Rock Paintings of Mirzapur*. Surveying the range of animals depicted, he noted fifteen rhino depictions, fifty-five figures of the elephant, and an overall absence of the members of the feline family with the exception of the leopard.

Radha Kant Varma (1996) chose the rock paintings of the Vindhyan region to understand the subsistence economy of mesolithic people. Based on a close study of the depictions, and reflecting upon the range of animals hunted as well as the methods employed to hunt them, he inferred that the subsistence economy of the mesolithic population in the region primarily relied on hunting, which was supplemented by gathering. Not just this, Varma also underlined the popularity of animals based on the frequency with which they were depicted.

The mention of rock art instantly brings to mind the fascinatingly illustrated works of Erwin Neumayer (1983, 1993, 2013). By undertaking extensive surveys in the various rock art regions of India, Neumayer effectively brought alive glimpses of early human interactions with the animal world in different cultural contexts. In fact, some of Neumayer's representations shed valuable light on issues (particularly of subsistence) not easily answered by archaeology.

Animals and animal hunts are undoubtedly a popular theme in early rock art. However, this is not to say that rock art studies projecting them along with other themes, are limited to the aforementioned contributions. The studies of Buddha Rashmi Mani (2000–1) and Thsangspa (2014) for Ladakh, V.H. Sonawane (2014), Murari Lal Sharma (2013; 2014), and A.K. Prasad (2014) for Gujarat, Rajasthan, Bihar, and Jharkhand respectively deserve equal emphasis.

Giriraj Kumar (2014) gave us an excellent sense of the evolution of the field, and also delineated the typical features of the rock art of different regions in the subcontinent. For Kumar,

> The theme of the figures in rock art generally was not a depiction of the day-to-day life activities or socio-cultural and natural environment as seen by their authors. Rather, it reflects the reality as perceived by their authors in particular and the related community in general, and also the human behaviour developed in the light of the wisdom so earned at different stages in human history. (2014, 342)

Beyond rock paintings, there are those quaint terracotta animal figurines as well as representations on seals, descriptions of which accompany almost every excavation report. While there is no dearth of writings on Indian terracotta art, I will confine myself to the ones which have at some length dealt with animal representations in the medium. The importance of such works lies in the fact that they help us chart cultural perceptions regarding the animals chosen for representation. Not just this, they also enable us to map ecologies when we keep in mind the context of the retrievals, and push the evidence to ask whether the animals, particularly wild ones, were a part of the local landscape, and if not, why at all would they be modelled?

For an immediate sense of the range of animals depicted in terracotta, one can turn to writings that have worked on different spatial and temporal canvases collating evidence which also helps us in tracing trends in the modelling of animal forms. Discussions of animal figurines have generally formed only a part of works dealing with terracotta art. Early beginnings were in the form of Charu Chandra Das Gupta's *Origin and Evolution of Indian Clay Sculpture* (1961), which was a panoramic survey of terracotta finds across time and space. Twelve years later, Rai Govind Chandra looked at animal figurines as part of his *Studies of Indus Valley Terracottas* (1973). At a time when the number of excavated Harappan sites was limited, Chandra brought together the evidence that was available, facilitating a sense of the animals which figured in the Harappan way of life.

Pratibha Prakash chose to focus on *Terracotta Animal Figurines in the Ganga–Yamuna Valley (600 B.C. to 600 A.D.)* (1985). Perhaps the only monograph dealing exclusively with terracotta animal figurines, the emphasis was on identifying elements of continuity and change by analysing the variety of animals modelled, the frequency of representations, techniques, and decorative devices employed.

Though fleeting, Nisha Verma's (1986) treatment of the terracotta animal figurines of Bihar from the pre-Mauryan till the Pāla periods serves to give us cursory pointers to the animal repertoire retrieved from sites in the region. Verma's emphasis was on understanding the religious and socio-economic importance of these representations.

B.M. Pande and Shubhangana Atre went beyond terracottas to look at Harappan seals. Pande (1984) made an interesting attempt to see how animal figures on Harappan seals looked 'in the round' when cast in plaster of Paris. The results showed not only the superiority of Harappan craftsmanship, but also a remarkable adherence to details and an understanding of the anatomy of the animals. Atre (1990), on the other hand, sought to make a case for the religious and mythological bearing of the motifs on the seals, which she argued were more often than not those of animals, real or fabulous.

Similarly, the writings of Elisabeth Christina Louisa During Caspers have extended the ways of perceiving and interpreting Indus iconography, suggesting and attributing magico-religious dimensions to the representations. 'Caricatures, Grotesques and Glamour in Indus Valley Art' (1979), 'Singular Aspects of Indus Valley Artistry' (1987), 'Magic Hunting Practices in Harappan Times' (1989), 'Rituals and Belief Sytems in the Indus Valley Civilization' (1992) are a few among her multifarious contributions. Particularly fascinating is her analysis of the famous *Paśupati* seal, particularly against the backdrop of the attention the seal has elicited since the time of its discovery at Mohenjodaro.

Vibha Tripathi and Ajeet K. Srivastava furnished us with a more detailed overview in *The Indus Terracottas* (1994). In the section dealing with animal figurines, they underlined the prominence assigned to bull figurines, and noted that the rest of the animals were confined to a limited number and variety. They also postulated that with the exception of the bull, the animal figurines had a secular character. More crucially, they argued that animals such as the rhinoceros, bison, antelope, crocodile, and buffalo being suggestive of a particular kind of ecology, have much to tell us about the ecological conditions of the time.

In the same year, Arundhati Banerji engaged with human and animal figurines in northern and western India between c. 2000 and 300 BCE. She sought to classify the figurines typologically and stylistically in their cultural context. Tracing developments in post-Harappan early iron age and early Northern Black Polished Ware (NBPW) cultures, she delineated trends in the tradition of depicting animals, and also explored possible reasons for cultural and artistic changes noticed during the period.

Similarly, Urmila Sant (1997) took us through Rajasthan in search of proto-historic, early historical, and historical terracotta figurines and artefacts. Briefly listing the terracotta animal finds from different excavated sites in the region, she went on to focus primarily on the religious and socio-economic significance of the terracotta repertoire.

Within the visual tradition, the focus has also often been on individual animals. The elephant, for instance, has been a distinct favourite. As early as 1946, the animal made its way into Heinrich Zimmer's *Myths and Symbols in Indian Art and Civilization.* Reiterating the special place of the pachyderm in the Indian subcontinent, and its overwhelming presence in visual representations, Vishwanath Shridhar Naravane (1965) set out to explore the curious association of the elephant and the lotus in literature, painting, and sculpture.

Less than a decade later, while the evolution of the elephant motif in ancient Indian art engaged Asis Sen (1972), Jeannine Auboyer (1972) underlined the exalted position of the animal throughout the course of Indian art. Beginning with Bharhut, she mapped phases in Indian artistic development up till the medieval epoch, coalescing them with changes in ways of representing the pachyderm. Subsequently, Bahadur Chand Chhabra (1973) took the Harappan civilization as a starting point to trail the 'Elephant in Indian Art'.

Calambur Sivaramamurti (1974) in his celebrated work *Birds and Animals in Indian Sculpture* underlined the extraordinary importance given to the elephant in sculptural representations spanning centuries. Similarly, K. Bharata Iyer (1977) demonstrated how literature and legend often translated into representations in sculpture. The elephant, he maintained, had 'a phenomenal career in which it very soon attained a pre-eminent position in Indian animal sculpture, and which it continued to retain. The Indian sculptor loved to depict it over and over again and it remains the most sculptured of animals, its images running into hundreds of thousands' (Iyer 1977, 48).

Subsequently, Sadashiv Gorakshar (1979) brought together animal motifs in paintings, sculptures, coins, seals, terracotta, and other miscellaneous objects spanning millennia. References to the pachyderm can be garnered from his listing of animals.

A fair sense of the interest generated by this mega herbivore also emerges from a quick overview of the range of writings (mostly from an art historical perspective) that have exclusively focused on it. Pratapaditya Pal's *Elephants and Ivories in South Asia* (1981) retrieves glimpses from mythology, literature, and art, while Vikramjit Ram's *Elephant Kingdom: Sculptures from Indian Architecture* (2007) was effectively written to drive home even to the lay reader the overwhelming presence of the pachyderm in visual narratives spread across millennia.

In 1983, S.K. Gupta's *Elephant in Indian Art and Mythology* sought to trace the evolution of the elephant symbol in Indian art over 5,000 years as also to demonstrate how intrinsically the animal has been associated with Indian history and art. Shortly thereafter, Kamal Shankar Srivastava's *The Elephant in Early Indian Art: From Indus Valley Civilization to A.D. 650* (1989) followed as a similar attempt.

Putting to use his experience as the incharge of the Department of Arms and Armour at the National Museum, New Delhi, G.N. Pant brought forth an exhaustive volume on *Horse and Elephant Armour* (1997). Pant used diverse sources such as museum collections of armour, sculptural reliefs, wall paintings, depictions on coins and seals, and corroborated them with literary evidence to trace cavalry and elephantry through the ages.

The rhinoceros and the tiger have been less fortunate in eliciting scholarly attention, though our one-horned hero does form the central theme of Joachim Karl Bautze's (1985) 'The Problem of the Khaḍga (*Rhinoceros unicornis*) in the Light of Archaeological Finds and Art'. Despite the animal's impressive size and power, what intrigued Bautze, an art historian, was its inferior role in the history of Indian animals. Though he marshals evidence from literature, the approach, unmistakably, is that of an art historian. The emphasis is on visual representations of the animal as he traces its occurrence on Indus seals, amulets, copper tablets as well as terracottas, archaeological finds from historical periods as also its frequent presence in wall paintings, Mughal miniatures, and animal carpets of the seventeenth century. What emerges is a clear sense of a strong visual presence of the animal in Indian art and archaeology. Yet, as Bautze points out, it is strange that the animal never became the mount of any god in the Indian pantheon. He speculates on the possibility of Indians sharing the European disdain for the animal and if that could explain its anomalous fate.

Joseph Manuel (2008a) recounts the journey of the animal from a common element in rock paintings, Harappan terracottas, seals, copper tablets, Mauryan, Jaina, and Buddhist art to one that came to be confined to art associated with the nobility by the time of the Mughals. What sets Manuel's analysis apart from that of Bautze's is his emphasis on the progressive dip in representations, which according to him, had more to do with changing cultural perceptions rather than the animal itself becoming a rarity. In a way then, Manuel's account is also a history of the perceptions that determined the fortunes of the animal.

Animals in Stone: Indian Mammals Sculptured through Time (2008) by Alexandra Anna Enrica van der Geer, a palaeontologist and an ethnozoologist, is a finely illustrated volume mapping the art history of animals that have captured the Indian sculptor's attention across time. The approach, a combination of zoology and art history, is essentially different, and scientifically brings alive each species before tracing its narrative history.

* * *

Texts as historical documents often offer windows into past sensibilities. While attempts to retrieve glimpses of the political, economic, social, and cultural from these written repositories is common, textual traditions have also been approached for insights regarding the natural world. Important within this framework then are works which devoted themselves to the study of fauna as seen in ancient Indian literature.

As in art, even within the written tradition, animals have been approached in various ways. They have been treated as a category, or the focus has been on particular animals or an animal. Again, while some prisms sought to catalogue them, others strove to culturally contextualize them. Both methods have their own merits since any understanding of the cultural and ecological importance of animals can only be based on a firm knowledge of the available faunal range. It is, therefore, crucial to mention some early and lesser-known writings which realized the importance of separately dealing with the natural world in the form of its plant and animal resources, and set out to do so either through panoramic surveys of ancient Indian literature, or within the confines of a particular corpus of the same.

As early as the last quarter of the nineteenth century, Valentine Ball found it worthwhile to engage with the identification of the animals and plants of India known to early Greek writers. An Irish geologist with diverse interests, Ball worked with the Geological Survey of India, and in the course of his attempt to identify minerals mentioned by the writers of yore, he also chanced upon allusions to India's plants and animals. Despite their 'apparently mythical character', he was certain that the descriptions rested on 'substantial bases of facts' which could be established by putting them to comparison with known information. Ball (1879–88, 302) disclaimed possessing philological qualifications for the project, but claimed familiarity with a few languages of India, and with the plants and animals of the country. He subsequently returned with 'Further Notes on the Identification of the Animals and Plants of India, Which Were Known to Early Greek Authors' (1889–91).

In 1906, Monmohan Chakravarti in a study on Aśokan edicts, set out in search of 'Animals in the Inscriptions of Piyadasi'. More than two decades later, the rhinoceros caught the attention of George W. Briggs (1931), who sought to trace the cultic importance of the animal since ancient times. Drawing loosely on archaeological finds of seals depicting the animal from Harappa and Mohenjodaro, references occurring in the pillar edict of Aśoka, the *Sutta Nipāta,* the *Manu-smṛti* and other texts, and contemporary customs and traditions centring around the animal, Briggs tried to establish the antiquity of the rhinoceros as a sacred animal in India as well as to understand why it was held sacred.

While Briggs's was a cultural perspective regarding the rhinoceros, the elephant found place in more strategic realms in V.R. Ramachandra Dikshitar's *War in Ancient India* (1944), which used ancient Indian literature along with sculptural and epigraphic evidence for glimpses of weapons used in war, the army, and its divisions within which elephants occupied a crucial place. Sibadas Chaudhari worked on a broader canvas listing all the animals referred to in the *Rāmāyaṇa*. Published as a series between 1952–4 in the *Indian Historical Quarterly*, 'Concordance of the Fauna in the Rāmāyaṇa' was a monumental effort based on the Calcutta Sanskrit Series, and comprised 237 entries substantiated with footnotes citing references to animals in Vedic literature, the *Mahābhārata*, and animal remains from Harappa and Mohenjodaro.

While Chaudhari brought alive the faunal world of the *Rāmāyaṇa*, B.P. Sinha used archaeology, textual evidence, and numismatics in 'Elephants in Ancient Indian Army' (1955) to offer a historical perspective to the use of the animals in the military. Baij Nath Puri (1963) on the other hand, put together (as a section in his book) an account of Indian fauna as described in Western classical literature. The theme of warfare once again received attention in Sarva Daman Singh's *Ancient Indian Warfare with Special Reference to the Vedic Period* (1965). Using archaeological evidence bearing on the taming of animals and the use of metals, arms, and armour, Singh in his own words, added 'material proof to literary testimony' (1965, 3) to reconstruct warfare during the Vedic period. Elephants elicited particular consideration, and Singh also drew upon Buddhist and epic material to substantiate the picture of the early period.

Madhusudan Madhavlal Pathak (1968) engaged with contexts where animals served as figures of speech, and juxtaposed the two epics in their usage of similes alluding to the animal world.

Howard Hayes Scullard (1974) introduced us to the *Elephant in the Greek and Roman World.* An authoritative and exhaustive account of the Western world view of the pachyderm, Scullard did not confine himself to the 'Indian elephant' as he set out with a discussion of the natural history of the animal, and relied primarily on accounts left by Greek and Roman writers to reconstruct the interactions of this animal with the Graeco-Roman world. Particularly engaging is his treatment of the animal in 'war and peace', where he crucially points out that while the Greeks used elephants mainly for war, the Romans used them mainly for public spectacle (Scullard 1974, 250).

The title of Balkrishna Govind Gokhale's (1974) essay 'Animal Symbolism in Early Buddhist Literature and Art' is self-explanatory. Through literary and sculptural evidence based on early Buddhist literature (specifically the Pali *Tipiṭaka*) and art, Gokhale, a Pali scholar, sought to address two questions: the raison d'etre, if any, of the early Buddhist preoccupation with animal symbolism, and if it could be interpreted as an integral part of the thought world of early Buddhism. Choosing the elephant, horse, bull, and lion for analysis, he set out to explore their roles in a specifically Buddhist context. It was argued that

> These animal symbols had become a part of the collective Buddhist 'unconscious'. They had also become convenient means as visible signs of something that was invisible but ever present, namely, the Buddha and his *dhamma*, as objects representing something generally and peculiarly sacred . . . The growth of these symbols reflects the development of a Buddhist 'mythology', an essential part of the evolution of Buddhism from a mere monastic movement into a religion with its own 'pantheon', metaphysics, ethics and ritual. (Gokhale 1974, 120)

Sures Chandra Banerji's (1980) *Flora and Fauna in Sanskrit Literature* is a compendium of references to plants and animals as gleaned from Sanskrit texts starting from the *Ṛgveda* down to classical times. Banerji asserted that the work was not meant to be a mere catalogue of plants and animals in ancient India, but was also to place them in their cultural and social contexts. Apart from spelling out the general characteristics of plants and animals in ancient Sanskrit literature, Banerji also dabbles at some length with elephant lore, equestrian science, and ornithology. In an earlier study, tracing the origin and evolution of the Dharmasūtras, Banerji (1962) had devoted an entire chapter to the flora and fauna of the Dharmasūtras.

In his celebrated essay 'Elephants and the Mauryas', Thomas R. Trautmann (1982) sought to put in historical perspective the reliance on the elephant as a strategic factor in Indian history. Culling evidence from the *Arthaśāstra* and classical Greek accounts of India, Trautmann brought out the centrality of the pachyderm to the political fabric of the Mauryan Empire, and attributed the ascendancy of Magadha in the sixth century BCE as well as the success of the Mauryan Empire in the fourth century BCE to the royal monopoly of horses and elephants. Not just this, Trautmann made some remarkable observations regarding the distribution and acquisition of horses and elephants in ancient and medieval India as well as the measures exercised by the Mauryas to manage and control the elephant population of the empire. Particularly outstanding was his mapping of the eight *gaja vanas* listed in the *Arthaśāstra* which ranked them for the quality of their elephants.

Like Madhusudan Madhavlal Pathak, G.V. Bapat (1985–7) examined the similes in the *Rāmāyaṇa*. The focus, however, was specifically on elephants, wherein Bapat noticed about 250 references to the pachyderm, 110 of which occurred in similes as *upamānas*. Arguing that the use of such a large number of elephant images in one single narrative is, perhaps, unique in literature, he sought to reason why the animal was chosen as an *upamāna* so often. He proceeded to group the similes under seventeen themes, which validate his contention that the sheer diversity of aspects associated with the pachyderm not only suggested the poet's familiarity with it, but also reflected an audience acquainted with the animal in the wild as well as in captivity.

'The Fauna in the Āraṇyakaparvan of the Mahābhārata' by Madhukar Anant Mehendale (1987) is the only such exhaustive study for the *Mahābhārata*. Along with alphabetically collating the fauna encountered in the *parvan*, Mehendale, an esteemed scholar of Sanskrit, lucidly provided the contexts of their occurrence. What we have are hints of cultural perceptions as well as glimpses of the behaviour and ecology of animals.

The Jungle and the Aroma of Meats: An Ecological Theme in Hindu Medicine by Francis Zimmermann (1987) is an outstanding departure from the contributions mentioned hitherto. Underlining the polarity of the *jāṅgala* (dry), and *ānūpa* (marshy) lands, the work is a careful analysis of three principal treatises of classical Indian medicine, the *Suśruta Saṃhitā*, *Caraka Saṃhitā*, and the *Aṣṭāṅgahṛdya Saṃhitā*, and probes animal classifications therein.

While James P. Mcdermott (1989) tapped the *Sutta* and *Vinaya Piṭaka* of the Pali canon to elucidate the 'Buddhist Concern for an Ethically Grounded Relationship between Humans and Animals', Brian K. Smith (1991) used ancient Indian texts to tell us how these classified animals as domesticated or wild, edible or inedible, and sacrifiable or not, based on foot and dental structure, and the form of procreation. He also went on to establish that these classification schemes had social ramifications.

Uma Chakravarti (1993, 53) sought to do justice to the Jātakas as historical narratives for reconstructing the everyday lives of people in early India. But before taking up the narratives, she reflected on the context of their composition, pertinently observing that at the time the Jātakas were finalized, men, women, and beasts continued to live in close proximity to each other given the settlement pattern reflected in the Buddhist texts. Between each settlement and the cultivated area surrounding it was what the *bhikkhus* called *aranna* or forest land teeming with animal and vegetable life. The *bhikkhus* who lived on its edges often traversed it in the course of their wanderings. The presence of the animal world was, thus, according to Chakravarti, integral to the experience of the early Buddhists, and this partially explained its prominence in the narratives where animals and humans often interacted.

Chakravarti's is a nuanced approach as she proceeds with the help of specific narratives to demonstrate how the animal stories are used to serve different functions within the tradition, sometimes to lend a voice to the mute lower classes or to act as an allegory for the rights of the voiceless.

What prompts me to draw upon Chakravarti is the way she underlines that these animal stories often transpire in spaces and contexts where animals and humans interact. It gives us the vital push to access these tales for the subtle messages (which, as I shall show in my chapters, transcend the social world central to Chakravarti's argument) they convey, and in the process one often stumbles upon at least two of the protagonists of my narrative, the elephant (which is a frequently employed motif) and the tiger. The rhinoceros is more of a chance encounter, yet put together with the other two, they certainly help us in stringing together individual animal histories.

With Julie Hilton (1996), we plunge into the darker realms of omens and prophecies determined by birds and animals that were considered good or evil by their physical appearance or by the threat they posed to humans. Culling evidence from a wide range of ancient Indian texts, Hilton showed how animals in ancient India were both adored and abhorred. Propelled by socio-economic factors, notions of beauty and ugliness, she argued, led to the development of a hierarchy of creatures and the conferring of auspicious and inauspicious status. Associated with priests and the elite, cattle, elephants, and horses were beautiful animals, deemed as auspicious, virtuous, and holy. Admired for their prowess and virility, lions and tigers possessed qualities required in rulers and warriors. On the other hand, birds and animals inferior in beauty and behaviour were associated with the lower classes.

Trained as a historical and Indo-European linguist, Stephanie W. Jamison (1998) rooted herself in the ancient dharma texts to investigate the strange inclusion of the rhinoceros in the list of five-nailed animals (*pañcanakha*) whose flesh could be eaten. Based on the textual configuration of the 'older' and 'younger' five-nailed passages, Jamison sought to prove that the rhinoceros was a later addition to the text as also to explain the presence of this three-toed mega mammal in the list.

Ancient dharma literature also elicited the interest of Patrick Olivelle, an Indologist and a scholar of Sanskrit. Olivelle's '*Abhakṣya* and *Abhojya*: An Exploration in Dietary Language' (2002a) focused on these two terms relating to food prohibitions within this corpus of ancient Indian literature. His diverse scholarly interests also manifested in 'Food for Thought: Dietary Rules and Social Organization in Ancient India' (2002b) as well as 'Talking Animals: Explorations in an Indian Literary Genre' (2013).

In the former, he used early Vedic texts, the dietary regulations in the later legal texts (the Dharmaśāstras), and in the medical treatises of Caraka and Suśruta to examine animal classifications and the prohibitions based on them. He then proceeded to relate these regulations to broader social categories and relationships, particularly those relating to social hierarchy and marriage. 'Talking Animals', on the other hand, traces the genesis of animal fables, and reflects on 'the religious and cultural backdrop within which the anthropomorphizing habit of Indian animal tales took place' in Sanskrit literature (Olivelle 2013, 17). Beginning with the contention that in general people have been fascinated by talking and thinking animals, Olivelle approaches a range of Indian literary genres in order to understand the beliefs and ideologies which possibly underlay animal fables.

Animals have also been used and studied to resolve knotty debates in Indian history. I refer here to the renowned Indian archaeologist Braj Basi Lal's (2003–4) engagement with *Ṛgvedic* flora and fauna, and the light they cast on the 'Aryan invasion' debate. Lal sought to analyse the plants and animals mentioned in the *Ṛgveda*, and on that basis, he argued that their distribution suggested a tropical ecology, and not a cold one (which formed the premise of theories advocating that the Aryans came to India from outside, particularly from Central Asia). Lal maintained that the *Ṛgveda* preserved no memory of any typically cold-climate tree/plant or animal/bird, and, hence, urged for a ceremonial burial to the theory of 'Aryan Invasion of India'. This line of enquiry was subsequently integrated with other kinds of scientific data, including archaeological and linguistic, in his monograph *The Homeland of the Aryans: Evidence of Rigvedic Flora and Fauna & Archaeology* (2005) to further challenge the age-old Aryan invasion theory.

Animals in Early Buddhism engaged Arvind Kumar Singh (2006), who drew upon early Buddhist literature to elaborate upon the position of animals therein, the emphasis on *ahiṃsā* and animal protection, and an analysis of such attitudes vis-à-vis the same enjoined in other religions.

What is outstanding is that most of these works engaging with the natural world came from people who were neither zoologists nor ecologists, but ranged from being archaeologists, linguists, historians, and art historians. Evidently then, the consciousness regarding the need to retrieve glimpses of the faunal wealth of ancient India has transcended disciplinary barriers, and goes further back than is generally known or acknowledged. The value of these contributions should, therefore, be judged not on the basis of their accuracy, but in view of the fact that they were crucial stepping stones for anyone attempting to map animal histories.

Moving beyond writings that have used animals to address a mosaic of issues, I approach the realm of narratives which set out to give animals their own histories. While Mahesh Rangarajan (2001) used archaeology along with a range of textual material to give us a sense of the past and present of India's wildlife, chronicling the history of a species now extinct is *The End of a Trail: The Cheetah in India* by Divyabhanusinh (1995). As the author himself puts it, 'The main thrust of this expanded work is to trace the cheetah through historical records and to recount the animal's brush with human society leading ultimately to its extinction in India' (Divyabhanusinh 1995, xix). What sets the work apart is its method, which dexterously strings together diverse sources to tell the compelling tale of an animal that now survives only in memory.

Similarly, 'In Times Past', Romila Thapar (2004) takes us on a tiger trail. Perhaps the only piece of its kind on the predator in ancient India, Thapar maps the history of the animal using rock art, archaeology, and literature. Though she briefly mentions tiger bones at archaeological sites, and also explores the human–tiger interface as reflected in rock art and Indus iconography, the highlight of the essay is the evidence she marshals from literary sources spanning millennia.

The canvas is further enriched by *The Story of Asia's Lions* where Divyabhanusinh (2005) retains his narrative approach steeped in historical and visual records to trace the interaction of this mega carnivore with humans across millennia in South Asia. The animal is skilfully salvaged from the annals of ancient, medieval, modern, and contemporary Indian history. As in the case of the cheetah, this book reinforces the power of the visual archive in defining as well as refining our notions of the past as well as the present.

This trajectory also finds extension in Raman Sukumar's *The Story of Asia's Elephants* (2011). Splendidly illustrated, Sukumar clarifies that the book is not an art history of the Asian elephant or of ivory, but rather an ecological and cultural history of the species. It is a historical journey tracing the human–elephant relationship from the Stone Age to the present. The sources he uses to tell his story are primarily literary, ranging in time from the Vedas to the *Ain-i-Akbari*, though he also delves into fossil finds, representations in rock paintings, and on Indus seals while discussing the prehistory and proto-history of the pachyderm. Bones found at Harappan sites along with elephant tusks and ivory artefacts are also briefly alluded to.

Paul J. Kosmin (2014) employed archaeology, epigraphy, and textual evidence to give us a fascinating perspective of the Seleucid Empire, arguing, 'To think Seleucid is to see elephants.' More significantly, it is the Indian elephant that he refers to. The great Seleucid Empire, he contends, 'founded at the end of the fourth century by Seleucus I Nicator, a king titled "elephant-commander" by his enemies, took as its dynastic blazon the Indian elephant. Marching elephants and elephant-chariots and elephant-scalps adorned official coins and seals . . . Elephant trophies commemorated Seleucid military victories . . . and like the beasts, the Seleucid empire, enormous and vulnerable, brutally powerful and self-defeating, was as much a work of the imagination as a creature of war' (Kosmin 2014, 3).

It is crucial to add to this survey Thomas Trautmann's *Elephants and Kings: An Environmental History* (2015), which probes human–elephant relations in India, focusing particularly on the relation of kingship to elephants. The hypothesis is that it was the institution of the war elephant that promoted the persistence of the animal in India and Southeast Asia. Trautmann works on a vast spatial and temporal canvas using India as an entry point to show how war as a political compulsion created a royal interest in protecting elephants and their habitats, and contrasts the situation with China, where the institution was never established, and where elephants have mostly retreated.

What emerges from this historiographical survey is a range of approaches to the study of human–animal interactions in the past. However, what unites these is the use of diverse kinds of evidence including archaeological, faunal, artistic, and textual in the reconstruction of this interface.

***

Having surveyed the ways in which animals have been approached, I embark on my megafaunal ride. I would, however, at the outset like to emphasize that this work does not claim to be an archaeozoologist's perspective or method of charting animal histories, nor is it exclusively based on faunal evidence. Bones tell their own tales, and it is those that I look for when I use animal remains as one of the many prisms that can be employed to map larger faunal histories. My study is a cultural and ecological enquiry whose method reiterates the importance of interdisciplinary approaches in reconstructing the past.

With this groundwork, I set out keeping in mind the limitations of the sources at hand. As far as the archaeological record in terms of bones is concerned, I have proceeded with the consciousness that the evidence is scattered, and at times disappointingly patchy. This predicament is compounded by the fact that a lot more is required in terms of archaeozoological findings, particularly at early historic and historic sites, and it is equally important for the results to reach the public domain. Another problem is that often the cultural context of retrievals (particularly in the case of earlier reports) is not specified, rendering inferences ineffective. There is also awareness regarding the fact that with the dawn of the historical period, one encounters a gap in the archaeological record, which can primarily be attributed to historical archaeology's preoccupation with prehistory and proto-history.

Though this drying up of archaeological evidence is made up for by the vivid imageries derived from texts and textual traditions, these have their own widely known problems. For instance, while reading a text from a historical point of view, it is important to keep in mind the age, authorship, and geographical provenance of the composition, as also the genre it represents. While chronology remains important, it is not always possible to put textual traditions within neat time brackets. Time boundaries for such traditions are far from precise, and overlaps are not uncommon, yet linguists and philologists have given us relative chronologies (again with more disagreements than otherwise) to work with.

While dealing with textual evidence, it is crucial to get in place some sort of rationale behind the sequencing of the literary sources used. The approach being historical, chronological margins, no matter how blurred, have been kept in mind. The intention is not to convey a linear progression from one corpus to the other, but to carve out a narrative of the fortunes of these animals within religious and secular traditions separated in time and space. As is also evident, this is not a study based entirely on texts; hence, the choice regarding what to include was one that though not easy, had to be made.

My foray in search of my three protagonists is based on a range of ancient Indian texts—in Sanskrit, Pali, and Prakrit. It begins with the Vedic corpus (second to the first millennium BCE); followed by early Buddhist literature based on the Pali *Tipiṭaka* (the core of which is placed between the fifth and third centuries BCE, the Jātakas being somewhat later—between the third century BCE and second century CE); and the non-canonical *Milindapañha* (first century BCE to first century CE); some early Jaina literature in the form of glimpses from the *Ācārāṅga Sūtra, Kalpa Sūtra, Uttarādhyayana*, and *Sūtrakṛtāṅga* (origins of the canon placed in the fifth or fourth centuries BCE); the Dharmaśāstra literature, including the early Dharmasūtras (600–300 BCE) and the *Manu-smṛti* (200 BCE–200 CE); early medical texts in the form of the *Caraka Saṃhitā* and *Suśruta Saṃhitā* (first half of the first millennium CE); the two epics, *Rāmāyaṇa* (fifth or fourth century BCE to third century CE) and *Mahābhārata* (400 BCE–400 CE); and the *Arthaśāstra* (300 BCE–200 CE). Also captured are glimpses from the classical Western accounts beginning with Ktesias in 400 BCE.

Diacritical marks have been used primarily in the section dealing with textual sources. They have also been used in the names of deities and while mentioning the dynasties of ancient India. Diacritics have been retained in the context of place names when they occur in texts. However, exceptions have been made in the case of archaeological sites, well-known rivers and mountains, scripts such as Kharoshthi, languages including Sanskrit, Pali, Prakrit, and words in common usage such as gharial, pipal, sal, stupa, swastika, and others.

It is also crucial to clarify that my work makes no claims about being encyclopedic in its approach. The aim is not to compile an inventory of references to the three mega mammals, but to cull representative samples that equip me to tease out stories rendering them visible in history. Further, it may be qualified that as far as representations in art are concerned, while rock paintings and terracottas have been dealt with in substantial detail, my journey with an animal such as the elephant with its ubiquitous presence in Indian sculpture did not render it possible to discuss representations therein. The exclusion was also determined by the consciousness that sculptures speak their own language, hence, it would be unjust to look at them perfunctorily.

Within the framework outlined, Chapter 2 chronicles the much-neglected saga of the rhinoceros in ancient India, while Chapter 3 maps the journey of the tiger. The sheer quantum of evidence necessitated a splitting of the elephant story into two chapters, where Chapter 4 looks for traces of the animal in archaeology, art, and iconography, while Chapter 5 explores its presence in textual traditions. With reference to the elephant, it may be additionally emphasized that the aim of weaving in glimpses of ivory working in archaeological contexts as well as in literature is only to drive home the utility humans have found in it as a raw material since prehistoric times, and to underline its continuity in ancient Indian history. Hence, while the use of ivory as a resource has been touched upon, the survey by no means takes into cognizance the entire ivory repertoire of the region under study. This is particularly the case when I deal with the early historic sites, nearly all of which yield ivory objects. Here, rather than going into the details of the ivory collection at each site, and making the discussion unwieldy, the narrative proceeds with the awareness of the continuing (and in some cases prolific) use of ivory objects.

Another point which needs some emphasis is the caution that has been exercised in using the words 'taming' and 'domestication' due to the technical difference between the two. Taming involves subduing a wild animal into adapting and submitting to human control, while domestication involves structural, physiological, and behavioural modifications in certain species. In view of the lack of evidence for the latter in the context of elephants in ancient India, I have mostly chosen to associate the word 'taming' with it.

The chapters that follow will attempt to reconstruct the histories of these mega species, and to give a sense of the implications these histories encompass for the cultural and ecological moorings of ancient India in the millennia surveyed.

Put by the rod for all that lives,
Nor harm thou any one thereof;
Long not for son—how then for friend?
Fare lonely as rhinoceros . . .

With friends one is at beck and call,
At home, abroad, on tour for alms:
Seeing the liberty none want,
Fare lonely as rhinoceros.

—*Sutta Nipāta*
(i, 3, 35, 40; Hare 1947, 6)

2

TRAILING THE ONE-HORNED WONDER

A MASSIVE BODY, STUMPY LEGS, and an armour-clad prehistoric look is the image that is evoked by the greater one-horned rhinoceros (*Rhinoceros unicornis*). Despite being one of the two greatest success stories in rhino conservation (the other being the southern white rhino in South Africa), the animal is still vulnerable on the International Union for the Conservation of Nature's (IUCN) Red List of threatened species. As a result of strict protection measures launched by Indian and Nepalese wildlife authorities, the population of greater one-horned rhinos has rebounded. The recovery of the species has, however, been precarious due to poaching pressure remaining high in both India and Nepal.

Concerns regarding its conservation are fairly recent, but the awe, wonder, and curiosity this mega mammal has inspired dates back to antiquity. Besides painstakingly reconstructing the evolutionary lineage of the rhinoceros, popular and scientific writings have unanimously acknowledged the multitude of prehistoric forms distributed all over the world. Of this original multitude, only five remain in Asia and Africa as testimonies to the fascinating saga of evolution against odds such as geological upheavals, climatic changes, and biotic factors. These are the African white or square-lipped rhinoceros (*Ceratotherium simum*), and the black rhinoceros (*Diceros bicornis*) together with the three Asian species: the greater one-horned rhinoceros (*Rhinoceros unicornis*), the Asiatic two-horned or Sumatran rhinoceros (*Didermoceros/Dicerorhinus sumatrensis*), and the lesser one-horned or Javan rhinoceros (*Rhinoceros sondaicus*). Most of these have been threatened with extinction, and almost all are caught in conflicts of various kinds with humans.

Of these, the Indian rhinoceros is known to have once extensively roamed the marshes of northern India from Sindh in the north-west to the valley of the Brahmaputra in the north-east. A repertoire of historical and hunting references testifies to the extensive occurrence of this animal in the alluvial floodplains of large rivers such as the Indus, Ganges, and the Brahmaputra, as well as the Terai regions of Nepal and Sikkim. This is not surprising given the animal's preference for moist and swampy terrains.

As late as the eighteenth century, a map of Mughal India showed the rhino in Awadh (Gole 1988, 27). In the same century, we are told that North Bengal and Assam were so rich in rhinoceroses that a French map of India describes that area as 'Contrée de Rhinoceros', and that late medieval temples in Bengal approximately from the same period as the map, are decorated with terracotta panels showing rhinoceros hunts (Bautze 1985, 415). In an evaluation of the status of the three Asiatic species, E.O. Shebbeare (1953, 142) contended that the Indian rhinoceros 'inhabited the sub-Himalayan tract during historic times, the western limits of its range retreating from Peshawar, in the days of the Emperor Baber (1505–1530), to Rohilkhand (the Bareilli district) in the mid-nineteenth century and the Nepal Terai during the present century'.

Reviews abound in reiterating the past distribution of the animal in northern India, Pakistan, and Afghanistan (Rookmaaker 1984, 1999, 2000, 2002). We, thus, have ample testimony to the existence of the rhinoceros outside areas in the North-East until not so long ago, and its unfortunate retreat thereafter. Rookmaaker (2002, 928) charts this retreat, arguing that the animal gradually disappeared, starting from the western regions around Delhi and Agra as illustrated by the latest dates when the animals were recorded: 1590—Sambhal (Uttar Pradesh), 1625—Aligarh (Uttar Pradesh), 1665—Kora (Uttar Pradesh), 1700—Mirzapur (Uttar Pradesh), 1800—Patna (Bihar), and 1850—Rajmahal hills (Bihar). The last rhino was shot in the

Pilibhit district of Uttar Pradesh in 1878, and by 1890, Indian rhinos had vanished from most areas barring southern Nepal, the Bhutan Duars, parts of West Bengal, and the Brahmaputra valley of Assam (Martin and Martin 1982, 29).

Evidently, things had gone wrong somewhere for this mega mammal, and the explanations were not far to seek. Laurie (1978, 9) reasons that as a result of demographic pressure since the fifteenth century, more and more land was brought under cultivation, particularly the lush, fertile floodplains—the favoured haunt of the Indian rhino. While he does not statistically support the mounting demographic pressure that he refers to, his argument that habitat destruction made hunting easier as rhinos were deprived of the shelter of grasslands and forests is certainly persuasive. The combination of habitat destruction and hunting sounded the death knell for the animal, pushing it to the verge of near extinction.

My story is about the animal's journey through the tapestries of time and space before it disappeared from most areas. Apart from underlining how the mega herbivore has been perceived as a marker of ecology, an important line of enquiry would also be to ascertain whether the environment of the study area is now different from the environment to which the species is primarily adapted.

RETRACING THE TRAIL: THE TESTIMONY OF ARCHAEOLOGY

The rhinoceros is an animal of impressive antiquity in India. Traversing back in time, fossilized fragments of rhinoceroses retrieved from Indian geological formations tell tales of numerous extinct species. Amidst a plethora of such forms, the recognized species of the genus *Rhinoceros* are *Rhinoceros sivalensis*, *Rhinoceros palaeindicus*, *Rhinoceros platyrhinus* from the Upper Siwaliks (Pinjor), and *Rhinoceros perimensis* from the Middle Siwaliks (Chinji, Nagri, Dhokpathan). In late Pleistocene contexts, we know of *Rhinoceros deccanensis* from the Krishna Valley near Gokak, Belgaum district on river Ghataprabha, and *Rhinoceros karnuliensis* from the Kurnool caves, Andhra Pradesh (Badam and Jayakaran 1993, 249).

It has been argued that though the genus itself can be traced back to the Pliocene of northern India, most known fossils of *Rhinoceros unicornis* seem to go back to the Pleistocene (Laurie 1978, 6; Laurie, Lang, and Groves 1983, 2). The presence of this species in Pleistocene contexts is testified by fossil and semi-fossil remains retrieved from the Narmada Valley in central India together with those from peninsular India. Though spatially as well as temporally beyond the purview of this book, it may be meaningful to mention some discoveries from these areas suggesting the past distribution of the species.

For instance, mention must be made of the retrieval of three horn cores of the species from the gravel deposits of the Sher River tributary in the central Narmada Valley, in a late Pleistocene context dated to 40,000 BP. This may be the only find of a fossilized horn core not only in India, but possibly also in South Asia (Badam, personal communication). The rhinoceros also figured in the checklist of animals identified in the late Pleistocene fossiliferous gravels near Harwadi, Latur district, Maharashtra (Sathe 2015, 1). Equally crucial is the find of the partial skull of *Rhinoceros unicornis* at Sathankulum in Tirunelveli district of Tamil Nadu in a late Pleistocene to Holocene context. This is the first record of fossil *Rhinoceros* from the southern tip of India (Badam and Jayakaran 1993, 244). Rhinoceroses, known to have totally disappeared from south India, were argued to have inhabited this area when the low hills were forested with swamps in short canyons surrounded by grass (Badam and Jayakaran 1993, 259).

Perhaps one of the earliest accounts documenting the presence of the rhinoceros in conjunction with human activity is that by John Cockburn. During a hunt in July 1881 in the ravines of the Ken River, 3.2 km south of the town of Banda in Uttar Pradesh, he chanced upon the fossil remains of a rhinoceros assigned to the species *Rhinoceros indicus*. The remains he reported included a large quantity of fragments of teeth along with some longitudinally split pieces of the shafts of long bones. The bones and teeth he detailed included the ascending ramus of the left inferior maxilla as far as the insertion of the last molar in four fragments, a fragment of the glenoid cavity of the right scapula, the nearly intact right incisive tusk and several lower molars, and an upper molar. Within four feet of the rhino bones, Cockburn also picked up chert and shell knives on the surface of the soil (Cockburn 1883, 56–7).

What is even more significant is that Cockburn simultaneously drew attention to a rock painting in the Ghormangar rock shelter in Mirzapur district, representing a rhino hunt (Cockburn 1883, 58). While he seemed to consider this painting to be of a recent date (not more than 300 years old), the fact that he found stone implements there suggests that this was likely to be an ancient site whose antiquity escaped Cockburn.

The cohabitation of ecosystems by humans and the rhinoceros can, however, be pushed back much further. While lithic artefacts strongly testify to the presence of humans in the Middle Son Valley during the terminal Pleistocene, Blumenschine and Chattopadhyaya (1983, 283) reported the presence of the mega mammal in the terminal Pleistocene fossil faunal assemblage of the region in the form of a maxillary fragment bearing three cheek teeth. The inference that the species was *Rhinoceros unicornis* was based on the presence of the animal in this region of India during the late Pleistocene, as revealed by Badam (1979). The assemblage represented surface collections from eighteen localities distributed along the Son between Baghor and Patpara that expose the coarse member of the Baghor Formation. Although material from each locality was not sorted, seven were found to be the most fossiliferous, including in order from the most to the least identifiable and identifiable pieces: Kharabara (Son locality), Ram Nagar, Baghor, Tariha Dhaba, Odara, Baliar, and Pawariah.

The taxonomic composition of the assemblage was found consistent with geological evidence for arid conditions during the terminal Pleistocene glacial maximum. Based on the same, a mosaic vegetation was envisioned with grassland occurring over much of the floodplain, and woodland and swamp probably being found along the river and its tributaries (Blumenschine and Chattopadhyaya 1983, 283).

Here then, were the beginnings of an interface which was to mature and manifest itself in the realms of subsistence as well as aesthetics. Archaeological research has brought to light various dimensions of the Holocene cultures of the subcontinent. Significantly, numerous stratified deposits unearthed from sites in the study area have yielded biological materials, particularly the faunal component in substantial quantities. As a result, considerable data for the reconstruction of the environment and human adaptations is now available.

As will be evident, I am documenting the presence of the animal in various cultural niches from hunter-gatherer societies to the first urban civilization of India and beyond. However, before one sets out in search of the rhino in antiquity, it is essential to clarify that though for the sake of convenience in discussion the evidence has been classified according to cultural periods, it will help to remember that since these cultures do not evolve uniformly in a neat pattern all over the subcontinent, an overlap and coexistence of cultures is not uncommon. Where details are available, I shall also try to reconstruct the vegetational landscape that harboured these animals.

Mesolithic/Microlithic Contexts

Microlith-bearing sites in India, several of which are mesolithic, have a wide spatial range, and a fairly long temporal span. Discovered in most parts of the Indian subcontinent, the tradition flourished from the early to the late Holocene, often cutting across Bronze Age or chalcolithic traditions (Chattopadhyaya 2002, 370). The relevant mesolithic/microlithic sites where the rhinoceros finds a presence fall within two geographical regions—the Central Ganga Plains of Uttar Pradesh in northern India, and Gujarat in western India. It is important to underline that the two regions represent different ecological niches and temporal contexts. While there is chronological clarity about the first region, the antiquity of microlithic levels in Gujarat has not been precisely dated. The contextual evidence in some cases even suggests contemporaneity with the chalcolithic cultures of the region.

In the context of northern India, faunal evidence pertaining to the rhinoceros comes from the mesolithic sites of Sarai Nahar Rai, Damdama, and Mahadaha in the Central Ganga Plains. However, before looking at this evidence, it would be useful to briefly understand the palaeoenvironment and palaeovegetation of the region, an endeavour which is aided by the investigations conducted by the University of Allahabad, and subsequently amplified by the University of Lucknow and the Birbal Sahni Institute of Palaeobotany, Lucknow. These include systematic geomorphic, geochemical-mineralogical studies, chronometry using luminescence method, radiocarbon dating, and palynological studies.

A key inference of such studies, and one that is particularly crucial in the context of any attempt to understand the ecological setting of the region, tells us that the area has been a grassland with few thickets and patches of forest, at least for the last 45 kyr (kiloyears, signifying a thousand years). The landscape marked by a large number of small and large water bodies (ponds and lakes) and minor channels had some high grounds represented by alluvial ridges and natural levees. In fact, there is a continuous record of palynological evidence of grassland for the last 15 kyr (Singh 2004–5, 5).

The Middle Ganga Plain (where the cluster of mesolithic sites to be discussed are situated) is characterized by a large number of horseshoe or oxbow lakes, most of which have now been filled up by natural and human agencies, and converted into farmlands. Their morphology, however, still allows them to be identified as extinct lakes (Pal 2002, 61).

The importance of the aforementioned water bodies needs special emphasis since not only did their vicinity serve as sites for early human habitation, but also because of the crucial role they play in our understanding of past ecological conditions.

For instance, deposits from Basaha Lake (Unnao), Misa Tal (Lucknow), Lahuradewa Lake (Sant Kabir Nagar district) have provided crucial multiproxy data for the reconstruction of palaeoenvironmental and palaeovegetational conditions in the Ganga Plain. This coupled with geochemical-mineralogical studies and stable isotopes in calcrete and teeth enamel have aided the reconstruction of the palaeoenvironment of the region for the last 80 kyr BP, of which the last 15 kyr BP are considered to be of high precision. Such investigations reveal that the period from 10.5–5kyr BP was a time of high rainfall, enlargement of lakes, wetlands, and aquatic plants. The period was also marked by the formation of numerous lakes and ponds with alluvial ridges and mounds due to tectonic activity. The period from 5–3.5 kyr BP was one of relatively dry climate, high siltation rates in the lakes together with a reduction in the number of lakes and ponds. A grassland-dominated landscape with water bodies and mounds prevailed. On the other hand, the period between 3.5 kyr BP till the present showed century-scale fluctuations in rainfall (Singh 2005, 23).

With regard to palaeovegetational trends, the data came from areas between Shahjehanpur in the west, and Sant Kabir Nagar in the east, and from the southern part of the Ganga Plain, and covered a time span of around 80 kyr BP. The reconstruction based on palynological studies of lake deposit profiles suggested that the Ganga Plain was open grassland with a few thickets. Locally there may have been some forested areas. The vegetation predominantly comprised C_4 type grasses (warm-season, tropical) with a small proportion of C_3 type trees (cool-season, temperate). The lakes supported aquatic and marshy vegetation. It was, thus, argued that during the late Quaternary, the Ganga Plain was grassland throughout its expanse with only few thickets of forests (Singh 2005, 23).

An earlier pollen analysis by H.P. Gupta (1976, 109–19) of the horseshoe lake (Khoalan Jhil) near the archaeological site of Sarai Nahar Rai also deserves mention. Palynological studies indicated that during the Holocene, in the area of Pratapgarh (where all the three mesolithic sites yielding rhinoceros remains are located), vegetation was grassland dominated with scattered stands of tree vegetation. Lakes which remained prominent during the middle to late Holocene were formed around 8,000 years BP, and had a prominent swampy zone (Singh 2005, 20).

It can then be postulated that grassy forests were certainly more widespread in the past, and gave shelter to a number of herbivores such as different species of deer, rhinoceros, and wild boar (Chattopadhyaya 1996, 464). Pollen analysis together with archaeozoological evidence testifying the presence of such species suggest a mosaic environment in the past—forest, grassland, and marshland (Chattopadhyaya 1999, 128). These forests may have been interspersed with thorny bushes and trees as may be inferred from the charred seed of jujube excavated from Damdama. Patches of *dhak* and *singhor* forest seen even today suggest a similar picture of such forests during the mesolithic (Pal 1994, 96).

The landscape that archaeology reconstructs for us, thus, seems to have been one ideal for a moisture-loving species like the Indian rhinoceros, which is rarely found more than 2 km away from water. We also know that oxbow lakes (which the region abounded in) and other water features are important for the animal in view of its need to wallow and feed on aquatic plants (Wilson and Mittermeier 2011, 179). It is then a matter of little surprise that archaeological sites in the region have yielded evidence for the presence of the animal in early contexts.

To now move on to an assessment of the faunal remains, one can begin with the cluster of mesolithic sites in the Middle Gangetic Plain. It may be relevant to mention that though the rhinoceros does not figure in the initial faunal surveys conducted by Alur (1980, 208, 211) for the sites of Sarai Nahar Rai and Mahadaha, subsequent investigations have much to add to our story in terms of the presence of the animal at all the three sites including Damdama.

Damdama: Perhaps the most ancient in this context is this site in Pratapgarh district of Uttar Pradesh. The antiquity of mesolithic Damdama is tentatively placed in the first half of the seventh millennium BCE. The two AMS ^{14}C dates suggesting an early Holocene antiquity for the site are 8,865±65 BP and 8,640±65 BP (Lukacs et al. 1996, 303). A detailed and systematic analysis of its faunal material was undertaken by P.K. Thomas, P.P. Joglekar, V.D. Mishra, J.N. Pandey, and J.N. Pal (1995, 29–36; 1996, 255–66).

Damdama has yielded bones of large mammals such as elephant, rhinoceros, gaur, wild buffalo, and possibly wild cattle constituting about 9.52 per cent in the total collection of identifiable bones at the site. Though the details of the bone remains of the species identified were not mentioned, there were 26 rhinoceros bones that accounted for 0.64 per cent of an assemblage, which comprised a total of 4,054 identified bone specimens occurring throughout the occupation at the site. Rhino bones, however, did not occur in all layers, and were found in greater numbers in the upper ones (Thomas, Joglekar, Mishra, Pandey, and Pal 1996, 257–9).

Thomas and his team, however, cautioned that the bones of large mammals need not be seen as evidence of their having been hunted for food. This, they argued, was unlikely in view of the technological level of the mesolithic population. This also perhaps explains why they were excluded from the list classifying species according to their live weight in order to look for a pattern of resource exploitation. Instead, it was conjectured that the carcasses or isolated bones were collected and utilized for making bone tools. Additionally, it was observed that the bones of large mammals were concentrated on the eastern part of the site (trenches SA–SD), and that these bones were well preserved without much charring or fragmentation. A majority of the bones also revealed cut marks leading to the inference that they were possibly cut off from the original carcass and dumped at one place near the settlement for further use (Thomas, Joglekar, Mishra, Pandey, and Pal 1995, 31–2; 1996, 257–8). These were, therefore, postulated to have been intentionally kept raw material for the preparation of bone tools and objects.

Sarai Nahar Rai: Our next site is situated about 15 km south-west of Pratapgarh on the bank of a now nearly dry horseshoe lake. A calcified bone obtained from the excavation yielded a radiocarbon date of c. 8395±110 BCE. Preliminary excavations conducted by G.R. Sharma of the University of Allahabad and his team at the site revealed hearths and floors yielding burnt clay lumps along with charred as well as uncharred animal bones and microliths. These bones indicated the presence of stag, bison, and rhinoceros (*Indian Archaeology 1971–72—A Review* [*IAR* hereafter], 49).

Mahadaha: Another mesolithic site in the Middle Ganga Valley, Mahadaha is situated 31 km to the north-east of Pratapgarh city in the Patti subdivision of Pratapgarh district, Uttar Pradesh. Excavations conducted over two seasons in 1977–8 and 1978–9 under the general supervision of G.R. Sharma revealed three complexes: the cemetery-cum-habitation complex, the butchering complex, and the lake complex. Unlike Sarai Nahar Rai and Damdama, the dating of Mahadaha is fraught with uncertainty. This is because the calibrated range of three radiocarbon dates obtained from there range from 1385–885 BCE to 2675–2515 BCE suggesting a comparatively late date for the site. At the same time, as some archaeologists have pointed out, the archaeological material from Mahadaha is identical to that of the other two sites. Hence, it may well be much older than what the radiocarbon dates suggest (Misra et al. 2003, 24).

A detailed examination of the faunal assemblage at Mahadaha by the Department of Ancient History, Culture, and Archaeology, University of Allahabad, in association with P.P. Joglekar of Deccan College, Pune, furnished fresh insights. The 'cemetery-cum-habitation' area was referred to as the Western Area by the excavators, the area designated as the 'butchering complex' was called the Eastern Area, and the area adjacent to the lake was termed the Lake Area.

Many bones of large mammals, including the elephant and the rhinoceros with live weight of over a ton, were reported in the mesolithic and post-mesolithic context at Mahadaha. In the case of some large mammal bones, it could not be determined whether they belonged to the rhinoceros or the elephant. Hence, they were treated at the second level of identification and taken to represent the rhinoceros, elephant, or hippopotamus (Joglekar et al. 2003, 75).

Of the clearly identifiable remains of the rhinoceros, teeth, mandible, and a vertebra were found from all the three areas at Mahadaha. A cervical vertebra of a senile (old) whose centrum plates were fused and fairly worn out was found from the Eastern Area. Two teeth found from the embankment of the canal were completely charred and broken. A mandible fragment devoid of teeth found from the disturbed deposit in the embankment area was fossilized and had a thick cortex. Two teeth fragments found from the Lake Area were neither charred nor fossilized, while one tooth fragment was found to be completely charred (Joglekar et al. 2003, 75).

Though the faunal component of the economy had a diverse array of species, the contribution of rats/rodents along with possibly the rhinoceros, hippopotamus, and elephant to the food economy was considered doubtful (Joglekar et al. 2003, 109). A statistical examination of the relative importance of different groups in the three areas at Mahadaha showed that the abundance of animal species varied considerably in different parts of the site, and in different layers (Joglekar et al. 2003, 111, 115). The reason for this differential distribution was postulated to be the result of localized activities of specialized hunting/scavenging groups inhabiting the site (Joglekar et al. 2003, 111).

Deer contributed to about half of the total mesolithic faunal assemblage. The two groups most scantily represented at the site included that of the canids, felids, and bears, and that comprising big mammals such as the rhinoceros, elephant, and hippopotamus. However, the latter fared better than the former both in terms of its distribution in the three areas as well as in the layers. While the evidence is scanty, what deserves attention is the fact that the biggest cluster of the remains of these mega mammals comes from the Eastern Area or the butchering complex. The observation, overall, was that a wide spectrum of animals were utilized during the mesolithic phase, and that animal resources were used to their maximum capacity. Animals were not only being consumed, but their bones, teeth, and antlers were also being used for making various tools. Big mammals such as the wild buffalo, gaur, rhinoceros, hippopotamus, and elephant, however, were not considered to have contributed to the food economy, and their carcasses were postulated to have been scavenged to obtain meat for consumption and bones for fashioning tools (Joglekar et al. 2003, 109, 115–16).

The faunal assemblages from Damdama, Sarai Nahar Rai, and Mahadaha were also studied by Chattopadhyaya (1991, 1996, 2002) who suggested a hunting-gathering-fishing economy for this area. Like others, he also identified a variety of animal taxa including various species of deer and antelopes, large bovids including the gaur (*Bos bibos gaurus*), the wild buffalo (*Bubalus* cf. *arnee*), and the now extinct *Bos* cf. *namadicus*, rhinoceros, elephant, wild boar, *Equus* cf. *namadicus*, a few carnivores, aquatic resources including several species of freshwater turtles, fish, molluscs, and birds. Though Chattopadhyaya (2002, 376) attempted to assess the relative dependence on various resources for the three mesolithic sites of the Ganga Valley, nowhere did he discuss the possibility of large mammals such as the rhinoceros and the elephant being considered for dietary purposes.

From the evidence available, it is clear that the economic exploitation of the rhino hearkens back to ancient times. However, reservations regarding the flesh of big mammals such as the elephant and the rhinoceros having been consumed need to be examined from a broader perspective. For this, we will need to take into account all the evidence, including that derived from rock paintings which will subsequently be discussed. It will also be pertinent to argue that if the technological level of the mesolithic population permitted the utilization of the bones of a dead rhino (scavenged or hunted), then one wonders as to how the same technological level could have prevented them from utilizing its meat as well. After all, a large mountain of meat offered by a rhino would not be easily forgone since it is known that despite its hard exterior, a rhino is not so difficult to hunt. In fact, notwithstanding its tough appearance, rhino hide is known to be quite tender at places, making the animal far more vulnerable than it looks (Divyabhanusinh, personal communication).

Moving west, Gujarat is another state where the mesolithic/microlithic culture is well understood on the basis of excavations, especially at Langhnaj in its northern part. However, it is crucial to note that the archetypal art of manufacturing microliths was not confined to northern Gujarat, but extended to other parts as well. Excavations at the rock shelter site of Tarsang in the Panch Mahals district—along with a mesolithic deposit consisting of microliths and animal bones found sealed on a mound known as Sai-no-Tekro at Kanewal (Kheda district)—underline that the mesolithic culture in Gujarat was widespread, and flourished even in the hilly area of eastern Gujarat as well as in central Gujarat (Sankalia 1987, 28).

I now move on to look at the sites of the region yielding traces of the rhinoceros in mesolithic/microlithic contexts. As the discussion of sites in and around Kanewal and at Valotri will reveal, these microlith-bearing sites are likely to be late, since they also contain chalcolithic pottery. Like the hunter-gatherers at Langhnaj, subsequently discussed in this section, these too appear to be hunting-gathering communities that are contemporary with chalcolithic cultures.

Explorations conducted by K.N. Momin and his colleagues at Maharaja Sayajirao University of Baroda between 1972–3 in Bhalbara (Cambay taluka) in Kaira district of Gujarat revealed many late Stone Age and chalcolithic sites. Most of the late Stone Age sites were discovered around Kanewal, a large natural lake covering an area of about 11 km, and situated at a distance of 20 km to the north-west of Cambay (Momin et al. 1973, 801).

Khaksar: Situated on the southern side of lake Kanewal, the site yielded microlithic tools, pottery, and semi-fossilized bones that included two pieces of the cervical vertebrae of the rhinoceros (Momin et al. 1973, 801).

Valotri: This village, 8 km north of Tarapur on the Anand–Cambay railway line, also yielded another cervical vertebra of the rhinoceros. The finds from both Khaksar and Valotri included chalcolithic pottery as well as microlithic tools (Momin et al. 1973, 801).

Kanewal: Situated in the Khambhat taluka in the Kheda district of central Gujarat, animal remains at the site including those of the rhinoceros in a fairly preserved state were examined by D.R. Shah. Regrettably, no details of the remains were provided, and little was said beyond that the rhinoceros was present in Gujarat in the period from 8000 BCE to 1200 BCE. Shah underlined the existence of the animal in Gujarat at least up to the sixteenth century, and attributed its extinction from this region to human intervention rather than natural factors.

Based on the faunal assemblage at the site comprising wild species such as the chital (*Axis axis Erxl*), barasingha (*Cervus duvauceli cuvier*), and nilgai (*Boselaphus tragocamelus Pallas)* along with the rhinoceros, the landscape envisioned included swampy areas and thickets in favourable places, and grasslands in the areas away from water as well as on and around the dunes. This evidence was used to suggest that environmental conditions were more or less the same in those days as they are now (Shah 1980, 70).

Langhnaj: Far more telling evidence comes from this site, situated at a distance of 59 km from Ahmedabad in northern Gujarat between the Rupen and Sabarmati Rivers on a fossil sand dune called Andhario Timbo. On the basis of circumstantial evidence, H.D. Sankalia proposed a date prior to 2500 BCE for the initial phase of the Langhnaj or north Gujarat microlithic culture, and a date of 2000 BCE for its later phase (Sankalia 1965, 9). Only a single ^{14}C date (TF–744, 2040±110 BCE) based on charred bone samples from the lower and middle depths is available (Misra 1973, 66).

Regarding the antiquity of the site, it is also necessary to note that sites of the semi-arid region of Gujarat and Rajasthan such as Langhnaj and Bagor differ from the Ganga Valley sites (Lekhahia, Mahadaha, and Sarai Nahar Rai) in their proximity to intensive agriculturists (Lukacs 1990, 183). This assertion was made in view of the contemporaneity between the late 'mesolithic' occupation of Langhnaj and the Harappan site of Lothal, 100 km to the south of Langhnaj. The occurrence at Langhnaj of a 98.12 per cent pure copper knife, black-and-red ware typologically similar to sherds from Lothal, and Harappan disk beads was said to strongly suggest interaction in the form of exchange between the occupants of these two sites. This interpretation viewed Langhnaj as a campsite of nomadic hunter-gatherers or pastoralists whose movements brought them into repeated contact with the urban agriculturalists (Lukacs 1990, 183). What is also interesting is that despite its contemporaneity and possible cultural contact with agricultural communities, the faunal spectrum at the site does not include domestic animals.

Rhinoceros remains at the site examined by Clutton-Brock (1965, 9–10) included a left scapula, a right humerus, a talus, and a fragment of molar tooth (Figures 2.1, 2.2, and 2.3). Another scapula after cleaning and mending was found to have been used as an anvil for the production of microliths. A comparison of the scapulae of the three surviving Asian species with the evidence from Langhnaj led to the two rhinoceros scapulae at the site being attributed to *Rhinoceros unicornis* (Clutton-Brock 1965, 10).

It was argued that though no longer found in western India, *unicornis* was the only species of the genus likely to be found in Gujarat. Its presence, together with other swamp-loving animals such as the wild boar (*Sus scrofa cristatus*), in the faunal assemblage at Langhnaj was considered to suggest that perennial water was available in northern Gujarat during this period. It was also pointed out that while the rhinoceros and the hog deer (*Axis porcinus*) may be found in high grass and forested country, the remaining mammals at Langhnaj were all inhabitants of open scrub or grasslands (Clutton-Brock 1965, 37). The landscape, thus, seems to have been a mosaic comprising patches of dense vegetation along with open country. While this was the topography Clutton-Brock reconstructed for us based on the faunal remains, Zeuner (1963a, 19–28) used the same

Figure 2.1 Rib of rhinoceros, other bones, and stones, Langhnaj
Source: After Sankalia 1965, plate XI a. ©Deccan College Post-graduate and Research Institute, Pune.

Figure 2.2 Shoulder blade, other bones, and stones, Langhnaj. These are rare images since it is not usual to encounter visuals showing the primary context of faunal retrievals

Source: After Sankalia 1965, plate XI b. ©Deccan College Post-graduate and Research Institute, Pune.

Figure 2.3 Three fragments of rhinoceros rib bones, Langhnaj

Source: After Sankalia 1965, plate XII a. ©Deccan College Post-graduate and Research Institute, Pune.

evidence to recreate the environment at Langhnaj with a somewhat different conclusion.

Given that some species are highly indicative of certain types of environment, Zeuner's approach focused on the ecological implications of the presence of the rhinoceros at Langhnaj. What interested him was the fact that though no longer found in western India, the frequency of depictions on seals suggested that the animal must have been present in the Indus region in Harappan times. He argued that there are reasons to believe that Langhnaj is contemporary with or slightly later than the Harappan period, and that osteological evidence proves the presence of the rhinoceros in a dry part of western India. For Zeuner, the presence of the rhinoceros in a low-rainfall area with dune formation posed a problem that threw up three possibilities: that the rhinoceros was in those days able to live in open country, or the country was more densely covered with vegetation, or that the country was dry and open at Langhnaj, while river valleys and nalas contained a dense reed jungle suitable for the animal.

What ensued in Zeuner's investigation is enthralling. He demonstrated how a study of the skull forms of Recent and Pleistocene rhinoceroses can help reconstruct the mode of life of the animal and the environment in which it survived. According to him, it can be shown that the Recent species are accustomed to carrying their heads in different ways in accordance with their usual manner of feeding. The 'average carriage of the head' is exhibited by individuals walking or standing at ease, but this is not the position in which the head is carried most frequently in the course of the animal's life. Nevertheless, the skull is constructed on the basis of the average carriage, which was demonstrated statistically (Zeuner 1963a, 26).

Zeuner illustrated how the Indian rhino (Recent) carried its head horizontally, and largely consumed the foliage of high-growing plants. He compared the outlines of the skulls of different species, all arranged on the vertical axis of the foramen magnum (the opening at the base of the skull through which the spinal cord connects with the brain by integrating with the medulla oblongata, its lowermost portion). The angle of the foramen magnum informs scientists on whether the animal's head was in a horizontal position, as is the case in most four-legged beings, or in a vertical position, like in bipedal animals.

Through figures, Zeuner showed the elevated skull of *Rhinoceros unicornis*, the Indian primeval forest form. In the case of the *Diceros bicornis*, an animal of the forest steppe that feeds on leaves as well as grass, the average carriage of the head is intermediate between the forest forms and the grazing ones. Similarly, since *Ceratotherium simum*, the African white rhinoceros, lives entirely on grass, its head is constructed in a way so as to stretch the facial part with the mouth towards the food. Thus, forest rhinoceroses were said to carry their heads more or less horizontally. The head is inclined in case of those of the steppe, and strongly inclined in those of the grass steppe, and the skulls are constructed accordingly. Zeuner regarded this rule to be generally applicable to Recent as well as fossil forms, enabling us to reconstruct the biotope and the climate in which they lived.

Thus, according to Zeuner, the Indian species was never a dweller in open country. He returned to the problem of Langhnaj to postulate that the animal must have required dense jungle in the past as it does today. Since the evidence afforded by the fossil soil of the dune of Langhnaj suggested only slight vegetation in a relatively dry climate, he was inclined to believe that in northern Gujarat as also in the Indus region, rhinos were confined to the river valleys and their nalas, where at that time, dense vegetation was available all the year round.

Drawing heavily on Zeuner, V.N. Misra (1973, 58–72) in an attempt to reconstruct the palaeoecology and palaeoclimate of the microlithic cultures of north-west India, suggested arid climatic conditions for Langhnaj around 3875±105 BP (calibrated to 2550–2185 BCE). This perspective was, however, contested in view of the argument that if the climate was indeed arid, the narrow stretch of the Luni River Valley would have been like a linear oasis, and, thus, unlikely to have supported rhinos (Chattopadhyaya 2002, 373). Very pertinently, Chattopadhyaya also drew attention to the fact that the occurrence of the rhinoceros at Langhnaj is not an isolated or regionally restricted phenomenon. It is attested by the retrieval of its remains from a number of contemporaneous Harappan sites such as Lothal, Surkotada. Kuntasi, Khanpur, Shikarpur, and also possibly Oriyo Timbo, covering the whole of Gujarat and Saurashtra. Moreover, he argued that if one accepts Zeuner's identification, then the presence of the Indian wild buffalo along with the rhinoceros at Langhnaj would certainly imply wetter conditions and greater carrying capacity during the period under consideration. The available faunal evidence, according to him, would indicate a mosaic topography of savanna and forest cover with interspersed wetlands in northern Gujarat during the period of occupation at Langhnaj (Chattopadhyaya 2002, 373).

The human exploitation of the rhino at Langhnaj was another aspect that received careful attention. Zeuner painstakingly demonstrated how the deliberate pits on the shoulder blade of the animal indicated its use as an anvil for making microlithic tools (Figure 2.4). Based on the predominance of game animals in the food debris at the site, he inferred that the economy must have been largely dependent on them. Whether the rhinoceros was a possible inclusion in the diet can be speculated since he mentions the animal as the most 'remarkable' of the game animals in the faunal composition at the site (Zeuner 1952, 129). Significantly, Clutton-Brock (1965, 37) also concluded that all species except the mongoose and the wolf were part of the food economy. In this, one wonders if she was reinforcing what Zeuner had ambiguously hinted at thirteen years earlier.

Osteological evidence testifying the exploitation of the animal can be further corroborated by turning to the rock shelters that have served as canvas to humans since upper palaeolithic times. Though the chronology of rock paintings has been problematic, these have invaluable clues to render when it comes to human–animal interactions since the subject matter of the earliest portrayals revolved around animals. Shown in different situations such as standing, grazing, running, or as part of hunting scenes, animals seem to have been uppermost in the minds of early humans because of the crucial role they played in subsistence (Varma 1996, 330).

That the rhinoceros captured the artist's imagination from early times is apparent in the way it found favour as a subject of rock paintings. The earliest representations divulge an association that unquestionably transcends mere acquaintance. Our evidence comes primarily from Mirzapur in Uttar Pradesh, the second-largest rock art centre in the country after Madhya Pradesh, and also from Rajasthan and Gujarat.

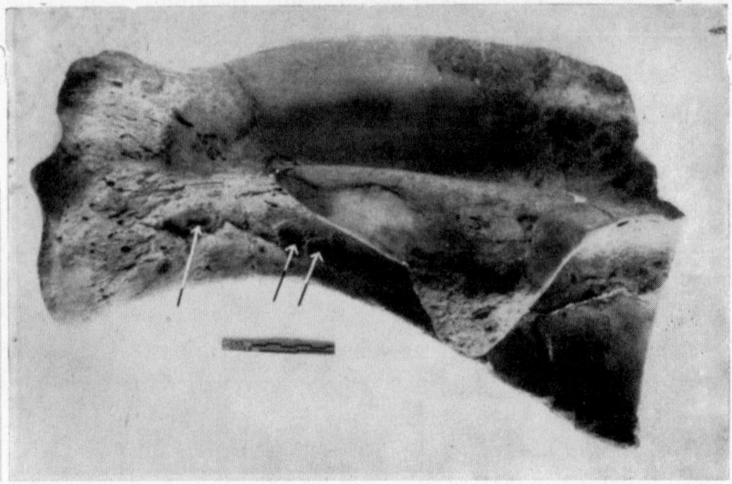

Figure 2.4 Shoulder blade of rhinoceros, Langhnaj
Source: After Sankalia 1965, plate XIII (1). ©Deccan College Post-graduate and Research Institute, Pune.

Cockburn's (1883) report was the first scientific paper on Indian rock paintings, wherein he described a painting of a rhinoceros (which he initially identified as a boar) hunt in a shelter near Ronp village in Mirzapur. He later retracted from his erroneous identification after encountering well-preserved rhinoceros hunt scenes at Ghormangar and Harni-Harna in the region. Cockburn (as mentioned earlier), however, erred in the dating of these paintings when he tried to prove that they were of more recent antiquity, being not more than 300 years old.

Notwithstanding his miscalculation, what is significant is Cockburn's sense of how intensely the landscape had changed since then. He remarks:

> It requires ... abstraction to conceive that this now semi-arid region largely productive of reh and usar was covered with forest so recently, but such was without doubt the case ... The change effected in the climate has undoubtedly been great and everywhere in the plains of N.W.P. dried watercourses and rivulets, barren

> ravines and saline efflorescence, attest to the slow but certain progress of aridity and exhaustion. As regards the precise locality where the drawing of the rhinoceros hunt was found, sal forests yet exist there in patches and the occurrence of numerous characteristic drawings of the Bison (*B. gaurus*), a forest loving animal, renders it nearly certain that primaeval forest existed at the time. In the swamps engendered by these forests I would suppose the rhinoceros depicted to have lived. (Cockburn 1883, 64)

The antiquity of the rock paintings of Mirzapur like those from other sites in the subcontinent has been a subject of debate. Rakesh Tewari's (1990, 37–43) investigations of the superimpositions revealed twelve stages for the rock paintings of the region. On the basis of general observations, the identified phases were divided into three major groups, where paintings represented a hunting-gathering society (mesolithic—phases I–IV), a pastoral life (neolithic/chalcolithic—phases V–VII), and scenes from the historical period (phases VIII–XII). A reasonable date for the earliest rock paintings of the region was considered to be 6000 BCE, while the entire group was placed between c. 6000 BCE and 1700 CE.

Explorations revealed more than fifteen figures of the one-horned rhino in the rock shelters of Gochara, Kerwa, Kauva-Khoh, Ghormangar, Panchmukhi, Soraho, Duara, and Karihawa. The sites were found restricted to a particular area of Mirzapur, roughly near the Son River in the southern region of the central part of the district. Rhino portrayals were encountered only in the paintings of earlier phases suggesting that the animal may have become extinct in the area in the later period. In most depictions, as at Matahawa, Panchmukhi, Ghormangar, and Kauva-Khoh, the animal is shown being hunted (Tewari 1990, 12–13).

In an illustration on the ceiling of the Matahawa rock shelter, two rhinos—one grown up and a young one—are shown being attacked from behind by hunters holding barbed spears or harpoons (Figure 2.5). The hunters seem to be approaching at a fast pace, so much that they almost seem to be flying in the air (Tewari 1990, 23).

Another portrayal from Panchmukhi Ronp shows a rhinoceros struck by a hunter (holding a bow) from behind (Figure 2.6). As a result of the impact of the shot, the animal seems to have come to

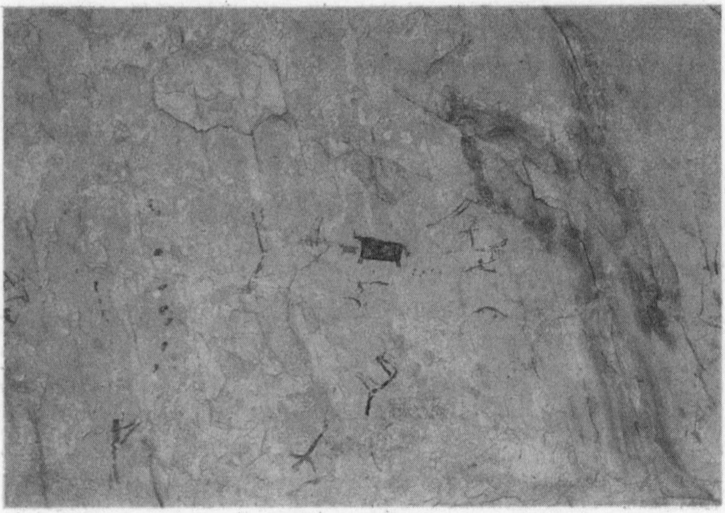

Figure 2.5 Rhino hunt, Matahawa, Mirzapur, Uttar Pradesh
Source: After Tewari 1985.

Figure 2.6 A mesolithic(?) rhino-hunt scene, Robertsganj (Panchmukhi Ronp, Mirzapur, Uttar Pradesh)
Source: © Indira Gandhi National Centre for the Arts (IGNCA), New Delhi.

a halt, also suggested by its lowered head. There is another animal within the frame in front of the rhinoceros.

It is, however, the rhino-hunt scene at Ghormangar in Mirzapur which has been considered the most natural and the rarest amongst such depictions found in the rock shelters and caves of India (Tewari 1987, 26). Tewari's (1987, 25–9) description of this famous scene painted in dull ochre with a thin but dark outline is a reappraisal of the scene published by Cockburn and others after him.

As can be seen, in the original portrayal, a wounded one-horned rhinoceros is surrounded by eleven hunters armed with multi-barbed harpoons (Figure 2.7). The harpoons appear to have been made with wooden shafts and stone blades leading Tewari (1987, 27) to attribute this scene to the late upper palaeolithic or mesolithic. The animal has been pierced with five harpoons in different parts of the body, three on the back, one on the hip, and one below the knee. Its open mouth communicates its anger at being attacked as well as its anguished plight because of the wounds inflicted. While the swift-attacking postures of the hunters effectively portray the desire to overpower the quarry, it also seems to suggest that hunting accidents where hunters often became victims

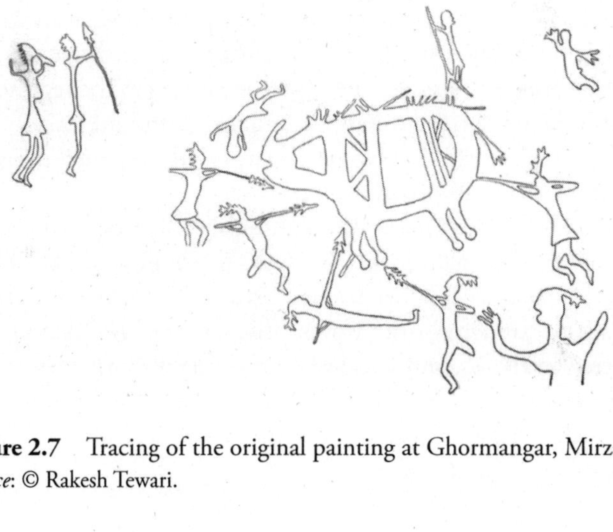

Figure 2.7 Tracing of the original painting at Ghormangar, Mirzapur
Source: © Rakesh Tewari.

of their chosen preys were anticipated. The perils of hunting are unambiguously conveyed when one hunter is shown tossed in the air by the horn of the rhino.

Writing on the rock art of southern Uttar Pradesh with special reference to Mirzapur, Varma (1984, 207) noted the popularity of the rhinoceros as a subject of portrayal, though according to him, single rhino figures were rare. Rather the animal was shown either with a rider or as part of a hunting scene, being attacked by men with bows, arrows, or spears. Subsequently, Varma (2012, 62) shed light on the animals represented in the region, and found the rhinoceros to be the second most popular animal in the painted rock shelters of the north Vindhyan region.

Neumayer (1993, 114) similarly draws our attention to the rock paintings of the northern Vindhyan range where rhino depictions are very common in contrast to regions further west in the Vindhyas, where rhinos are rare. While there are several extensive scenes showing hunts of rhinos, far more telling are the scenes from Kerwaghat in Mirzapur (Figure 2.8). Perhaps the only representation of its kind, a remarkable portrayal presumably belonging to the mesolithic period shows the butchering of a big stylized rhinoceros in the vicinity of what appear to be huts. The human figures seem to have brought vessels and containers to carry the meat away from the site. The next illustration again shows two stylized rhinos of varying sizes (the smaller one seems to have been left unfinished). Equally significant is the depiction of the incomplete rhino figure next to what looks like a doorway (Neumayer 1993, 115). It has been argued that while it may not be possible to assign conclusive meanings to the representation itself, what can certainly be postulated is the presence of the animal in the vicinity of human settlements in the region during this period (Bose 2018, 40).

Closely resembling the rock paintings of Mirzapur in Uttar Pradesh are the rock paintings of Kaimur. The similarity is attributed to the fact that the Vindhyan range extends right from Mirzapur district up to Sasaram in the Rohtas district, and the Kaimur hills and plateau in the Kaimur and Rohtas districts of Bihar form a part of the same range. Even in the matter of antiquity and relative dating, the rock paintings of Kaimur run parallel to the Mirzapur cluster (Anand 1991–2, 63–4).

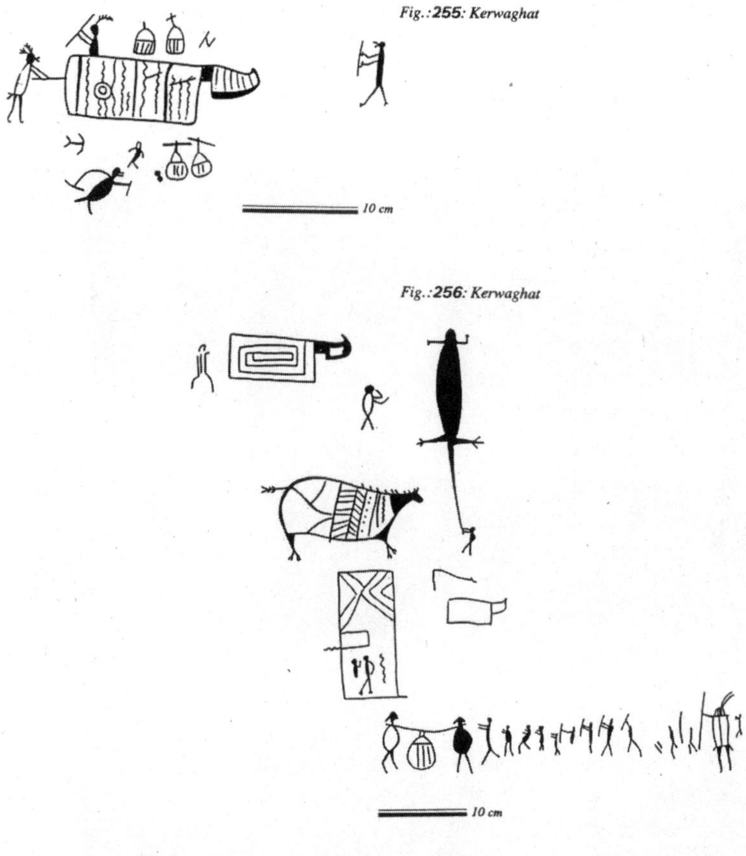

Figure 2.8 Stylized mesolithic(?) rhino depictions, Kerwaghat, Mirzapur, Uttar Pradesh
Source: After Neumayer 1993, 115. © Erwin Neumayer.

Rhino figures were noticed in the rock shelter in Badki Goriya, while a big and a baby rhino were depicted in the Mithaiya Mand rock shelter. These portrayals were taken to suggest the presence of the animal in the region during the period in which the paintings were executed (Anand 1991–2, 60).

In and around the area of the Kaimur hills and the plateau which extends along the southern boundary of the Kaimur district forming the eastern limit of the Vindhyan range, rock paintings document

Figure 2.9 Mesolithic-Chalcolithic animal figures and designs, Isco, Hazaribagh, Jharkhand. The rhinoceros with its prominent horn is faintly outlined in white at the bottom of the panel
Source: © Indira Gandhi National Centre for the Arts (IGNCA), New Delhi.

human activities right from the mesolithic period till later times though their exact date is yet to be determined (Figure 2.9). A brief report on prehistoric rock paintings in Bihar observed that animals seem to have been the most obvious concern of the early artists, and precisely those which were hunted, and around which their lives revolved. Wild cattle, bison, deer, boar, rhinoceros, bear, birds, reptiles, and other animals are depicted running, leaping, and grazing (Prasad 1995–6, 88).

Moving west, explorations along the banks of river Bilas, district Kota, Rajasthan, also brought to light thirty-six rock shelters that contained paintings. While some were assigned a prehistoric antiquity, others were attributed to historical times. Drawn on the wall surface and projected ceilings of the rock shelters, the paintings were executed in monochrome, in red and dark red colour, and depicted animals like the bear, monkey, deer, rhinoceros, dog, ox, buffalo, horse, fox, scorpion, camel, antelope, elephant, humped bull, and peacock (*IAR 1981–82*, 56). The paintings depicting wild animals were attributed to the prehistoric period (Misra 2007, 348).

Explorations in the Panchmahals district of Gujarat by Sonawane brought to light several rock shelters, but only the ones found near Tarsang village, situated about 30 km to the north-west of Godhra, the district headquarters of the Panchmahals, revealed rock paintings together with associated material remains of the mesolithic and historic periods. The rock shelters located within the Maheshwari Hill as well as in the north-eastern small granite hillock known as Nano Dungar housed rock paintings (Sonawane 2002, 71–2).

At Tarsang, probably the earliest rock painting is housed in one of the south-eastern shelters located on the lower slopes of the Maheshwari Hill, and consists of line drawings of several deer, two hunters, and an animal like a rhinoceros (Figure 2.10). The deer representations consisting of the neck and head portions were considered reminiscent of the mesolithic representations of the animal in the rock shelters at Bhaldaria, Bhimbetka, Kharwai, and Raisen in central India. Significantly, one of the rock shelters at Tarsang yielded charred and uncharred bones of deer from the mesolithic level along with microlithic artefacts, but unfortunately,

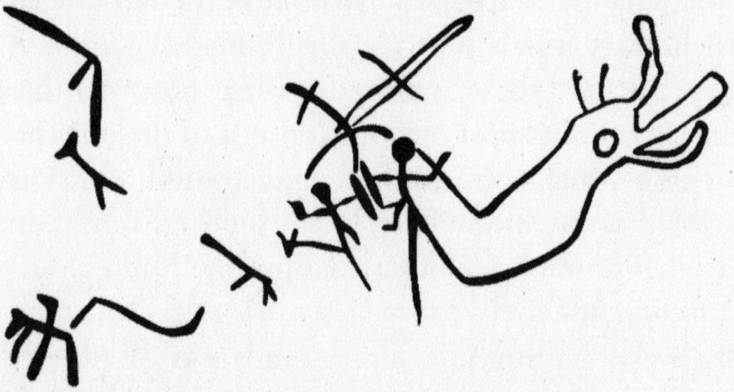

Figure 2.10 Mesolithic hunting scene, Tarsang, Gujarat
Source: After Sonawane 2002, 71.

the depiction of the rhinoceros is not corroborated by the retrieval of its material remains indicating its existence at the site (Sonawane 2002, 72). However, it is imperative to remind ourselves that rhinoceros remains found in mesolithic/microlithic contexts at Langhnaj and Kanewal, and at several sites in the Harappan context clearly establish Gujarat as a rhino habitat from the earliest times till about the sixteenth century CE.

To reflect a little on the painting itself, it can be safely said that the rather peculiar portrayal of the rhinoceros dominates the scene. That it is indeed one can be argued on the basis of the depiction of a horn-like projection. What seems noteworthy in this illustration, which seems to be a hunting scene, are the two figures that appear to be holding shield-like objects, possibly in their defence. What is intriguing, however, is the absence of any weapons of offence. The animal seems upturned, and the human figures in motion.

The Neolithic

Chirand: This site in Saran district lies in the vast alluvial tract of north Bihar near the confluence of the rivers Ghagra and Ganga. Lala Aditya Narain's studies (1970, 1972) reconstructing the neolithic way of life at the site attempted to offer an environmental analysis of the area. It was postulated that the environmental setting of Chirand was such that the monsoon type of climate in the area helped the growth of luxuriant jungle in which animals of open country must have thrived. Having the Ganga in its vicinity, the area was always susceptible to flood. The extra water in the river possibly led to the formation of water pools facilitating the growth and survival of aquatic species as well as of other animals requiring moist habitats. It is in these environs then that we must go in search of the rhino.

The presence of the animal at the site was confirmed by Nath and Biswas (1980, 122) through the finds of the proximal fragment of ulna with semilunar notch and broken olecranon process, the left tibia without proximal and distal epiphysis, fragment of upper molar tooth, and a complete left humerus. A moist and swampy past climate as compared to the present-day dry conditions was inferred from the presence of the rhinoceros at the site (Nath and Biswas 1980, 122). Though Nath and Biswas (1980, 115) assigned an antiquity of c. 1700 BCE to the neolithic phase at Chirand, the beginning of occupation at the site has been suggested to have been even earlier than the middle of the third millennium BCE (Chakrabarti 1999, 244).

The Harappan Civilization

Before I set out to collate the evidence telling us about Harappan interactions with the one-horned rhino, it is important to have an overview of the environmental milieu in the third and the second millennia BCE.

Given the conflicting strands within the debate regarding the issue of environment during the Harappan civilization, a consensus seems elusive. What has been postulated, however, is that the Indus region experienced a series of climatic changes during the Holocene. The questions that have essentially engaged archaeologists as well as Quaternary scientists concern the extent, duration, timing, and cultural responses to these climatic changes (Possehl cited in Shaffer and Lichtenstein 1989, 120). It has, however, been cautioned that climate-based reconstructions can prove inadequate 'when the evidence is brought to bear on the environmental repercussions generally and in local settings specifically', that is to say, there may be some disparities between local evidence and the more general reconstructions of the palaeoclimate (Wright 2010, 38–9, 41). Against this backdrop, I look at some positions that have emerged from palaeoenvironmental investigations undertaken for regions of the subcontinent relevant to the study area.

Ecology and Environment in the Third and Second Millennia BCE

The complexity in reconstructing ancient environments becomes apparent when we look at how the presence or absence of particular animal species has been widely treated as ecological markers. In the case of the Harappan civilization, for instance, the preponderance of the rhinoceros, and compared to it, the relative paucity of the horse can scarcely be missed. Manuel (2004–5, 21–6) set out to solve this puzzle, which according to him is intriguing given the fact that in any habitation, horses are likely to be more common than the rhinoceros. After illustrating how the animal 'covered a large ground from association with deity, to entertainment and food' (though apart from mentioning the retrieval of rhino bones from Harappan sites, he does not cite specific evidence for the animal being used as food), and in view of the numerous graphic depictions, Manuel argued that the animal could not have pervaded all these realms of Harappan life unless the Harappans were at least occasionally seeing it. The preponderance of the rhinoceros and the paucity of horses in the larger part of the Mature Harappan period was then attributed primarily to the humid Harappan environment which nurtured a 'terra firma' interspersed with patches of swamps and ponds besides grasslands and forests.

On the other hand, Chitalwala (1990, 80) has argued that though it is generally assumed that rhinos favour grassy lands with swamps, and do not prefer a dry environment, the animals actually exhibit a lot of flexibility when it comes to adapting to a particular ecological backdrop.

It is in view of such conflicting interpretations that it becomes crucial to turn to palaeoclimatic records in order to understand the factors that made for favourable rhino habitats in areas where the animal is now extinct. The available data, we know, has been used to support arguments both for and against climatic change within the zone of the Harappan civilization. While a detailed survey of the theories supporting either position would be out of context here, my emphasis will be on some of the evidence which points to an already arid climate during the Mature Harappan period. The attempt will be to reconstruct a landscape permitting the survival of the moisture-loving Indian rhinoceros in Harappan times.

We know that while older theories suggested wetter conditions during Harappan times (Marshall 1931; Vats 1940), these were effectively critiqued by later archaeologists, hydrologists, and botanical experts (Raikes and Dyson 1961; Chowdhury and Ghosh 1951).

Nevertheless, the theory of a wetter climate was revived by Gurdip Singh (1970–1, 70–6), a plant palaeoecologist, when his sedimentological investigations of the northern salt lakes of Rajasthan (Lunkaransar, Didwana, and Sambar) showed significant fluctuations in precipitation based on pollen sequences dated by radiocarbon. These fluctuations were then correlated with the origin, efflorescence, and decline of the Harappan civilization. As per Singh's pollen graph, the environmental sequence was grouped into six phases, of which we are concerned with Phase IV ascribed to c. 3000–1000 BCE. While sub-phase IVa (3000–1800 BCE) was associated with a sudden change to wetter climatic conditions, the onset of aridity during sub-phase IVb between c.1800–1500 BCE was argued to have resulted in the weakening of Harappan culture in the arid and semi-arid parts of north-west India.

Much has come along since then in terms of theories arguing for a wetter Harappan climate as well as an arid one. Michel Danino (2015, 40–2) gives an overview of the theories supporting either position. However, since my aim here is to juxtapose the evidence for the presence of the rhinoceros at Harappan sites with theories postulating an arid Harappan climate, the focus will be on the latter.

Based on a recalibration of the dates offered by Singh, Jim G. Shaffer and Diane A. Lichtenstein (1989, 120–1), for instance, argued that sub-phase IVa actually began at c. 4000 BCE and reached its mid-point by c. 3100 BCE, or 600 years before the generally accepted recalibrated date for the appearance of Harappan sites,

c. 2500 BCE. The Harappan phenomena, therefore, according to them, appeared during a period of increasing aridity, coinciding with Singh's sub-phase IVb, and not during a period of increased rainfall.

A later sedimentological study of lake Lunkaransar in the Thar Desert also argued that the environment and climate of the past 5000 ^{14}C years were similar to those of the present (Enzel et al. 2000, 234). The data showed that the lake had completely dessicated around 4800 ^{14}C yr BP, and that the Harappan civilization began and flourished in this region 1,000 years after the dessication of the lake during an arid climate (Enzel et al. 2000, 226). This chronology clearly defied any relation between the proposed drought that caused the dessication of the lakes and the collapse of the Indus culture, as the lakes dried out more than 1,500 years earlier. It was emphasized that the Indus civilization flourished mainly along rivers during times when north-western India experienced semi-arid climatic conditions similar to those at present (Enzel et al. 2000, 235).

Similarly, M.B. McKean, who studied the palynology of Balakot, also inferred that 'there is nothing in the Balakot pollen data, which might suggest that the climate during the proto-historic period in Las Bela was decidedly wetter than at present'. This, therefore, serves as reinforcement for the evidence from Rajasthan when dates are calibrated, indicating that the mid-Holocene wet phase was more or less over by the time of the establishment of the Mature Harappan civilization of the second half of the third millennium BCE (cited in Madella and Fuller 2006, 1292).

From pollen sequences we can also turn to the study of soil stratigraphies. Ronald Amundson and Elise Pendall's (1991, 13–27) study of the same from Harappa, for instance, is a case in point. As an indicator of pre-Harappan environmental conditions, basic conclusions of the laboratory tests revealed either a very arid, sparsely vegetated site (matching presumed latest Pleistocene conditions) or a nearly pure C_4 flora, indicative of a tropical grassland (presumed early Holocene conditions).

Again, soil stratigraphies of the Upper Indus region showed that moister climates prevailed around 8000 to 5000 BCE, followed by a period of diminished precipitation and drier conditions

(c. 5000–3000 BCE), the latter coinciding with the time when increasing numbers of settlements were present on the alluvial plains. The data for the period from c. 3300 to 1500/1300 BCE, and the peak period of occupation for settlement at Harappa and along the Beas (c. 3300–1900 BCE), suggested frequent fluctuations in climate and larger-scale stream migrations and precipitation levels than at other times. This was followed by a surge in precipitation, but by then the region had already been abandoned (Wright 2010, 39–40).

In view of these theories, it is possible to say that a substantial volume of evidence points to a dry climate during the pre-Harappan and Harappan phases. However, it is crucial to underline that by virtue of their riverine situation, several cities of the Indus civilization were in floodplain environments where rainfall was of minimal importance (Thapar 1977, 69). Cities such as Mohenjodaro, Chanhudaro, and Kot Diji prospered with Indus in their vicinity as did Harappa on the Ravi, Ropar on the banks of the Sutlej, Kalibangan on the Ghaggar/Hakra, while Rangpur and Lothal benefitted from the Bhadar and Sabarmati (Lahiri 2000, 12). Hence, it would serve us well to remember that the majority of Harappan settlements flourished in alluvial terrains created by these mighty rivers.

In such a situation, the landscape must have been able to carry marshy-swampy patches for a species like the rhinoceros. Thus, despite climatic and tectonic fluctuations (the timings of which have been a matter of contention), it is not difficult to envision conditions permitting the survival of the animal at least along river valleys and in their vicinity.

Faunal Remains

With this sense of the environmental milieu, I now move on to look at Harappan sites that have yielded rhinoceros bones. The presence of the mega herbivore is widely testified through the retrieval of its remains from the sites of Nausharo in Baluchistan, Harappa in Punjab, Pakistan, Kalibangan and Karanpura in Rajasthan, and from a number of sites in Gujarat such as Lothal, Surkotada, Shikarpur, Khanpur, Kuntasi, and possibly also Oriyo Timbo (Chattopadhyaya 2002, 393, 395). More recently, the animal has also been reported from the site of Madina in Haryana (Joglekar and Sharda, 2016). If we consider Meadow's (1993, 197) argument that the bones of animals not normally kept or not normally eaten are likely to be rare in archaeological deposits, then the presence (and in some cases like Kalibangan, the profusion) of rhinoceros remains at some of the major urban sites of the civilization along with the evident popularity of the animal as a subject of portrayal on seals as well as terracottas raises numerous possibilities which can be explored. These range from the possible exploitation of the animal for its meat and bones to the role of the animal in the religious life of the people. Before proceeding with the narrative, it is important to note that while dealing with early faunal reports like those of Mohenjodaro and Harappa, a limitation to be kept in mind is that it is not always clear from what kind of context or even from what phase of occupation particular faunal specimens are derived (Meadow 1993, 196).

Nausharo: I begin with this site near Mehrgarh in district Kachi, Baluchistan. During the Harappan phase at the site, the rhinoceros along with other wild species such as the onager, wild boar, and gazelle was hunted, but the remains accounted for less than 10 per cent of the faunal assemblage (Meadow 1989, 68). Rhinoceros bones were found in a wide hollow along with some trash and misfired sherds in front of a staircase of baked bricks leading from the edge of a monument to its internal part (Jarrige 1987–8, 187).

Amri: Discovered in 1929 by N.G. Majumdar, this site in the Dadu district of Pakistan is about 160 km from Mohenjodaro, further downstream. On the map it appears to be facing Chanhudaro, on the other side of the river. The presence of the rhinoceros was documented in sub-period C (corresponding with what is termed 'the Late Mohenjodaro' phase) of Period III representing the Indus civilization (Casal 1964, 164–9).

Harappa: At this site in Sahiwal district (earlier known as Montgomery), Punjab, Pakistan, the remnants of the animal occurred in the form of 'an almost complete right shoulder girdle'. This led Prashad (1936, 30–1), who examined the faunal remains at the site, to postulate that the distribution of the species must have been more extensive previously in the Punjab suggesting the presence of marshy forest environs in the vicinity of Harappa where the animal was found. However, the scapula being the only remain of the animal, Prashad (1936, 3) contended, like he did in the context of the jackal and the wolf, that the remains of these animals may have been accidentally introduced, or could be the result of the hunting of these animals by the Harappans.

Madina: Situated about 90 km north-west of Delhi, the village of Madina in district Rohtas, Haryana, revealed seven (Madina-1, Madina-2, Madina-3, Madina-4, Madina-5, Madina-6, and Madina-7) archaeological sites within its revenue jurisdiction. Excavations at Madina-3, located about 3 km south-west of the village revealed eight habitational phases and nine layers at the site, which yielded Painted Grey Ware (PGW) and Black Slipped Ware (BSW) along with sherds of Late Harappan pottery and antiquities (Kumar 2016, 1–2). A rhinoceros astragalus retrieved from Phase III (layer 7) did not show any marks of utilization, and was suggested to have been collected and brought within the settlement by the inhabitants (Joglekar and Sharda 2016, 228).

Kalibangan: The presence of the rhinoceros in the western state of Rajasthan was documented in an early report by Banerjee and Chakraborty (1973, 430). They argued that the occurrence of this species at Kalibangan (Hanumangarh district) supported the idea of its extensive distribution in the past. More importantly, they held that since *Rhinoceros unicornis* generally inhabits swampy land mixed with forest, its occurrence in Rajasthan strengthens the geological evidence that the aridity of this area is of recent origin.

The remains reported included the distal fragment of left tibia, distal fragment of right humerus, first phalanx of the fourth metatarsal of both the right and left feet, and the third metatarsal of the right foot. All the fragments were found to be structurally similar to the bones of the full-grown modern female specimen from West Bengal present in the Zoological Survey of India, but were slightly larger in size. Banerjee and Chakraborty (1973, 430) also found the fragment of rhinoceros scapula from Harappa and that of the mandible from Lothal to be larger than those of the modern specimen. However, in the absence of adequate comparative material it could not be ascertained whether *Rhinoceros unicornis* found in the western part of the Indian subcontinent during ancient times was larger than the modern one of the eastern part.

In an unpublished report on Kalibangan, Nath (cited in Joshi et al. 2003, 18–19) also mentioned the existence of a large number of animal remains, and observed that amongst wild animals, the remains of rhinoceros (*Rhinoceros unicornis* L.), shells of turtles, the barasingha, sambar, and spotted deer were found. These remains, he argued, showed that the climate at that time was more humid than the arid climate of the present day.

Subsequent excavations at the site revealed an Early Harappan Period I placed between c. 3000 and 2700 BCE, and a Mature Harappan Period II dated from c. 2550 BCE to c. 2000 BCE with a margin of 50 years on the earlier side, and of about 100 years on the later (Lal 2003, 26). In the detailed report of animal remains, the rhinoceros was documented either from the middle or late levels of Period II. The remains retrieved in substantial quantity included the distal fragment of radius, one broken medial, and one broken lateral condyle of the left humerus, fragment of the shaft of left ulan-3, broken right ramus of a mandible, second phalanx of right hind limb, fragment of right pelvic girdle with acetabulum, broken rib-7, left lunate, left trapezoid magnum, left external cuneiform, right metacarpal, right unciform, distal portion of left femur, lumbar vertebra, right astragalus, proximal portion of left ulna, proximal end of right radius, fragment of a rib, and broken left pelvic girdle with pubis (Banerjee et al. 2003, 267–339).

Karanpura: Subsequently, rhinoceros remains were also identified at the site of Karanpura at a distance of 102 km from Kalibangan. Four complete rhino bones were found from the Mature Harappan levels at the site. Preliminary information tells us that these bones were found in a street deposit along the northern fortification wall. The remains included a shoulder bone, a part of pelvis, one radius and ulna among the complete ones (V.N. Prabhakar, personal communication). These finds are significant since along with the evidence from Kalibangan, they affirm the presence of the animal in proto-historic times in this now arid zone.

I now move towards Gujarat where rhinoceros bones have been identified from a large number of Harappan sites. Known to have inhabited a major part of the Gujarat plains in the proto-historic period, the animal came to be restricted to certain ecological niches by medieval times. It has been postulated that strips of grassland along the banks of rivers must have formed the habitat of animals like the rhino and wild buffalo during Harappan times (Thomas 2000–1, 82). Let us begin with sites in the district of Kutch.

Surkotada: At this site, 160 km to the north-east of Bhuj, the entire habitational deposit revealed three sub-periods: IA, IB, and IC, primarily based on ceramic evidence. Radiocarbon dating placed Period IA between c. 2055 and 1970/1940 BCE indicating the earlier to mature phases of the Harappan culture. Period IB was chronologically placed between c. 1940 and 1790 BCE, while Period IC placed between c. 1790 BCE and 1660 BCE was Late Harappan in context (Joshi 1990, 62–6).

Our evidence comes from Period IC, where bones of the one-horned rhinoceros constituted a meager 0.21 per cent of the faunal assemblage (Sharma 1990, 377). The details of the remains were, however, not mentioned.

Based on the faunal assemblage, it was argued that the terrain was not very different from the present one, but climatic conditions were more favourable, and fresh water available in plenty permitted the growth of comparatively thicker vegetation in the vicinity. The marshy character of the land and tidal marshes stretching well into the interiors were also underlined (Sharma 1990, 383).

Shikarpur: Locally known as Valamiya Timbo, Shikarpur lies 5 km south-west of the present-day village in Bhachu taluka in Kachchh district of Gujarat. First excavated during 1987–90, the site revealed Harappan cultural material spread across nineteen layers, with layers 1–9 representing the Mature Harappan, and layers 10–19, the Early Harappan period (Thomas, Joglekar, Deshpande-Mukherjee, and Pawankar, 1995, 33).

The presence of the rhinoceros was documented in the Mature Harappan layers (Thomas, Joglekar, Deshpande-Mukherjee, and Pawankar, 1995, 38). Though the details of the remains were not mentioned in the report, the tabulation of the 'Number of Identified Specimens' (NISP) put the count at 12 for the rhinoceros, which constituted 0.173 per cent of the total faunal assemblage of the Mature Harappan period (Thomas, Joglekar, Deshpande-Mukherjee, and Pawankar, 1995, 38).

Kuntasi: Moving on to Harappan sites in Saurashtra, Kuntasi in Rajkot district is located almost at the boundary between Kutch and Saurashtra on the right bank of river Phulki. Excavations brought to light mainly two cultural periods at the site. Period I representing the Mature Harappan dated from about c. 2400 BCE to 1900 BCE, and Period II representing the Late Harappan was placed between c.1900 BCE and 1700 BCE. Based on structural remains, three phases were identified, with Period I being divided into phases A and B, while the last phase C belonged to Period II representing the Late Harappan period (Thomas, Matsushima, and Deshpande 1996, 297).

Receiving less than 600 mm average rainfall per annum, and having little forest cover, the area once harboured the rhinoceros. Among the twenty-three wild species identified by Thomas and his colleagues, the animal was recognized by a solitary bone, the third phalanx belonging to the Late Harappan period.

The fauna in the two cultural periods at Kuntasi also revealed some interesting trends. While only a few animal species were represented between layers 14–19, and a moderate increase was discernible between layers 9–13, the maximum representation of animal species was between layers 1–8. The report maintained that on the basis of archaeozoological evidence for subsistence strategies, the whole cultural deposit at Kuntasi can be divided into the mentioned three major units, wherein the maximum exploitation of wild fauna was noticed in the Late Harappan period. A similar pattern of faunal exploitation at Rangpur, Surkotada, Nageswar, Mohenjodaro, and other sites was explained by a failure in agriculture due to environmental degradation caused by prolonged human interference with nature as also due to erratic monsoon rains. The over-exploitation of animal resources in the Late Harappan period was also attributed to demographic pressure compelling greater reliance on animal food for sustenance (Thomas, Matsushima, and Deshpande 1996, 305).

Climatically though, not much was considered to have changed since the bones of the blackbuck, nilgai, and gazelle, which are typically semi-arid species, are found abundantly at Kuntasi. As for animals such as the wild buffalo, rhinoceros, and probably wild cattle, it was presumed that the grassland along the river banks must have suited these animals, suggesting conditions more congenial for faunal life in proto-historic times (Thomas, Matsushima, and Deshpande 1996, 305).

Lothal: Another well-excavated Harappan site in Saurashtra, the ancient mound of Lothal, is set in a flat featureless alluvial lowland called *bhal*, approximately 83 km from the state capital, Ahmedabad. Two cultural periods were identified at the site: Period A (c. 2450–1900 BCE) representing the Mature Harappan phase and Period B (c. 1900–1600 BCE) representing the Late Harappan. Based on structural evidence, Period A was subdivided into four phases (Rao 1979, I: 24, 28).

Overall, it is conceded that though the plains today are slightly more arid owing to biotic factors, denudation, and erosion, the climate and vegetation have not changed much during the last 3,000 years except for the fact that there were more swamps allowing a profusion of tall grass and reeds. It was around these swamps that animals such as the rhinoceros must have found congenial conditions for survival (Rao 1979, I: 20).

Now that we have a sense of the landscape, let us look at the traces of the animal at the site. Nath and Rao (1985, II: 637, 642) reported the solitary find of a fragment of rhinoceros mandible comprising the right horizontal ramus and some portion of vertical ramus. The find bore close similarity with the modern specimen in the Zoological Survey of India collection. A more humid and moist ancient climate was inferred from the presence of the animal in this region, where it is now extinct. The faunal report, however, does not mention the cultural period to which the remains belonged.

Interestingly, a later study gives the rhinoceros a miss in its account of the animal remains excavated from Lothal (Saha et al. 2004, 1–162). The discrepancy between the two reports is perplexing, and according to Joglekar and Sathe (personal communication) can be explained by the possibility of the remains having gone missing before they reached Saha and his colleagues, or the latter not identifying the species in the assemblage handed to them.

Khanpur: Situated near the village of the same name in the Morbi taluka of Rajkot district in Saurashtra, the site revealed two chronological horizons: Late Harappan and post-Harappan. The former on the basis of pottery shapes was assigned to Rangpur IIB, and the latter to Rangpur IIC (Chitalwala and Thomas 1977–8: 11–12). It may be mentioned that the dates assigned to Rangpur IIB are c. 1500–1100 BCE and c. 1100–1000 BCE respectively (Rao 1962 and 1963, 27).

Our evidence comes from the Late Harappan horizon, where the rhinoceros was represented by a single bone—a complete and well-preserved humerus (Thomas 1977, 138). The identification of the animal at the site was significant particularly in view of its wide distribution in Gujarat in early times testified by sites such as Lothal and Langhnaj. Its presence was also considered indicative of a moist tropical climate in Gujarat during the post-Pleistocene period (Thomas 1977, 196). The evidence from Khanpur also attested to the extent of the penetration of the eastern half of Saurashtra by the rhinoceros (Chitalwala 1990, 80).

In view of the widespread distribution of the animal in Saurashtra, Chitalwala (1990, 79–82) sought to reconstruct the ecological variables that formed a rhino habitat in the region about 4,000 years ago. His enquiry regarding its subsequent disappearance from the region points out that the floral record from archaeological sites in Saurashtra indicates the existence of xerophytic vegetation (plants adapted to living in dry arid habitats) there. Evidence of *Acacia*, *Zizyphus*, and bajra from the Harappan settlement of Rangpur underline semi-arid conditions. Sites such as Chiroda, Rojdi, and Lothal have yielded the seeds of different varieties of grasses and sedges. Chitalwala argues that given that climatic conditions over Saurashtra have remained more or less stable in the last four millennia with the exception of a higher and slightly less erratic rainfall, and the overall faunal spectrum also suggesting an ecological and climatic set-up not too different from conditions obtaining today, the disappearance of the rhino from this region needs to be investigated since other animals that shared the rhino habitats survive today. This paradox is explained by putting forth some hypotheses which could explain the exit of the rhino from this region. The contention is that by the time the Harappans arrived in Saurashtra, rhinos were already struggling for existence in the eastern part (since thick forests in the western part hindered rhino penetration there), and the location of Harappan settlements near watercourses close to rhino habitats brought biotic pressures to bear on the animal's existence.

Chitalwala further maintains that as long as the Harappans of the Mature phase (2300–1750 BCE) were given to trade and commerce, there was no organized and intensive pattern of land use. But by Late Harappan times, there was a tremendous increase in population entailing an intensive pattern of land use based on dry farming. Also, the location of Late Harappan settlements along watercourses began to overlap rhino habitats as a result of expanding catchment areas due to more and more land being brought under cultivation. Additionally, the intrusion of cattle into grasslands that formed rhino habitat also greatly reduced the chances of rhino survival in Saurashtra. Given that rhinos become asocial when faced with dwindling food supplies, and death rates progressively mount, Chitalwala argues that it was the Late Harappans who gave the 'coup de grâce' to the already embattled animals that were surviving in the face of heavy environmental and ecological odds.

Having reviewed the Harappan faunal records, it can be safely said that there is sufficient evidence for the presence of our moisture-loving protagonist in proto-historic times in areas now known to be arid and semi-arid. Not just bones, even visual representations amply testify to the animal having been seen around frequently.

Harappan Art and Iconography

Returning to our search for the rhino in the archaeological records of the Harappan period, apart from bones, the creature also made its way into the realm of art in the form of representations in terracotta and steatite as well as portrayals on seals. What calls for particular attention is the faithfulness of these depictions revealing an interaction close enough to have facilitated a careful observation of the anatomical features of the animal. A closer look at some of these representations from different sites will serve to reinforce this contention.

Though rhinoceros remains are not reported from Mohenjodaro, the animal occurs frequently in the animal figurines mostly made of terracotta as well as on seals at the site, and, hence, seems to have been found within close vicinity. This situation, where forms have been identified from iconography but not yet in bones and vice versa, suggests that economic significance need not equate with social, ideological, or conceptual importance (Meadow 1993, 196).

In the rhino figurines at Mohenjodaro, the wrinkled skin is realistically portrayed by hatching or pitting in some cases, while in others, strips of clay were used to show the folds in its hide. The models are usually roughly made, and were described by Mackay (1931a, I: 348) as being in every case 'a child's handiwork'. However, it is unlikely that in an urban context like that of Mohenjodaro, children would have had intimate and extensive access to wild habitats where rhinos roamed, facilitating such detailed depictions. Nevertheless, the frequency of representations does suggest that the animal was well known.

The animal also appears on seals though not as frequently. Curiously, it is shown standing over a manger-like object. Marshall was emphatic about the fact that these depictions bore no relation to domestication. His own sense was that these troughs were meant to symbolize food offerings, and that their presence implied that the animals to which these offerings were made, whether in captivity or in the wild, were objects of worship (Marshall 1931, I: 70).

As can be seen from the illustrations which follow, there seemed to be no ambiguity regarding the fact that in every case it was the *Rhinoceros unicornis* that was depicted on the seals (Mackay 1931b, II: 387). This unequivocal assertion has much to convey about the extreme fidelity, even to the 'wicked pig-like eye' with which the animal was rendered. Illustrating the details of one such seal, Mackay noted the well-represented thick hide with the nearly real wrinkles and folds of the skin. Rough excrescences on the skin were indicated in some cases with holes made with a fine drill, while in others, hatched lines were employed.

In another seal, only a fragment of which survives, the animal at first glance seems to be a rhinoceros as the fore and hindquarters bear drill-marks arranged to represent the roughness seen on other rhinoceros seals. However, a closer look rules out the possibility of the legs (resembling those of a bull or an ox) being those of a rhinoceros (Mackay 1931b, II: 387). Could this then be a composite animal, the hind part of which was taken from the rhinoceros?

This could be a possibility since though not a common part of the hybrid compositions one encounters in Indus imagery, there is a composite animal on a copper tablet from Mohenjodaro which has the hindquarters of a rhinoceros and the forequarters of a leopard or tiger. It also has the unicorn's horn, and a manger in front of it (Mackay 1931b, II: 399).

Further excavations at Mohenjodaro also served to underline the popularity of the animal as a subject of depiction on seals and amulets as well as for modelling in clay, particularly during the later occupations. The collection was a curious mix of graphic representations of the animal and gross misrepresentations of the same. While Mackay (1938, I: 290) described a faithfully rendered model where the folds of the skin were represented by strips of clay pitted all over to show the great horny bosses characteristic of the actual animal, he also reported examples from the lower levels which were not so realistically modelled. In one, the armour-plated hide of the animal was not indicated, emphasis being placed only on the tubercles, which were erroneously represented as covering the whole of the body. In the other, there seemed to be a complete disregard of the animal's bodily characteristics. However, in the absence of a sufficient number in the lower strata, it could not be inferred whether or not they were in general as elaborately modelled as those from the upper levels. On the seals, however, the animal was delineated with remarkable faithfulness, leading at times to an overelaboration of detail (Mackay 1938, I: 331).

Another interesting observation emerges when we compare the position of the heads of the rhinoceroses with mangers in front on the seals published in Marshall's report on Mohenjodaro (1931) with those published later by Mackay (1938). While in the earlier set of seals, the animal seems to be looking straight ahead despite the trough in front, in the later set, two of which show the animal with a trough in front, the head is inclined in a way as to suggest that the animal is eating out of it.

In one of the five seals Mackay (1938, I: 330–1) documented, the usual dish-like manger in front of the creature was substituted with the cult object exclusively associated with the urus bull. With the exception of the ears and the horn, the head was found to be similar to that of the urus particularly in its upright position, and so was the body with the exception of the warty excrescences characteristic of the rhinoceros. The feet were those of a rhino. What emerged then, was a new composite animal, whose association with the urus bull permitted the use of the cult object peculiar to it. The possibility of this representing the mingling of two cults was suggested (Mackay 1938, I: 330–1). However, in the absence of a deciphered script, it would perhaps be best to attribute such representations to a more complex world view which sought to invest its own rationale in such portrayals.

A survey of the seals and seal impressions also revealed the trend of depicting animals in a file. This motif, common on archaic Mesopotamian and Susian seals (showing antelopes and lions), occurs on the seals of both Mohenjodaro and Harappa, which show the unicorn, rhinoceros, gharial, short-horned bull, elephant, and tiger (Mackay 1931b, II: 398).

In this context, an interesting find at Mohenjodaro is a perforated baked clay triangular prism which was described as having on one face from left to right an elephant, a rhinoceros, a tiger or leopard, and another cat-like animal (Figure 2.11). Above these animals, which were arranged *en file*, was a fish on the left, followed by a gharial with a fish in its mouth. The other face showed from left to right another file of animals including a unicorn, what seems to be a cow, a short-horned bull, and a rhinoceros. Above the file, were the indistinct figures of what appeared to be a jungle-fowl and a gharial. The third face showed from left to right, two goats eating from a tree, a jungle-fowl, a man seemingly pushing a goat or similar animal along, and an antelope-like animal with two heads. A hole at each end of this object, suggested that it was used as a kind of revolving bezel (Mackay 1931b, II: 395–6).

A number of amulets also depict the animal often as part of a procession with other animals. Mention may be made of a particularly interesting one documented by Mackay (1938, I: 352), and retrieved from the upper levels of Mohenjodaro. It is an amulet of baked clay, pressed somewhat out of shape. On one side, a man is seated in a tree with a cat-like animal looking back over its shoulder at him, a scene which recurs on seals as well as sealings. On

Figure 2.11 Magnified view of the face of a terracotta prism showing a tiger(?), a rhinoceros, an elephant, and a gharial, Mohenjodaro

Source: After Blakiston 1927, plate XXII (c). © Archaeological Survey of India.

the opposite side, a row of animals though quite rubbed, show a rhinoceros, an elephant, and an urus bull in file. Above them on the left is a gharial with a fish in its jaws, and on the right there seems to be a bird. A third and narrower side is the most interesting, and though badly abraded, shows a tree on the left beneath which is a small truncated pyramid-like object, perhaps a shrine, surmounted by a standard. A goat-like animal with long horns and a short tail turned upwards stands on each side of the pyramid with its forelegs resting on its top. A number of objects representing either leaves or fruits hang from the branches of the tree. In the middle of this side, a kneeling man holds in both hands what may have been a small sapling that he is about to plant, while a woman standing before him bent down with what looks like a basket in her hands. Behind the woman is a badly weathered device of which only the lower portion remained. Further to the right were some indefinite details, one of which, according to Mackay, may have been a dancing figure. If this was so, the middle and right hand of the scene may have depicted some religious rite, but this is clearly a speculation. Given that little remained of the fourth segment, not much could be said regarding the interpretation of this side or the amulet overall.

A steatite rhinoceros seal was also retrieved from the Harappan site of Allahdino, approximately 40 km north-east of Karachi, Pakistan. The find, a part of the collection of eight seals obtained from one level of the site's occupancy, was found in the 'alley' close to the north-east corner of Building III, and on the highest part of the site. The importance of the collection was said to lie in the fact that it is proof that even small Harappan settlements had their quota of seal bearers and users (Fairservis 1976, 7, 13). As for the animal itself, on the mentioned rhinoceros seal, it is a clearly identifiable one-horned rhinoceros with a lowered head, and the folds of the skin delineated by criss-cross patterns. The animal seems to be facing a roughly etched object which could be the oft-depicted trough.

The rhino seems to have enjoyed a privileged position at Harappa. The largest number of terracotta toys at the site were those of bulls, followed by the rhinoceros, the goat, and ram together with a range of other animals such as the tiger (and perhaps lion?), elephant, pig, dog, monkey, cat, rodents, reptiles, aquatic animals, and birds (Vats 1940, I: 301).

An admirable find at Harappa was a faithfully executed sitting rhinoceros in burnt steatite with a prominent snout, short horns, and rough hide. A pinhole in the underside was used to fix it to a table or some flat surface. Another find was a roughly modelled version in terracotta, where again the portrayal of the rough and wrinkled hide is rendered realistic by incised lines and a pitted patch over the beast's hindquarters. The eyes consisted of two holed pellets (Figure 2.12). Significant in this and two other figures was a collar of two bands perceived as suggesting domestication (Vats 1940, I: 307–8). Vats also points out that on seals, however, the animal was never shown with a collar, but portrayed with a feeding trough before it.

We know we are on firm ground as far as identification is concerned when in a preliminary report on the fourth season of research

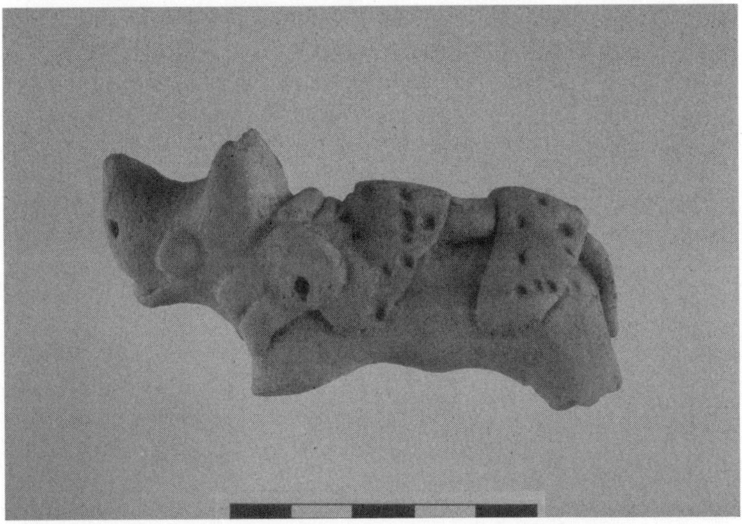

Figure 2.12 A rhinoceros figurine from Harappa. The animal is strangely portrayed with a collar. The strips on the back probably suggest the tuberculed hide

Source: © Richard H. Meadow, Harappa.com, courtesy Department of Archaeology and Museums, Government of Pakistan.

at the site, George F. Dales (1989, 20) observed that amongst all the animals in terracotta at Harappa, the rhinoceros, the humped bull, and the dog stand out in terms of their typical iconographic features. According to the tentative statistics furnished by him, the thirty-four depictions of the rhinoceros comprise 6.3 per cent of the total terracotta collection at the site, which is more than double the representations of sheep and goat (fifteen representations comprising 2.8 per cent of the collection). Cattle outnumber sheep/goat by 10:1, water buffalo by 3:1, and rhinos by 5:1. The figures though tentative, are certainly enough to convey the popularity of the animal as an object of portrayal among the masses.

Vats (1940, I: 323) reported two seals carrying the mega herbivore. While in one, only the hindquarters of the beast were left, the other was considered the best of the animal found at either Mohenjodaro or Harappa, with the anatomical detailing of the rough hide with its folds and excrescences clearly giving away the intense familiarity of the engraver with the animal (Figure 2.13). Vats also concurred with Marshall's contention that the troughs in front of the tiger, rhinoceros, bison, and in one case, an elephant probably symbolized only the offering of food to these animals, implying that the animals were objects of worship.

The clay rhinoceros at Chanhudaro (retrieved from Mound II) seems like a bit of an interloper amidst the faithfully rendered models of the animal retrieved from other sites (Figure 2.14). The missing realism is conspicuous, as is the artist's lack of acquaintance with the real animal. It is a quaint-looking animal with a rather short body, a narrow, tubular muzzle, and the horn curved in the wrong way. A deep groove at the end of the snout indicated the mouth, above which there were two deep holes for the nostrils. Some care was taken with the modelling, though without the horn, identification would have been difficult (Mackay 1943, 159).

The rhinoceros also occurs on a seal of the Jhukar period at the site. On what forms one side of a seal, there is an apparently composite animal, a combination of a humped bull and a rhinoceros with a tail (which does not seem to belong to either of these animals) bent over its back. On the other side was an endless coil pattern. The seal with its faces rounded, was very carefully cut, and had a milled edge through which it was pierced for suspension (Mackay 1943, 142).

Figure 2.13 Seal from Harappa showing a rhinoceros with a trough in front. Note the prominent horn and the folds on the skin with tubercles.
Source: National Museum, New Delhi, Collection Acc no.: HR 5992/119.

Unlike at Mohenjodaro and Harappa, the rhinoceros does not figure on seals and amulets at Lothal. But it does capture the terracotta artist's imagination as is revealed by a head of the animal where the detailing of anatomical features such as the thick folds of the hide around the neck, the short horn on the snout, beady eyes, and nostrils clearly reveal an interaction close enough to have facilitated a careful study of the animal. The thick eyebrow is indicated by a curved incised line, and the tongue with a pellet inserted in the mouth. Though the ears are damaged, and one of the pellet

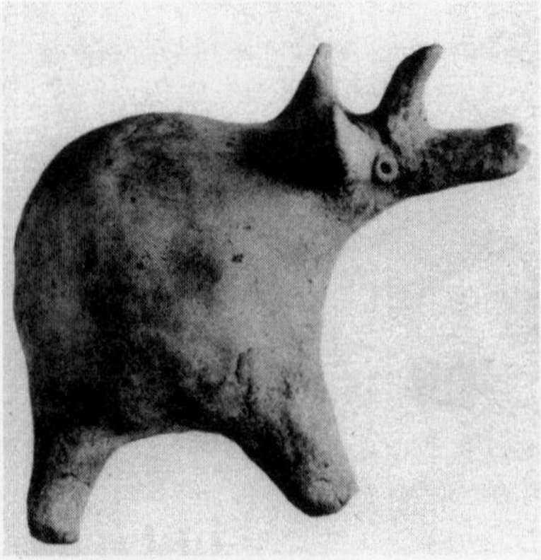

Figure 2.14 This quaint-looking clay rhinoceros from Chanhudaro is a departure from the otherwise faithfully rendered representations we encounter
Source: After Mackay 1943, plate LVI, 8. © American Oriental Society.

eyes has fallen off, the realism of the depiction is unmistakable. A smaller model has a sturdy body, thick short legs, and short prick ears, and the mouth is indicated by a nail punch. Both the models are evidently indicative of *Rhinoceros unicornis*, which is postulated to have inhabited the swamps and marshes around Lothal in the proto-historic past (Rao 1985, II: 485).

It may not be out of context to point out how archaeological excavations have demonstrated that the animal's popularity transcended spatial and temporal barriers by drawing attention to distant sites such as Shortughai and Tell Asmar that have yielded Harappan or Harappan-influenced seals with rhino depictions.

In this context, what deserves mention is a cylinder seal of glazed steatite with an engraving of an elephant and a rhinoceros in file with a gharial above at Tell Asmar in Iraq (Figure 2.15). The peculiarities of design as well as the subject showed such resemblances to the seals of the Indus region that its Indian origin was considered certain. The rendering of the skin of the rhinoceros (closely resembling plate armour), and the sloping back and bulbous forehead of the elephant were evidently carved by an artist familiar with the

Figure 2.15 Seal from Tell Asmar
Source: Courtesy of the Oriental Institute of the University of Chicago.

animals. H. Frankfort observed certain other peculiarities of style which connected the seal with the Indus civilization.

> Such is the convention by which the feet of the elephant are rendered and the network of lines, in other Indian seals mostly confined to the ears, but extending here over the whole of his head and trunk. The setting of the ears of the rhinoceros on two little stems is also a feature connecting this cylinder with the Indus valley seals. (Frankfort 1939, 305)

One is in this context also reminded of a seal impression from Mohenjodaro which has a similar procession of animals with a gharial above them.

Similarly, a square-shape Indus seal depicting a rhinoceros along with Indus pictograms also comes from the Harappan settlement of Shortughai, located at the confluence of the Kokcha and Amu Darya Rivers in north-east Afghanistan. The Harappan material recovered from the site belonged to the Mature (Shortughai Period I) and Late phases (Shortughai Period II), ^{14}C analyses placing Period I at the end of the third millennium BCE. On the basis of the finds, it was asserted that all the artefacts and technology from the first period of Shortughai clearly originated in the Indus region, and that nothing in those levels could be attributed to another culture or civilization (Francfort 1984, 303).

On the basis of the evidence collated so far, it can be said that the animal occupied a significant place in the Indus belief system. Not only this, the abundance of portrayals on seals and terracotta objects certainly conveys that the rhinoceros was an animal well known to the Harappans. Rookmaaker (2000, 66) maintains that it is possible that the animals were encountered near Indus settlements either in the lower part of the valley, or perhaps slightly more northwards of where Harappa is located.

A Waning Presence in Later Art and Iconography

The rhino tale, however, takes a curious turn as we approach the early historical period. Manuel (2008a, 36) rightly argues that something went amiss after the Harappans. Though depictions persisted, portrayals progressively lessened. It is indeed perplexing that despite being the second-largest land mammal after the elephant, the animal figures but rarely in later Indian art. Bautze (1995, 28) observes that the animal is shown mainly in the earlier periods, but its importance could never rival that of the elephant.

The paucity of depictions after the Harappans is intriguing, and the reasons can best be speculated. What is discernible is a gradual distancing of the animal from popular imagination, which now gets captured with imageries of the mightier elephant and the faster horse (Manuel 2008a, 36). Whether this had to do with the regularity with which the former as well as the latter were encountered, as also their potential to be tamed, controlled, and used vis-à-vis the rhino, is worth considering. Given the evidence for the existence of the rhinoceros outside areas in the North-East until not so long ago, the dwindling of the species itself does not seem to have propelled this shift. It would perhaps be more plausible to argue that the wavering fortune of the animal had to do with changing social and cultural perceptions, conditioned perhaps by new forms of human settlement and production. With the growing importance of agriculture and the clearing of forests, horses and elephants possibly found greater use for harnessing and other purposes. Rhinos, on the other hand, were neither encountered nor engaged with regularly. Though it could be tamed, the animals had no draught or domestic use (Divyabhanusinh, personal communication).

The rhinoceros once again figures in popular imagination during the NBPW period. The realistic modelling of the terracotta figurine of the animal in deep red ochre paint retrieved from sub-period IB of Period II (c. 600 BCE to second century BCE) at Kausambi in Allahabad district of Uttar Pradesh, clearly indicates the familiarity of the potter with this creature. The rendering of the mouth, the eyes, the nose, and the head call for particular attention. The climate and topography of the region must have been such as to attract this animal. Though at present this area is more or less arid, it is contended that much of it was covered with forest in the third century BCE, and received a larger amount of rainfall than today. This inference seemed plausible in view of the references to jungles in the vicinity of Kausambi in the time of Buddha. Even during the times of the Chinese pilgrims Fa Hien (fifth century CE) and Hiuen Tsang (seventh century CE), the whole area was infested with dense forest (Sharma 1969, 70, 73).

We meet the rhinoceros again during a survey of the gamut of Mauryan and late Mauryan art objects. Although largely confined to the third and second centuries BCE, some of them were believed to be from a century or so later (Gupta 1980, 53).

A red and grey soapstone seal dated to around third century BCE from Bhita in Allahabad, Uttar Pradesh showing a rhinoceros is preserved in the Allahabad Museum (Figure 2.16). Pramod Chandra graphically writes, 'The animal, in what seems to be a flying gallop, moves to the right. The bulky body is divided into two globular parts, each enclosed by a ridged border. The small tail hangs close to the back and a horn is visible at the end of the snout' (1970, 36).

The mega mammal also figures on a stone disc of the Murtaziganj-Patna group (Figure 2.17). This disc has a fourteen-petalled lotus in the centre. The flower is encircled by a line. The outer frieze has seven animals and two birds: an elephant facing right, a peacock looking behind, a stag facing right, a stag (?) facing right, a rhinoceros facing right, a stag looking behind, a stag facing right, a horse facing right, and a peacock facing right. The reverse has a roughly engraved ladder symbol (Gupta 1980, 58).

Figure 2.16 Red and grey soapstone seal, Bhita, Allahabad, c. 3rd century BCE
Source: Courtesy of the American Institute of Indian Studies. AIIS: 86.13. Acc. No. 84070.

Figure 2.17 Stone disc depicting an array of animals, including the rhinoceros with a prominent horn from the Murtaziganj–Patna group
Source: After Gupta 1980, plate 25a. © B.R. Publishing Corporation.

IN SEARCH OF THE RHINO IN ANCIENT INDIAN LITERATURE

The Vedic Corpus

With the dawn of recorded history, literature casts additional light on this enigmatic creature. In this context, the Vedas furnish us with the earliest glimpses of the human–rhino interface. However, before one sets set out to reconstruct aspects of this interface using this corpus, it is crucial to keep in mind the essentially ritualistic nature of the texts and proceed accordingly.

One can begin with an ambiguous reference in the *Ṛgveda* to a creature called the *parasvant.* Though there are no definite markers that enable us to assertively identify this creature, the word deserves attention since we have at least one instance where it has been interpreted as the rhinoceros. Vṛṣākapi is said to have found a slain wild animal (*parasvant*), a dresser, new-made pan, knife, and wagon with a load of wood (*Ṛgveda* X, 86, 18; Griffith 1963a, II: 509). In his translation of the *Ṛgveda*, Ralph Thomas Hotchkin Griffith reasons that though the speech is difficult to understand, here Indrāṇī seems to be speaking depreciatingly of a sacrifice offered by Vṛṣākapi comprising an unsuitable victim, prepared with instruments and means he has encountered by chance.

Significantly, Sāyaṇa (the celebrated commentator of the Vedas) interpreted the word as a 'wild animal' (Suryanarayan Nanda, personal communication). However, that the term itself has been more specifically construed is evident from lexicons listing it. Böhtlingk and Roth (1990, IV: 497), for instance, specifically conjectured the *parasvant* to be the 'wild ass', and so did Monier-Williams (1963, 589).

We encounter the *parasvant* more than once, for instance, in the *Atharvaveda* (VI, 72, 2–3; Whitney 1962, I: 335), but its identity remains far from certain in the absence of the mention of any definite physical markers except its large 'member' (apparently its genitals), which is invoked for virile power. This, however, is hardly any indicator since it is a feature common to many big animals. Nevertheless, Bautze (1985, 409) cites Heinrich Lüders' assertion that the *parasvant* was the oldest word used for the rhinoceros, and accordingly infers that the Vṛṣākapi passage of the *Ṛgveda* indicated that rhinoceros meat was edible. Significantly, there is a reference in the *Baudhāyana Śrautasūtra* (2, 5) where the *khaḍga* is mentioned along with the *parasvat*: 'in the *parasvat* my failure, in the *khaḍga* my misfortune' (cited in Bautze 1985, 410). Despite the ambiguity, what is evident is that whatever it stood for, the *parasvat* was distinct from the *khaḍga*, the word most commonly used for the rhinoceros. Hence, though it is possible to cull other references to the word as Bautze does, the endeavour is clearly not worthwhile in the absence of any specific physical traits aiding its identification as a rhinoceros.

We tread on more certain ground in our search for the armoured giant when we encounter the *khaḍga*. That it is indeed the rhinoceros is evident from the fact that most middle and modern Indic descendants of Sanskrit *khaḍga* seem to indicate the animal or its parts (Jamison 1998, 252). Another pointer to the identity of the animal is that several Vedic passages 'situate the *khaḍga* in the realm of fierce wild beasts and suggest that its hide is armour-like', an observation that accurately describes the Indian rhino (Jamison 1998, 252).

The quest for the *khaḍga* leads us to the *aśvamedha* sacrifice where animals dedicated to different deities are tied to the twenty-one *yūpas* (sacrificial stakes) and in the intermediate spaces. While domestic animals are bound to the stakes, in the spaces between them are confined wild animals, including the elephant and the rhinoceros. However, not all the animals are killed, some including the rhinoceros being only confined till the culmination of the ceremony (Griffith 1899, 218). It may be argued that this may have had to do with the way animals were classified in ancient India, which distinguished between village animals (*grāmya*) with the generic term *paśu*, and wild animals (*āraṇyaka*) with the generic term *mṛga*, and considered only the former as *medhya* or fit to be sacrificed. Hence, it has been premised that though included in the list of animals at a horse sacrifice, wild animals are eventually released, and that their inclusion was possibly to ensure the fulfilment of the rite by integrating the sacrifice of all beings (Olivelle 2002b, 9). Returning specifically to the rhinoceros, in the enumeration of the animals assigned to different deities, the animal is dedicated to the Vaiśvadevas (All Gods) (*Vājasaneyī Saṃhitā* XXIV, 40; *Maitrāyaṇī Saṃhitā* III, 14, 21).

Since these contexts do not suggest anything beyond a ritual significance of the rhinoceros, we can perhaps turn to other allusions to the *khaḍga* (which reflect an interest in the use of the animal) for more telling clues.

We are told about the Vedic use of rhinoceros-hide (*khaḍgakavaca*) in a ritual *dakṣiṇā* or priestly gift at the one-day Soma rite known as the Apaciti (Jamison 1998, 255). The *Śāṅkhāyana Śrautasūtra* (14,33,20) mentions, 'The sacrificial fee is a horse-chariot, coated with rhinoceros-hide, covered with tiger fell, with a quiver boar-hide, with a bow-case of panther-hide, drawn by brown horses' (cited in Bautze 1985, 410). Similarly, the *Jaiminīya Brāhmaṇa* (II, 103) expounds, 'The *dakṣiṇā* for this (ritual) is a horse chariot, yoked with four (horses) . . . Its covering is made of tiger(skin), its bow-case of leopard(skin), its quiver of bear(skin). There is a mounted warrior, with armor of rhinoceros(-hide), girded (for battle), along with a girded charioteer' (cited in Jamison 1998, 255). What stands out is the nature of the animals prescribed for use, and it seems reasonable to concur with the view that their combined ferocity was possibly employed to compel the 'respect' for which the ritual was undertaken (Jamison 1998, 255).

Glimpses from the Pali *Tipiṭaka* and Beyond

In this corpus, animals occur most frequently in similes and metaphors, and the ones that lead the day are clearly the lion, the elephant, the horse, the bull, and the monkey (Gokhale 1974, 111). The pursuit of our protagonist leads us to the *Khaggavisāṇa Sutta* or the rhinoceros *sūtra* of the *Sutta Nipāta*, which immortalized the animal by extolling its solitary character.

The verses recurrently urge one to wander alone like a rhinoceros—*eko care khaggavisāṇakappo* (Hare 1947, 6–11). It may be relevant to delve upon the compound *khaggavisāṇa.* T.W. Rhys Davids and William Stede (1975, 230) point out that *khagga* in this context means the rhinoceros, *visāṇa* stands for the horn of a rhinoceros, and *kappa* as a figure of speech means like or resembling (Davids and Stede 1975, 187). *Khagga-visāṇa-kappa* is then interpreted as meaning 'like the horn of the rhinoceros'. On the other hand, Franklin Edgerton (1970, II: 202) reasons that since Sanskrit *khaḍga* and Pali 'khagga' mean the rhinoceros, the Pali commentary on the *Sutta Nipāta* paraphrases the term by 'rhinoceros horn'. However, it is argued that the compound means rhinoceros, which is equivalent to the Sanskrit *khaḍgin*, originally having a sword(-like) horn. The comparison, therefore, it is asserted, alludes to the animal, and not to its horn.

Though allegorical, this *sutta* shows remarkable familiarity with animal behaviour. The thrust unmistakably is on acquiring the resilience of a rhinoceros in wandering alone, but the mega herbivore shares this space with the elephant and the lion whose virtues are also considered worth abiding by.

As large and full-grown elephant,
Shapely as lotus, leaves the herd
When as he lists for forest haunts
Fare lonely as rhinoceros . . .

Like lion fearful not of sounds,
Like wind not caught within a net,
Like lotus not by water soiled,
Fare lonely as rhinoceros.

As lion, mighty-jawed and king
Of beasts, fares conquering, so thou;
Taking thy bed and seat remote,
Fare lonely as rhinoceros.

(*Sutta Nipāta* i, 3, 53, 71–2; Hare 1947, 8, 11)

Ekacara or *ekacarin* seems to be a rhino's characteristic because a prototype of Buddha is called *khaḍga* as he wanders alone (M.A. Mehendale's correspondence with Divyabhanusinh). Similarly, in early Buddhist sources, 'Pratyeka Buddhas, those who attain enlightenment for themselves alone', are also likened to the animal, and given the epithet *khaḍga/khaḍgin[a]/pratyeka-khaḍgin*. The creature is 'thus identified as the archetypal solitary beast whose behaviour solitary sages should emulate' (Jamison 1998, 252–3). Significantly, this ancient characterization of rhino behaviour fits in perfectly with modern studies testifying the solitary nature of the animal.

The Jātaka Tales

These parables narrating the stories of the Buddha's former births abound in references to his coming to life in the form of many animals.

The list is significant, and includes the fish, crab, cock, woodpecker, quail, pigeon, goose, crow, zebu, buffalo, monkey, elephant, antelope, deer, and horse (Auboyer 1972, 117). Strikingly, the rhinoceros is missing from this list though it does put in an appearance in the *Sudhābhojana Jātaka* (535) as part of the setting of a hermitage.

> Thus Honour, glorious nymph, at his behest
> In Kosiya's home was welcomed as a guest:
> Fruits and perennial streams therein abound,
> And thronging saints are in its precincts found.
>
> Here flowering shrubs in a dense mass we see,
> The mango, piyal, bread-fruit, Judas-tree;
> Here sal and bright rose-apple deck the glade,
> There fig and banyan cast their holy shade.
>
> Here many a flower with fragrance scents the wind,
> Here peas and beans, panic and rice we find:
> Bananas everywhere rich clusters show,
> And bamboo reeds in thickest tangle grow.
>
> On the north side, hemmed in by smooth and level bank,
> And fed by purest streams, behold a sacred tank . . .
> Hither do lions, tigers, boars resort their thirst to slake.
> This bears, hyenas, wolves are wont their drinking-place to make.
>
> The buffalo, rhinoceros and gayal too are here,
> With antelope, elk, herds of swine, and red and other deer,
> And cats with ears like to a hare's in numbers vast appear.
>
> (Francis 1969, V: 215–16)

Similarly, an enchanting landscape rich in fauna is envisioned in the *Vidhurapaṇḍita Jātaka* (545) when it mentions a magic jewel through which the entire world could be seen. On the one hand, are the splendours of the material world, on the other, is the untainted natural world.

> See on the slopes of the mountains troops of various deer, lions, tigers, boars, bears, wolves, and hyenas; rhinoceroses, gayals, buffaloes, red deer, rurus, antelopes, wild boars, niṁkas and hogs, spotted kadalī-deer, cats, rabbits, all kinds of hosts of beasts, created in the jewel.
>
> Rivers well-situated, paved with golden sand, clear with flowing waters and filled with quantities of fishes; crocodiles, sea-monsters are here and porpoises and tortoises . . .

> See too lakes well-distributed in the four quarters, filled with quantities of birds and abounding with fish with broad scales. See the earth surrounded by the sea, abounding with water everywhere, and diversified with trees,—all created in the jewel. (Cowell and Rouse 1969, VI: 135–6)

The term employed for the rhinoceros in the two tales is *palāsāda* and *palasata* respectively. Trenckner (cited in Hultzsch 1969, 127) points out that *palasata* (=Vedic *parasvat*) is the original of Pali *palāsāda*, 'a rhinoceros', and the latter, which literally means 'eating leaves', is an etymologizing corruption of the former.

In the *Vessantara Jātaka* (547), the rhinoceros and the buffalo are said to make the woodland ring (Cowell and Rouse 1969, VI: 258). In the same narrative, while the one-horned creature is a part of the canopy on the banks of lake Mucalinda (Cowell and Rouse 1969, VI: 278), when it comes to encamping there, it is also one of the forces to be kept at bay along with dangerous and wild beasts such as lions and tigers (Cowell and Rouse 1969, VI: 299). Such contexts give an inkling of the ecological sensibilities of ancient India by situating the rhino in environs typically conducive as habitats.

Milindapañha or *The Questions of King Milinda*

The narrative here centres around a chain of discourses between King Milinda and Nāgasena the Elder regarding numerous points of Buddhist doctrine. The aim is said to be didactic since the king makes his enquiries, and the monk addresses them with either a tale about the Buddha in a previous life or an analogy (Davids 1890, xvii). Having originated in north-west India, the work is assigned to the first century CE, and stands out amidst the great bulk of non-canonical Pali literature (Winternitz 1977, II: 174–5).

The text alludes to 'an elephant hemmed in by rhinoceroses' (Davids 1890, 38). The context is clearly metaphorical, but it may be worthwhile to mention that the elephant–rhinoceros animosity (not conclusively proved in the wild) is often referred to in later popular writings.

There is also a passing reference to the solitary character of the animal when Pacceka Buddhas are defined as those who are self-reliant, needing no teacher, and dwelling 'alone like the solitary horn of the rhinoceros' (Davids 1890, 158). I have already discussed the compound *khaggavisāṇa*, and given the context, would stand by the argument that the allusion in such cases seems to have been to the animal rather than to its horn since the interpretation that one should wander alone like a rhinoceros horn 'conjures up an unintentionally comic picture' (Jamison 1998, 253). After all, it is the animal which is known to be solitary in character, the horn being a mere appendage.

The Jaina World

The *Kalpa Sūtra* extols the fortitude of Mahāvīra by saying that he was single and alone like the horn of a rhinoceros (Jacobi 1884, 261). In this sense, the natural behaviour of the animal makes for an image which cuts across traditions.

The Dharmaśāstras

Within this corpus of the 'legal texts' of early India—a phrase used by Olivelle (2002b, 7), though he admits that their scope is much broader than the merely legal—we meet the mega herbivore in the context of prescriptions and proscriptions regarding food. Dharma literature makes a clear distinction between *abhakṣya* (forbidden) and *abhojya* (unfit) foods. While the former are prohibited because of their very nature, the latter include those that have become unfit for reasons such as having been given by an unfit individual, been touched by an impure person or animal, having got contaminated by an impure substance, or gone stale or bad (Olivelle 2006, 278). Thus, while the category of *abhakṣya* refers to the physical and biological world, the category of *abhojya* has to do with social boundaries, and with the pure/impure distinctions governing social relationships (Olivelle 2002a, 352–3).

Regarding the genesis of these lists, Olivelle (2002a, 346) underlines their absence in Vedic literature and in the Śrauta- and Gṛhyasūtras, and notes their first appearance in the Dharmasūtras. However, despite encountering them for the first time in the Dharmasūtras, the possibility of the existence of these lists, at least in informal ways before them has also been put forth (Olivelle 2002a, 346).

The Dharmasūtra texts widely forbid the eating of the flesh of 'five-nailed' (*pañcanakha*) or 'five-toed' animals, except for a restricted list comprising the porcupine, hedgehog, monitor lizard, hare, tortoise, and very often the rhinoceros, a strange inclusion, since the animals have only three toes.

However, before one sets out to explore the rather anomalous position of the rhinoceros in these texts of early India, it is necessary to have a sense of the parameters used for classifying animals in ancient India. Early Indian texts have employed different principles for classifying animals, where they have been distinguished by their anatomical characteristics or their mode of procreation; as either domestic (*grāmya*, 'of the village') or wild (*āraṇya*, 'of the jungle'); those suitable for sacrifice, and those which were not; and those which were edible or inedible (Smith 1991, 527–8). With reference to the last mode of classification, being 'five-clawed', the human foot served as an anatomical paradigm for identifying animals that were not to be eaten (Smith 1991, 529). Anatomical, cultural, ritualistic, and religious parameters could, thus, be employed to omit animals from the category of 'food' (Smith 1991, 538).

Within these classifications, the ones pertaining to pedal or foot structures concern us here since they lead to two categories of animals: those which have five toes with nails (*pañcanakha*), and those which have hooves, and are divided into double-hoofed (*dviśapha/dvikhura*) and single-hoofed (*ekaśapha/ekakhura*) animals. In the context of dietary regulations, both five-nailed (*pañcanakha*) and single-hoofed (*ekaśapha*) animals were prohibited. Five animals falling within the former category are, however, permitted (the refrain being *pañca pañcanakhā bhakṣyāḥ*): the hare, hedgehog, porcupine, tortoise, and monitor lizard. This list of exceptions often also includes the rhinoceros, clearly an oddity amidst those that accompany it, owing to its three toes and enormous size.

Let us proceed to examine some of these lists. In the section dealing with forbidden food, the *Āpastamba Dharmasūtra* (1:17.37; Olivelle 2003, 55) enjoins that animals with five claws are forbidden with the exception of the Godhā monitor lizard, tortoise, porcupine, hedgehog, rhinoceros, hare, and Pūtikhaṣa. A similar injunction occurs in the *Gautama Dharmasūtra*, which also forbids the eating of five-clawed animals with the same exceptions spelt out by Āpastamba excluding the latter's Pūtikhaṣa (17.27; Olivelle 2003, 165). The *Baudhāyana Dharmasūtra* permits the eating of the porcupine, Godhā monitor lizard, hare, hedgehog, tortoise, and the rhinoceros, specifying, however, that with the exception of the rhinoceros, these are the five five-clawed animals (1:12.5; Olivelle 2003, 223). The *Vasiṣṭha Dharmasūtra* excludes the rhinoceros from the list of animals with five claws (14.39; Olivelle 2003, 409) qualifying that there are conflicting opinions regarding the rhinoceros and the wild pig (14.47; Olivelle 2003, 409). In the same context, the *Mānava-Dharmaśāstra* enjoins,

> He must never eat those that wander alone; unknown animals or birds, even if they are listed among those that are permitted; as also all animals with five nails. Among animals with five nails, they say, the porcupine, the hedgehog, the monitor lizard, the rhinoceros, the tortoise, and the rabbit may be eaten; as also animals with incisors in only one jaw, with the exception of the camel. (5.17–18; Olivelle 2006, 139)

Overall what manifests here, are attempts to sanction the eating of rhino meat, which is further amplified when Āpastamba, for instance, tells us that the meat of a rhinoceros (*khaḍgamāṃsa*) offered on a rhinoceros skin (*khaḍgopastaraṇe*) is said to gratify ancestors for an unlimited time (2.17.1; Olivelle 2003, 99). Similarly, the *Gautama Dharmasūtra* details the periods for which offerings made to ancestors satisfied them, asserting that ancestors are satisfied for an unlimited time if amongst other things, meat of the rhinoceros mixed with honey is offered to them (15.15; Olivelle 2003, 157). The *Mānava Dharmaśāstra* (3.272; Olivelle 2006, 122) also spells out the periods for which the flesh of animals offered at ancestral rites satisfies them, and rhinoceros meat is listed amongst the items that are 'efficacious in perpetuity'.

Certain observations emerge from these injunctions. One is an obvious attempt to sanction the use of rhinoceros meat and skin. What is more subtle are the glimpses of early India's sense of the animal world. For instance, while the rhinoceros stands out as a definite oddity amongst the five-toed animals, one pauses to ponder over Baudhāyana's inclusion of the mega herbivore and his subsequent qualification that the ones mentioned with the exception of the rhinoceros were the five-clawed animals. So while the text first includes the rhinoceros in the edible *pañcanakha* animals, it goes on to set it apart from the group based on a recognition of the fact that it is not five-toed.

Similarly noticeable is the *Mānava Dharmaśāstra*'s (5.17) injunction against the eating of animals which wander alone even if they are a part of the permitted list. What ensues in the next provision (5.18) are the edible *pañcanakha* animals amidst which the rhinoceros is the *ekacara* (solitary) animal. In view of the structure of Manu's 5.17–18, it has been argued 'that the edible rhinoceros in Manu 5.18 (and elsewhere) was originally listed as an exception not to the ban on eating five-toed animals, but the ban on eating solitary, *ekacara*, animals' (Jamison 1998, 253). For whatever the arguments put forth may be, it is remarkable that ancient Indian jurists while dealing with legal and other issues were also closely and more or less accurately grappling with the faunal world, taking note of pedal structures and animal behaviour.

There have been attempts to explain why in the first place the *pañca pañcanakha* were singled out as edible, while other five-nailed animals were forbidden (Zimmermann 1987, 174). One is, however, more interested in understanding why ancient jurists would include an anomaly like the rhinoceros. Why despite an awareness of the incongruity of the inclusion (as seen in Baudhāyana, and to an extent in Manu) and voices of dissent (as seen in Vasiṣṭha) would the rhinoceros still be added to the original list of the five five-nailed animals? It evidently is a misfit since it is neither single- nor double-hoofed but triple-hoofed, and also wanders alone, and, thereby, falls into the prohibited category of solitary animals (Olivelle 2002b, 24).

Could this inclusion possibly be an extension of a tradition which had firmly embedded itself in early human consciousness? It may be worthwhile to remind ourselves of the importance assigned to the rhinoceros even in Vedic ritual contexts. What also deserves to be noted is how an interest in the skin of the animal (as seen in the Śrautasūtras) had over time graduated to an interest in its meat. As seen, the dharma texts unanimously reinforce the pre-eminence of rhinoceros meat in appeasing ancestors.

A significant and nuanced intervention in this context is made by Jamison, wherein she cites historical and linguistic reasons regarding how and why the rhinoceros came to be added to the list. The rhino, she argues, was clearly a later inclusion since 'older, rhinoceros-free versions of the five-nailed provision are found almost exclusively in *non*-dharma texts, while the updated version with rhinoceros is found across the *dharma sūtras* and *śāstras* but hardly elsewhere' (Jamison 1998, 250). This, according to her, is perplexing, but can be explained by the way most dharma texts of the time extolled rhino meat as the best food to be served to ancestors. She contends that this exalted status of rhino flesh in the food chain is rooted partly in Vedic ritual, and partly in textually unpreserved lore about the animal. Rhino meat, thus, came to top what Jamison (1998, 255) refers to as 'a particularly important food hierarchy' that was remembered and commended later even in the medical treatise of *Suśruta*.

Not just this, she points out that the skin and the horn of the animal were also accorded importance, and several dharma texts require that implements be made out of them (Jamison 1998, 255). These passages have then been perceived 'as transitional between the Vedic use of rhinoceros *hide* in the *dakṣiṇā* [customary fee given to the officiating priest] at the Apaciti ritual and the rhinoceros *meat* of the *śrāddha* [ceremony performed in honour of departed ancestors] dinner proper: from the artifact created from the formidable exterior of the rhino, its armour-like skin and sword-like horn, to the consumption of its actual flesh, presumably to internalize this same power' (Jamison 1998, 256).

In a similar vein, Zimmermann (1987, 183) reasons that rhino meat was consumed primarily within the ritual context of an ancestral offering, and that clearly it was not an animal one could have had for dinner on a regular basis. He also argues that the sacrifice of the rhinoceros or other wild animals for a *śrāddha* presupposed catching the victim in the forest, suggesting possible links between hunting and sacrifice though the two social activities in principle remained separate.

What clearly emerges from this maze of injunctions and the attempts to explain them is the interest ancient India had in the skin, horn, and meat of the rhinoceros. Though the contexts of use in Brāhmaṇical sources are primarily ritual, we shall see how this extended to strategic and medicinal realms as well. Not just this, perhaps the legends surrounding the efficacy of the horn (commonly written about in classical Western accounts regarding the animal) had their genesis here.

Ecology, Therapeutics, and Rhino Meat—The Legacies of Caraka and Suśruta

The theme of animal classification and a sense of the realms it extended to would be incomplete without a reference to the two early medical texts, the *Caraka Saṃhitā* and the *Suśruta Saṃhitā* (dated to the first half of the first millennium CE). Olivelle (2002b, 11) makes a crucial distinction between the injunctions in these texts and those spelt out in the Dharmaśāstras, arguing that the regulations in the former relate to the health benefits of various meats rather than to socio-religious prescriptions. He further observes that these medical treatises not only sanction the eating of many of the animals prohibited in the legal texts, but also attribute definite health benefits to them (Olivelle 2002b, 13).

Let us now proceed to look at some of these regulations. I begin with the *Caraka Saṃhitā*, considered to be the older of the two texts (Olivelle 2002b, 11). In an eightfold classification based primarily on feeding habits and habitats, the rhinoceros belongs to the *ānūpa* class of animals, or those which are the inhabitants of marshy lands (*Caraka Saṃhitā* I.27.39; Sharma 1994, I: 197). As for the meat of the animal, it is said to impart strength, and alleviate *vāta* (wind). It is sweet, unctuous (oily), nourishing, beneficial for complexion, and relieves fatigue (*Caraka Saṃhitā* I.27.84; Sharma 1994, I: 201). Olivelle (2002b, 25) contends that strength here may well suggest sexual potency. Elsewhere, the physician is urged to give the well-spiced meat of the animal in order to cure emaciation (*Caraka Saṃhitā* VI. 8. 154; Sharma 1994, II: 156).

Apart from an interest in the flesh of the animal, the text also mentions the use of its horn when it stipulates that amulets to be worn by a child were to be made of the tip of the 'right horns' of a living rhinoceros, deer, gayal or bull (*Caraka Saṃhitā* IV.8.62; Sharma 1994, I: 486). The allusion is ambiguous since it mentions the horn in plural, and that too a 'right' one. Nonetheless, it does give an inkling of the beliefs associated with the animal and its body parts.

Moving on to a more complex classificatory system in the *Suśruta Saṃhitā*, we encounter the *ānūpa* category again in our search for the rhinoceros, but here the category is divided into five subgroups where the rhinoceros along with elephants, boars, and other animals appears amongst those living on riverbanks (*kūlacara*) (*Suśruta Saṃhitā* I.46.49–50; Bhishagratna 1963, I: 487–8).

Significantly, the flesh of *kūlacara* animals is said to be spermatopoietic (sperm-producing), and eliminating deranged *vāyu* (wind) and *kapha* (phlegm). It is sweet in taste and cooling (*Suśruta Saṃhitā* I.46.51; Bhishagratna 1963, I: 488). Among the *kūlacara* animals, the flesh of the rhinoceros is said to have an astringent taste. The meat is pleasing to ancestors. It is sacred, imparts longevity, tends to suppress the discharge of urine, is dry, and pacifies *vāyu* and *kapha* (*Suśruta Saṃhitā* I. 46.53; Bhishagratna 1963, I: 489).

For fear of digressing from my focus on the rhino, I have not gone into the details of the classification in these texts since they involve a range of animals including birds and fishes, but what emerges from a look at these classificatory systems is an intense awareness of the ecologies of the animals mentioned. Also noteworthy is the preservation of the lore recommending rhino meat.

The *Khaḍga* in the Epics

The *khaḍga* occurs frequently in the epics, but in most cases the word denotes a sword. 'The "sword" sense is extremely common—especially in texts with a preoccupation with weaponry, like the battle books of the *Mahābhārata*' (Jamison 1998, 254). Hence, though the word is encountered quite often, my narrative will be confined only to contexts engaging with the animal in the two epics.

The Rāmāyaṇa

I begin with the *Rāmāyaṇa*, where Valmīkī recounts how Hanumān's arrival in Laṅkā is followed by his frantic search for Sītā. As part of this quest, Hanumān explores Rāvaṇa's palace, searches the harem as well as the drinking hall with all its splendour and the aroma of delicacies. We are told that

> Hanumān saw boars and *vārdhrāṇasakas*, prepared with curd and *sauvarcala* salt . . . porcupine, deer and peacocks; as well as various kinds of *kṛkaras* and half-eaten *cakoras*, buffalo, *ekaśalya* fish, and goats, all well-seasoned. And there were soft delicacies, all manner of drink, and

> every sort of food, along with various kinds of condiments: sour, salty and pungent.(*Sundarakāṇḍa* 5.9.13–15ab; Goldman and Goldman 1996, 139–40)

J.L. Brockington (1984, 93) emphasizes as an oddity the sole occurrence of the rhinoceros, *vārdhrāṇasa*, in the text, as one of the items in the banquet that Hanumān sees spread in Rāvaṇa's dining hall.

It may, however, be pointed out that the interpretation of the term '*vārdhrāṇasakas*' as rhinoceros does not appear in the translations of the *Rāmāyaṇa* either by Griffith (Canto XI, 1963b, 402) or by Shastri (1957, II: 362). Even the translation by Robert Goldman and Sally Sutherland Goldman (1996, 377) refrains from interpreting the term, and uses it verbatim, expounding that no commentator could specify with certainty what sort of a creature it was, most offering three alternative explanations, based on a purāṇic, lexical, or other source. Suggestions ranged from construing it as a bird with a black neck, red head, and white wings—possibly the black-necked stork; an old goat; and the rhinoceros (*khagamṛga*). Notwithstanding the thrill of finding the rhinoceros amidst the gastronomic spread for Rāvaṇa, I would argue (in view of the uncertainty regarding the term '*vārdhrāṇasakas*') that the presence of the animal in the text in general, and in the banquet hall of Rāvaṇa in particular, should best be considered a possibility rather than as a certainty.

Significantly also, the rhinoceros does not feature in the list of the five five-nailed animals whose flesh the epic says may be enjoyed by brāhmaṇs and kṣatriyas. The animals mentioned are the hedgehog, porcupine, lizard, rabbit, and turtle (*Kiṣkindhākāṇḍa* 4.17.34; Lefeber 1994, 90).

The Mahābhārata

Turning to the *Mahābhārata*, the *Ādiparvan* seems to preserve a rather quaint reference to the animal. Vasiṣṭha consecrates the Paurava as the sovereign of all baronage 'to become the one horn (*viṣāṇabhūtam*) of the entire wide earth' (*Ādiparvan* 1.89.39; Buitenen 1973, I: 212). The expression lacks clarity, but has been interpreted as a reference to the horn of the rhinoceros, a symbol of uniqueness and solitude (Buitenen 1973, I: 455). However, it is the *Karṇaparvan* (8.11.6; Bowles 2006, I: 157) that unambiguously establishes the identity of the animal as our one-horned hero when it describes a grim contest between Bhīma and Aśvatthāman, where the latter strikes the Pāṇḍava on the forehead with an iron arrow. Bhīma is then said to have borne the arrow protruding from his forehead like a 'proud rhinoceros bears his horn in the forest'. Capturing the burly giant is no mean feat since Bharata's prowess as a child is demonstrated by an account enumerating the dangerous species he overpowered in the wilderness. The rhinoceros, along with leopards, tigers, elephants, and lions, were the ones he is said to have subjugated and earned himself the epithet of the Great Tamer (*Droṇaparvan* 7.68.5; Pilikian 2009, II: 49). Elsewhere, a certain sage is reported to have lived in a forest, and among the animals terrifying to look at which gathered around him are lions, tigers, and rhinos. They came to meet him considering his good nature (*sadbhāva*) and behaved like his pupils. They inquired about his well-being and then went different ways (*Śāntiparvan* 12.117.3–8; M.A. Mehendale's communication with Divyabhanusinh).

Does the epic saga evince an interest in the utilization of the animal itself? Saṃjaya, the charioteer and confidant of Dhṛtarāṣṭra, narrates the magnificent scene of the consecration of Karṇa (according to scriptural prescriptions) on a seat made from *udumbara* wood and covered with linen, with sanctified golden and earthen pots, with water-filled horns of elephants, rhinoceroses, and great bulls, and others filled with jewels and pearls and pleasant-smelling herbs (*Karṇaparvan* 8.6.37; Bowles 2006, I: 125). Yudhiṣṭhira probes Bhīṣma regarding the duration of offerings that gratify the ancestors, and is told that the gratification received from the flesh of the rhinoceros (*khaḍgamāṃsa*) was inexhaustible (*Anuśāsanaparvan* 13.88.10; Ganguli 1970, XI: 145).

The *Khaḍga* and the Mauryas

In an unprecedented attempt to exert control over forest as well as wildlife, the Mauryan state adopted a strict policy towards both. Emblematic of this attitude is the *Arthaśāstra* of Kauṭilīya, which spells out the following as forest produce to the Director of Forest Produce: 'Skin, bones, bile, tendons, eyes, teeth, horns, hooves and tails of the lizard, *seraka*, leopard, bear, dolphin, lion, tiger, elephant, buffalo, *camara*, *sṛmara*, rhinoceros, bison and *gavaya*, and also of other deer, beasts, birds and wild animals' (*Arthaśāstra* 2.17.13, Kangle 1963, II: 149). Similarly, the Superintendent of the Armoury is instructed to arrange for 'machines for use in battles, for the defence of forts and for assault on the enemies' cities' (*Arthaśāstra* 2.18.1, Kangle 1963, II: 150). *Nistriṁśa*, *maṇḍalāgra* and *asiyaṣṭi* are swords whose hilts are formed by the horn of the rhinoceros and buffalo, the tusk of the elephant, wood and bamboo-roots (*Arthaśāstra* 2.18.12, Kangle 1963, II: 152). It is also said that 'a coat of mail of metal rings or metal plates, an armour of fabrics and combinations of skin, hooves and horn of dolphin, rhinoceros, *dhenuka*, elephant and bull are armours' (*Arthaśāstra* 2.18.16, Kangle 1963, II: 152).

The text, thus, reflects an unambiguous interest in the skin and horn of the rhinoceros, and this seems absolutely credible in view of Abul Fazl's testimony to the use of breastplates and shields made of rhinoceros skin, and finger-guards for bow strings from its horn even during the reign of Akbar (cited in Ali 1983, 16).

Beyond its economic and political uses, the animal seems to have been used for contests to entertain the sovereign as well as a befitting item of gift for him. Drawing on classical Western accounts, which shall subsequently be examined in detail, Radha Kumud Mookerji (1960, 61–2), in his reconstruction of the times of Candragupta Maurya, cites the testimony of Claudius Aelianus (henceforth referred to as Aelian) to the king having enjoyed seeing animal fights in his arenas. These included wild bulls, tame rams, rhinos (Aelian refers to them as 'unicorn asses'), tusked elephants, and more. Aelian is once again tapped to tell us that the animals brought to the king as gifts included stags, antelopes, gazelles, oryxes, and rhinos.

For our next encounter with the mega herbivore, we turn to the edicts of Emperor Aśoka (268–232 BCE), which mention a range of animals. Chakravarti (1906, 361–74) lists the fauna named in rock edict I and in pillar edict V. These two edicts, he asserts, are the well-known *ahiṃsā* orders of the emperor, where the former forbade the general destruction of life both in his own kitchen and in his empire, while the latter specified a number of animals which should not be killed or cruelly dealt with (Chakravarti 1906, 363).

Amplifying the general edict about *ahiṃsā* or non-destruction of life, it is in the fifth pillar edict that we meet the rhinoceros. The proclamation clearly indicates that human depredations on wildlife had begun as it decrees: '(When I had been) anointed twenty-six years, the following animals were declared by me inviolable, viz. parrots, *mainas*, . . . the rhinoceros, white doves, domestic doves, (and) all the quadrupeds which are neither useful nor edible' (Hultzsch 1969, 127). Aśoka's word for the rhino here is *palasata.*

Westerly Encounters

Turning to the earliest Western accounts of the animal, we have the fabulously magnified classical accounts hovering between legend and history. I begin with the account of Ktesias, a native of Knidos, an ancient Greek city in southwest Asia Minor (modern Turkey). Ktesias earned himself the distinction of being the first writer to give to the Greeks the only systematic account of India till the time of the Macedonian invasion.

Having spent seventeen years of his life serving as royal physician in the court of Persia, Ktesias's own understanding of India was based on reports of Persian officials who had visited the country, as also on reports of Indians who visited the Persian court either as merchants or tribute bearers from the princes of northern India, then subject to Persian rule. Traditional verdicts condemned him as a writer of 'unscrupulous mendacity' fabricating 'tales of wonder' though more moderate views credit his writing with certain elements of truth, and attribute the fallacies therein to 'misconceptions which were perhaps less wilful than unavoidable' (McCrindle 1882, 1–5). Similar judicious approaches have reminded us that hardly anything that can be attributed directly to Ktesias has survived—the overwhelming majority of information attributed to him having been handed over to us by other authors who did not always exercise caution while using their source. The same note suggests that we should no longer regard Ktesias primarily as a historian but as a precursor of a new literary genre combining historical fact with fictitious elements (Stronk 2007, 25).

What remains of the *Indika* of Ktesias (dated c. 400 BCE) is its abridged version by Photios, the Patriarch of Constantinople in 858 CE, and fragments available in the works of other writers. Ktesias is invaluable to our narrative, and it is worth quoting him at some length because of his elaborate description of a creature which seems to be the rhinoceros, though not all the details furnished by him correspond with the animal. Notwithstanding the garbled nature of the references, the importance of the text lies in its being the first allusion to the Indian rhino by a Western author. It would also serve us well to remember that subsequent accounts of rhinos were usually flavoured strongly with the 'original, rather fanciful, description' furnished by Ktesias (Laurie 1978, 3).

Coming to what Ktesias has to say about the rhinoceros, it is imperative to begin with his well-known though rather lengthy description of the creature which to him was a horned wild ass.

> Among the Indians, there are wild asses as large as horses, some being even larger. Their head is of a dark red colour, their eyes blue, and the rest of their body white. They have a horn on their forehead, a cubit in length [the filings of this horn, if given in a potion, are an antidote to poisonous drugs]. This horn for about two palm-breadths upwards from the base is of the purest white, where it tapers to a sharp point of a flaming crimson, and, in the middle, is black. These horns are made into drinking cups, and such as drink from them are attacked neither by convulsions nor by the sacred disease (epilepsy). Nay, they are not even affected by poisons, if either before or after swallowing them they drink from these cups wine, water or anything else. While other asses moreover, whether wild or tame, and indeed all other solid-hoofed animals have neither huckle-bones, nor gall in the liver, these *one-horned asses* have both. Their huckle-bone is the most beautiful of all I have ever seen . . . It is exceedingly fleet and strong, and no creature that pursues it, not even the horse, can overtake it.
>
> On first starting it scampers off somewhat leisurely, but the longer it runs, it gallops faster and faster till the pace becomes most furious. These animals therefore can only be caught at one time—that is when they lead out their little foals to the pastures in which they roam. They are then hemmed in on all sides by a vast number of hunters mounted on horseback, and being unwilling to escape while leaving their young to perish, stand their ground and fight, and by butting with their horns and kicking and biting kill many horses and men. But they are in the end taken, pierced to death with arrows and spears, for to take them alive is in no way possible. Their flesh being bitter is unfit for food, and they are hunted merely for the sake of their horns and their huckle-bones. (McCrindle 1882, 26–7)

Another fragment of the *Indika* of Ktesias is extracted from Aelian, who attributes the following observation to him:

> These particoloured horns are used, I understand, as drinking cups by the Indians, not indeed by people of all ranks, but only by the magnates, who rim them at intervals with circlets of gold . . . They roam about in the most desolate tracts of the Indian plain . . . Their mode of attack is to charge the horsemen, using the horn as the weapon of assault, and this is so powerful, that nothing can withstand the blow it gives . . . They sometimes even fall upon the horses, and so cruelly rip up their sides with the horn that their very entrails gush out. The riders, it may well be imagined, dread to encounter them at close quarters, since the penalty of approaching them is a miserable death both to man and horse. And not only do they butt, but they also kick most viciously and bite . . . they must be dispatched with such missiles as the spear and the arrow. This done, the Indians despoil them of their horns, which they ornament in the manner already described. The flesh is so very bitter that the Indians cannot use it for food. (McCrindle 1882, 54–6)

How precise or vague Ktesias was in his description of the actual Indian rhino can be ascertained when we closely scrutinize his observations in the light of what we know about the animal. Disconnecting fact from fiction, Lassen (cited in McCrindle 1882, 75–6) points out that the wild ass was distinguished by Ktesias by his horn, by the gall on his liver, and by his ankle bone, but reasons that while a large gall bladder is something that the rhinoceros possesses, ankle bones are common to all quadrupeds. This, he argues was, therefore, an error of the author, though one that was surprising since he was a physician, and had himself seen such ankle bones. Lassen is likewise sceptical about the red colour attributed to the animal by Ktesias, as also about the great swiftness he credited it with (the rhinoceros though is known to charge with speed and force for short distances).

Nevertheless, Lassen (cited in McCrindle 1882, 76) infers that by piecing these remarks together, it seems probable that by the wild ass Ktesias meant the rhinoceros, because no other Indian animal suits the description better. As against its actual grey-brown colour, the red head and white body attributed to the animal by Ktesias, he reasons, could have been because he had been so informed.

A departure from this position regarding the colours assigned to the animal by Ktesias comes from Ball, wherein he rather amusingly reasons that the horned ass of Ktesias had probably been 'white-washed', and had had his horn painted blue and scarlet by his owner 'who little foresaw what food for discussion and comment he was affording, by that simple act, to twenty centuries of philosophers and historians' (1879–88, 318).

Ball's (1879–88, 317–18) own contention was that the rhinoceros from which the description of Ktesias emanated was a tamed one which had been painted according to native tastes to take part in some pageant. He substantiates this line of reasoning by citing from Rouselet's work on the *Native Courts of India* where he found an account of a rhino fight at Baroda which took place before the Gaikowar. Chained at opposite sides of the arena, one of the animals was painted black, and the other red so that they could be distinguished from each other. Ball argued that at the time when he was writing, tamed rhinoceroses were still kept by many natives, and were often trained like elephants to carry howdahs with riders in them. He also recounted how he had once encountered a native dealer in animals who had taken with him over a long distance through the jungles, a rhinoceros which he ultimately sold to the ruler of Jaipur, and had also told Ball how he had driven the animal before him as if it were a cow.

Reports of Indian rhinos being tamed and trained by humans also lend credence to Ball's contention. Laurie (1978, 4), for instance, quotes early accounts of rhinos being used to pull ploughs in Assam. He also cited recent experiences in Indian zoos confirming that though dangerous and unpredictable, Indian rhinos could be tamed and trained.

The original treatise being lost, there is no way we can establish what inspired Ktesias to generate the image of a colourful rhinoceros, if at all it was his own construct, or whether it was based on what he had been told, or merely the consequence of misinterpretations and distortions in the works of later writers who serve as the only means of accessing Ktesias. Nevertheless, Ball's speculative stance regarding decorated rhinos being part of pageants or public spectacles comes as no surprise if we also turn to Francois Bernier's *Travels in the Mogul Empire*, which is an account of his travels in India in the seventeenth century. In describing a march made with the camp of Emperor Aureng-Zebe to Kachemire, he describes as part of the retinue, 'the lions and the rhinoceroses, brought merely for parade' (Constable and Smith 1914, 364).

A brief reference to the one-horned Indian ass by Aristotle (384–322 BCE), a contemporary of Alexander of Macedon, is also a likely allusion to the Indian rhinoceros when he says, 'We have never seen a solid-hoofed animal with two horns, and there are only a few of them that have one horn, as the Indian ass and the oryx' (cited in McCrindle 1896, 187).

The knowledge of India among the nations of the West would perhaps have advanced little beyond where Ktesias left it if Alexander had not marched into the plains of India spanning the Indus. The invasion in 326 BCE, 'drew aside the veil which had till then shrouded India' from the rest of the world by bringing with it a number of Alexander's officers and contemporaries who were recording their impressions of India in the memoirs they composed (McCrindle 1896, 6). Regrettably since these writings have not survived, glimpses of their contents are only available to us from the histories of Alexander compiled several centuries after his death by writers such as Diodorus the Sicilian (100 BCE–100 CE), Quintus Curtius Rufus (100 CE), Plutarch (46–120 CE), Arrian (200 CE), and Justinus Frontinus (around 500 CE).

Turning to these accounts in search of the rhino, it is the narrative of Q. Curtius Rufus which comes to our aid. As mentioned earlier, it is one of the five accounts of Alexander's Indian campaigns compiled several centuries after his death on the basis of the works of writers who either witnessed the events they described or lived at the time they unfolded. Notwithstanding the uncertainty surrounding his life, and the time in which he lived, this historian stands out by virtue of his eloquent writing style as one of the most popular of the classical authors. In his description of India, he affirms, 'The same country yields fit food for the rhinoceros, but this animal is not indigenous' (McCrindle 1896, 186). Curtius is clearly erring here, as McCrindle (1896, 186) also points out, arguing that not only is the rhinoceros bred in India, but that the brāhmaṇs allowed its flesh to be eaten though most other kinds of animal food were prohibited.

Similarly, it is through Curtius that we know that following the arduous yet memorable triumph at Hydaspes, when Alexander is rousing his soldiers for the conquest of the east, they are told that the region was abundant in timber, and also had the rhinoceros, an animal rarely found elsewhere (McCrindle 1896, 216).

As we sift through images of the animal in these early Western writings, a treatise that comes our way is the *Indika* of Megasthenes, parts of which can also be retrieved only through the works of later writers. Nevertheless, being the most complete first-hand account of India then known to the Greek world (despite its occasional inaccuracies), glimpses from the *Indika* can be perceived as far more reliable than the mosaic of images culled from writers who never visited India themselves, and were writing on the basis of wisdom received from diverse sources.

Strabo (c. 60 BCE–19 CE), for instance, cites Megasthenes when he mentions one-horned horses with heads like those of deer (McCrindle 1901, 59). This, McCrindle (1901, 59) remarks, is a rather curt summary of the observations of Megasthenes, found at considerable length in the work of Aelian who goes on to describe a one-horned animal called the *Kartazon*. The description though inaccurate, unmistakably relates to the rhinoceros.

> It is of the size of a full-grown horse, and has a crest, and yellow hair soft as wool. It is furnished with very good legs and is very fleet. Its legs are jointless and formed like those of the elephant, and it has a tail like a swine's. A horn sprouts out from between its eyebrows, and this is not straight, but curved into the most natural wreaths, and is of a black colour. It is said to be extremely sharp, this horn . . . The males are reported to have a natural propensity not only to fight among themselves, by butting with their horns, but to display a like animosity against the female, and to be so obstinate in their quarrels that they will not desist till a worsted rival is killed outright. But, again, not only is every member of the body of this animal endued with great strength, but such is the potency of its horn that nothing can withstand it. It loves to feed in secluded pastures, and wanders about alone, but at the rutting season it seeks the society of the female, and is then gentle towards her,—nay, the two even feed in company. The season being over and the female pregnant, the Indian *Kartazon* again becomes ferocious and seeks solitude. The foals, it is said, are taken when quite young to the king of the Prasii, and are

set to fight each other at the great public spectacles. No full-grown specimen is remembered to have ever been caught. (McCrindle 1877, 59–60)

The horn of the rhinoceros, which seems to have enthralled without exception all classical writers, also figures frequently in the items of trade mentioned in *The Periplus of the Erythrean Sea* attributed to an anonymous writer of the first century CE. An extract mentions a Chinese account referring to the ninth year of the Yen-hsi period, during Emperor Huan-ti's reign when the king of Ta-ts'in, An-tun sent an embassy which from the frontier of Jih-nan (Annam) brought ivory, rhinoceros horns, and tortoise shell. Ta-ts'in is argued to have been the Chinese name for the Roman Empire, while An-tun is Antonius, the family name of Marcus Aurelius, and the date is put at c. 166 CE. The embassy seems to have comprised a group of Western merchants trying to buy silk directly from the Chinese instead of through Indian middlemen. What they offered in exchange—ivory, rhino horn, and tortoise shell were all available in India (Casson 1989, 27).

Living around the middle of the second century CE, and writing *On the Peculiarities of Animals*, Aelian also has much to tell us about the animals of India. A teacher of rhetoric, settled in Rome, his book was regarded as a standard work on zoology where he mentions a horn brought to Ptolemy the Second from India which held three amphorae (about 26 gallons), and presumes it to have been from an ox 'which grew a horn so prodigious'. He reports being told about the breeding of one-horned horses and one-horned asses in India, and affirms that drinking cups were fashioned from their horns, and if a deadly poison was thrown in, the drinker would escape unharmed since the horn both of the horse and of the ass is an antidote against poison (McCrindle 1901, 136). More significant is the fragment where he recounts how the 'great King of the Indians' chose a day every year for fighting between men as also between brute animals which were horned.

> These butt each other, and with a natural ferocity that excites astonishment, strive for victory, just like athletes straining every nerve whether for the highest prize, or for proud distinction, or for fair renown. Now these combatants are brute animals—wild bulls, tame rams, those called

> *mesoi,* unicorn asses, and hyaenas, an animal said to be smaller than the antelope, much bolder than the stag, and to butt furiously with its horns. (McCrindle 1901, 145)

The wonder with which the Western world perceived the one-horned rhinoceros, along with the fact that many of the classical writers had not seen the animal themselves, perhaps accounts for some of the curious imageries we encounter. Yet, not all can be attributed to imagination since the mention of the single horn, the habitat of the animal, its solitary behaviour as well as the reference to the tradition extolling the properties of rhino horn assure us of being on apt terrain in our quest for the armoured giant.

Come back, O Tigers! to the wood again,
And let it not be levelled with the plain;
For, without you, the axe will lay it low;
You, without it, for ever homeless go.

—*Vyaggha Jātaka*
(No. 272; Rouse 1969, II: 244–6)

3

BEYOND THE GLITTERING EYE
TIGER TALES FROM ANCIENT INDIA

THE EYES EVOKE MORTAL FEAR, the gait exudes elegance, and the stripes on the orange coat symbolize magnificence and beauty. The biggest and one of the most charismatic of the big cats, the tiger stealthily makes its way through the shadows of the day and the stillness of the night. Typically an animal of the forest and jungle, its adaptability enables it to occupy varied habitats ranging from sea level to elevations of almost 3,000 m. Also remarkable is the predator's ability to adapt to the low temperatures of the Russian Far East as well as the heat of northern India (Karanth 2002, 31–2).

Cultural fascination with this mega carnivore has transcended spatial and temporal barriers, yet wildlife histories have done little to salvage it from the shadows of the past, particularly with reference to ancient India. Any attempt to reconstruct the tiger trail would, therefore, provide valuable glimpses into India's ancient ecological past.

TRACES IN ARCHAEOLOGY

Before I embark on my quest for the animal, it is important to underline the dearth of fossil evidence for it within the Indian subcontinent in general, and the study area in particular. For instance, though explorations in the Manjra Valley have led to the discovery of fossil remains of the species in the Late Pleistocene fossiliferous gravels near Harwadi, Latur district, Maharashtra, it is imperative to point out that this is only the second recorded occurrence of the species in peninsular India after its appearance in the Hunsgi Baichbal Valley (Sathe 2015, 1). The scarcity of carnivores such as the tiger in Indian Quaternary deposits is explained by the fact that these animals occupy the topmost position in an ecological pyramid, feeding easily on scavengers and herbivores which have a higher rate of reproduction and are, hence, more in number than these carnivores (Badam, personal communication).

The archaeozoological record also renders patchy evidence when it comes to telling the story of this daunting carnivore. Within the study area, the faunal remains of the animal, though sparse, have been retrieved from diverse ecological niches and cultural and temporal contexts. Its rarity in archaeological deposits can possibly be explained by the infrequency of its interaction with humans, as also the fact that the animal did not feature significantly in the human scheme of affairs. Scattered as the evidence may be, it needs to be collated in order to give us a sense of the heterogeneous ecologies that harboured the animal. The majestic carnivore, however, makes up for its negligible presence in faunal records. Depictions in rock paintings being rare in the study area, I will explore how the animal can also be looked at through terracotta seals and figurines.

The Mesolithic

Mahadaha: I begin with the Middle Ganga Valley where the mesolithic faunal assemblage at the site of Mahadaha revealed the presence of at least three felid species: the jungle cat (*Felis chaus*), the tiger (*Panthera tigris*), and the panther (*Panthera pardus*). The Western Area (cemetery-cum-habitation area) yielded a complete third phalanx of a tiger. The bone belonged to layer 1, and had no modification marks of any kind. Another phalanx of the animal interestingly came from grave VIII sealed by layer 2. Though completely charred, it showed no evidence of having been cut. This also reminds us of evidence of human–carnivore association from grave no. XXI, where a mandible attributed to a jackal or young wolf was found near the ninth thoracic vertebra. Layer 9 in the Western Area yielded another bone fragment of a large felid. It was, however, not possible to definitely identify whether this charred orbital bone belonged to a tiger or a panther (Joglekar et al. 2003, 102, 105).

The Harappan Testimony

The archaeozoological record falls silent when we probe the published faunal reports of Harappan sites in search of the striped predator. Despite the absence of actual skeletal remains of the animal in the Harappan context, representations on seals, copper tablets, and terracottas suggest definite familiarity with it in Sindh, Punjab, and beyond at this time. Iravatham Mahadevan's (1977, 793) tabulation of the frequency of animal devices on Indus seals lists sixteen seals with tiger motifs, the depictions being generally with a trough in front, and five representations of a horned tiger. I will attempt to bring together a few such retrievals (with an emphasis on the seals) from individual sites. The aim will be to shed more light on the context and significance of such portrayals, as also to highlight the centrality attributed to the human–tiger liaison in the Indus belief system.

At Mohenjodaro in Sindh, for instance, the tiger occurs frequently on seals. What distinguishes the animal on them are its stripes. In what has been considered as the finest and one of the most realistic representations of the animal at the site, the carnivore is shown with an open mouth and a protruding tongue. Double wavy lines broken at intervals to relieve the monotony were employed to depict the broad stripes on the body (Figure 3.1). The engraver's acquaintance with the animal comes forth in the realism employed while portraying the folds of skin in the upper jaw caused by the opening of the mouth (Mackay 1938, I: 330, 658).

In another two, the animal is peculiarly represented with a manger before it as in the case of the short-horned bull and the rhinoceros (Mackay 1931b, II: 387). The presence of the trough in front of the animal seems strange since we know that it cannot be easily tamed. I have discussed in Chapter 2 Marshall's (1931, I: 70) surmise about the troughs symbolizing food offerings, and their presence implying that the animals to which these offerings were made, whether in captivity or in the wild, were objects of worship.

Figure 3.1 Broken tiger seal, Mohenjodaro. Note the wavy lines indicating the stripes
Source: National Museum, New Delhi, Collection Acc no.: DK 12245/117.

Shubhangana Atre (1990, 44) points out that there are no signs of these animals being bound by any rope or chain, and that it is absolutely unlikely that animals such as the tiger and the rhinoceros could be kept in a captive state without either heavy shackles or a cage. Underlining the religious and mythological bearing of the motifs on the seals and their artistic quality, she argues that an analysis of seals from Mohenjodaro and Harappa (which have yielded the maximum number of seals) suggests that the motifs, in all probability, whether 'theriopic', 'anthropic', 'therianthropic' (combining the form of an animal with that of a man), or narrative, formed part of 'a single continuous chain of, in all likelihood, mythological episodes'. Each single motif, she argues, was picked up as a 'still' forming important links in that chain (Atre 1990, 43). The evidence I attempt to put together seems to support Atre's hypothesis, though one is uncertain about their being parts of a single continuous chain of mythological episodes as she has argued. However, the depictions certainly seem to suggest an association with the realm of ritual, legend, and mythology.

Let us begin with the most celebrated and intensely debated iconographic representation on seal no. 420 (Mackay 1938, II, Plate XCIV) at Mohenjodaro (Figure 3.2). The seal, which has been a subject of scholarly scrutiny since the time of its discovery, shows a half-human and half-animal horned figure seated on a dais in a

Figure 3.2 The celebrated *Paśupati* seal, Mohenjodaro
Source: © John Marshall/Harappa.com, reproduced with permission.

meditative posture. Famously and variously perceived as *Paśupati* or the prototype of the historic Śiva (Marshall 1931, I: 54–5), the 'Mistress of the Animals' (Sullivan 1964, 122), or the 'Lady of the Beasts' (Atre 1985–6, 9), the appellations male or female, in either case, evidently have to do with the animals surrounding the therianthropic deity, and it is these which are primarily of interest to us. The animals include the elephant, tiger, rhinoceros, and buffalo. Beneath the dais on which the figure is seated, are what Marshall identified as deer or ibexes (Marshall 1931, I: 55), though Atre (1990, 44) perceived them as composite animals.

In a close study of the iconographic detailing of the animals on the seal under consideration, Alexandra van der Geer (2008, 372) observes how the stripes on the forepart and on the front limb of the tiger run oblique, those on the middle part of the trunk vertical, and those on the hind part horizontal. Though she does not qualify as to how she makes her observation regarding the hind part, which is barely visible due to the damaged condition of the seal, overall, she argues that the pattern of the stripes gives a more realistic impression. The circle around the eye, she contends, is, however, exaggerated since tigers have unmarked eyes.

It may be worthwhile to bring together some observations that have served to shape our understanding of this representation. Since my purpose here is to engage primarily with the animals on this seal, and to assess their relation with the horned figure, I shall steer clear of positions that have debated and deliberated upon the iconography and sex of the figure, unless relevant to the enquiry.

Atre (1990, 45), for instance, underlines some of the obvious details differentiating the animal categories on the seal. The first is that while some of the animals consistently appear along with the manger, others never do so. Second, two of the ones (the rhinoceros and the tiger) that are portrayed with the manger are among the ones surrounding the 'goddess', though they sometimes occur even independently without it. Additionally, she argues that these two are placed at opposite ends of the same diagonal axis, whereas the other two animals, the elephant and the buffalo, which are never associated with the manger, are placed on the other diagonal axis cutting the first one.

A fundamental problem with Atre's classification of the animals represented comes forth when contrary to her assertion we encounter seals where the buffalo as well as the elephant are depicted with feeding troughs (Marshall 1931, I: 70). Again, Atre (1990, 45) does not push this observation enough to clarify its implications, but she goes on to argue that if the association of wild animals such as the tiger and the rhinoceros with the manger, as also their subsequent 'pacifity', indicate the prowess of the goddess over them, animals without it would certainly suggest a different status quo.

I would argue that Atre's position perhaps needs reconsideration, since divorced from her interpretation of the seal, what emerges from a careful look at it immediately seems to unsettle views espousing that the animals depicted appear to be at peace (Atre 1987, 195). While that maybe the case with the elephant (significantly, it is the only animal which seems to be moving away from the central figure), the rhinoceros, and the buffalo, the tiger with its open mouth and claws is perceptibly an exception. The seal though mutilated from the corner where the tiger is depicted, makes it clear that while all the other animals are on their fours, the tiger stands on its hind legs as if rearing to pounce on the figure.

This seal has also received considerable attention in the writings of During Caspers (1987, 1989, 1992), who associates it and similar representations on another seal and some 'sealing amulets' with hunting magic. However, my focus will be on her analysis of seal no. 420 since this alone carries the four animals surrounding the central figure. During Caspers' interpretation becomes particularly relevant to our discussion in view of the centrality she attributes to the human–tiger liaison in the Indus belief system.

Emphasizing that the face itself is exceptional, During Caspers argues that the entire frontal composition seems to suggest a human figure 'wearing the enshrouding mask of a horned tiger' (1987, 221–2). The positioning of the comparatively large figure amidst the surrounding wild animals, she asserts, can be construed as bearing a connection with hunting magic. She further expounds that the attitude of the figure with its extended arms may be interpreted as one of dominance and subordination as well as of protection. Also,

> If the tiger be regarded as the animal *par excellence* in this part of the ancient world, then it is but a short step to see it virtually as 'King of the Beasts'... The magic connotation relating the seated man ... playing the dimly glimpsed part of hunting sorcery, may well also represent

> a magical transference of the power of the animal to that of the man, thus creating the man-god as 'Lord of the Beasts', and strengthening the suggestions of Marshall, Wheeler, Rao and others, that we are, indeed, contemplating an early, proto-form in the evolution of Śiva. (During Caspers 1987, 221–2)

The four wild animals, including the tiger, which surround the 'magic hunter' can be seen as the hunter's prospective game which he brings under his spell in a ritual dance, and so can be the goat or gazelle-like animals under the hunter's stool (During Caspers 1989, 235).

A closer look at the seal tells us that while During Caspers' suggestion regarding a feline character of the face cannot be ruled out, her notions of 'dominance', 'subordination', and 'protection' describing the attitude of the central figure call for reconsideration. It seems reasonable to argue that seated amidst the four wild animals there is nothing to suggest that the central figure is in any way inhibited by its surroundings. Neither does the ready-to-charge image of the tiger allow us to infer a hegemonic hold of the central figure over the setting. Similarly questionable is During Caspers' contention regarding the 'magic hunter' having cast a spell on the animals within the frame of the seal. If that is the case, then the tiger clearly stands out since the spell seems to have failed to have had its effect in subduing the charging animal.

Having said this, I, however, concur with the contention that there existed a special relationship between man and tiger in Indus beliefs, since the intaglios repeatedly show the animals looking backwards at human figures perched on trees, or present composite forms with a frontally shown human being wearing horns and being attached to the body and hindquarters of a tiger, or a semi-bovine, semi-human figure fighting a tiger with tall incurving horns. In fact, the horned mask comprising the frontal aspect of a tiger's face with the side faces being those of a human also seem to be 'an ingenious solution for expressing this interdependence' (During Caspers 1989, 235).

The representations mentioned by During Caspers can be taken up individually. Of special interest at Mohenjodaro are seals on which a man is seated on a tree with a tiger below watching him attentively (Figure 3.3). The man is holding on to the tree with one hand, while the other he holds out as if 'apostrophizing' the animal. Since the scene occurred on two seals, the representation was not considered an unusual one (Mackay 1931b, II: 387–8). Marshall (1931, I: 71)

Figure 3.3 The oft-repeated theme of a man on a tree and a tiger below, Mohenjodaro
Source: National Museum, New Delhi, Collection Acc no.: DK 7033/146.

was emphatic in arguing that such seals cannot be classed with the great majority of seals on which the animal is represented alone, his own contention being that they were possibly used as protective amulets against tigers and other jungle animals.

Another four seals bear an unusual scene which portrays a muscular hero or a deity gripping a tiger on either side of him by the throat. The animals with clearly delineated striped bodies are shown on their hind legs with open mouths and claws. The nude figure was paralleled with certain figures of Gilgamesh, and a definite Sumerian or Elamite influence was inferred from these four seals. It was also argued that tigers were substituted for lions only to correspond to Indian fauna (Mackay 1938, I: 337).

Yet another seal at Mohenjodaro depicts a partly human form struggling with a fabulous beast whose stripes suggest that it is a tiger (Figure 3.4). There are, however, horns on its head without which the representation can be perceived as an exceptional one of the carnivore with its mouth open, and the canine teeth showing. The scene was considered a reminder of seals and other objects from Sumer, where Enkidu is shown combating a lion, and Gilgamesh a bull (Mackay 1931b, II: 389).

A far more extraordinary depiction was what apparently seemed like a combination of a woman with a tiger's body but the forelegs of a human (Figure 3.5). The prominent pigtail was considered the distinguishing mark of a deity, while the close association of the figure with a tiger was perceived as an indication of it not being of a benign nature. The horns, on the other hand, suggested association with an animal remarkably opposite in nature to the tiger. There

Figure 3.4 A partly human form combating a horned tiger, Mohenjodaro
Source: National Museum, New Delhi, Collection Acc no.: VS 1574/144.

was also an attempt to associate the figure with either Durgā or Śiva depending on whether the human part of the composite figure was construed to be male or female (Mackay 1938, I: 339).

One would, however, steer clear of such interpretations, and explore alternative ways of perceiving such depictions. It is significant that just at the point where the body of the predator should have carried its head, it merges with the body of the woman. Could this possibly symbolize an amalgamation of human and animal power involving the assimilation of the prowess of the carnivore, or can the motif of the figure with the elaborate head gear and an arm extended over the body of the tiger be perceived as an indicator of a person of some importance within the Harappan social realm? Kenoyer (2010, 43) contends that in such representations it is not possible to determine if the animal or the human component is more important, but the prevalence of both suggests that they were important symbols, and that the line between human and animal was quite fluid in terms of ideological representation.

Figure 3.5 An enigmatic representation from Mohenjodaro. Significantly, the same motif is also encountered at Kalibangan
Source: National Museum, New Delhi, Collection Acc no.: DK 8203/141.

Additionally, the animal is employed quite frequently in the depiction of composite and fabulous animals. We can think, for instance, of the horned tiger at Mohenjodaro (see Figure 3.4). The stripes unambiguously reveal the identity of the animal, yet the head carries tall incurving horns.

Also, in a complex design on a broken seal at Mohenjodaro (Figure 3.6), the animals appear to be tigers and their three bodies cross one another in the centre (Mackay 1931b, II: 390).

From a ring-like motif on another broken seal at the site, the heads and necks of six animals radiate outwards (Figure 3.7). Of the four unbroken heads, one belonged to a unicorn, another to a short-horned bull, the third to an antelope, and the fourth to a tiger. The remaining two heads were attributed to a rhinoceros and an elephant (Mackay 1931b, II: 390).

Figure 3.6 A spectacular motif on a broken seal showing three intertwined tigers
Source: National Museum, New Delhi, Collection Acc no.: C2896/138.

In a set of seals retrieved from the site, the animal appears to be a ram (indicated by long locks of wool on its forequarters), with the horns of a bull, a human face, and the trunk and tusks of an elephant (Figure 3.8). The hindquarters and hind legs are those of a tiger. On

Figure 3.7 Note the head of the tiger radiating from the ring-like motif on this broken seal, Mohenjodaro
Source: National Museum, New Delhi, Collection Acc no.: E1886/75.

Figure 3.8 Seal from Mohenjodaro depicting an enigmatic composite creature
Source: National Museum, New Delhi, Collection Acc no.: DK 5935/139.

seals where the face is visible, it appears to be human with elongated eyes. Finally, the tail is depicted as an upraised cobra. The linking of the ram, elephant, tiger, bull, and human form was postulated to be an attempt at the fusion of several deities and a possible step towards monotheism (Mackay 1931b, II: 389–90).

Numerous attempts have been made to interpret and unravel these composite and fabulous creatures. An analysis of Harappan chimaeras by Dennys Frenez and Massimo Vidale (2012, 107), for instance, not only revealed a basic set of regular combinations, and their probable evolution over time, but also suggested the possibility of treating the structure of the images as an early form of 'hypertext' where selected parts of the chimaera's body could be interpreted at different levels. It was postulated that the chimaeras may have been linked to a specialized role within the urban administrative system. The assembling of the parts of different creatures within an outline may have been a metaphor for 'social inclusion'. While this was put forth as a possible interpretation, it was also agreed that the depictions represented a more complex system of ideas (Frenez and Vidale 2012, 120).

However, it seems most plausible to concur with Kenoyer's (2006–7, 18) view regarding these composite animals representing complex philosophical or religious ideas. Even more persuasive is his contention that the attempt to represent these concepts visually is remarkable, and since we cannot unravel the exact meanings of these images without deciphering the Indus script, it is best to simply describe them.

I now move on to representations on copper tablets, which are few and primarily attributed to the later periods at Mohenjodaro. Represented twice, the best example came from Chamber 27, Block 4 of L Area (Mackay 1931b, II: 400). A composite animal on one tablet had the hindquarters of a rhinoceros, and the forequarters of a leopard or a tiger (Mackay 1931b, II: 399). Another representation came in the form of a fabulous beast combining the forequarters of a tiger with the hindquarters of a bull or similar animal (Mackay 1938, I: 366).

It has been argued that these tablets were amulets often carried on the person. Also interesting is the observation that unlike the inscriptions on the seals, which were apparently their owners' names, the inscriptions on the copper tablets seem to be associated with the animals portrayed on them. This argument was made on the basis of the retrieval of tablets bearing the same inscription with the same animal (Mackay 1938, I: 363).

Interesting glimpses of the animal also emerge from a survey of the amulets at the site. For instance, a pottery amulet with a three-sided prism from the upper levels of SD Area carried on the extreme right a composite animal with its tail held straight up in the air. On the left of this creature was the familiar scene of a human figure being watched by a tiger. Beyond was a framed swastika, while on the extreme left was an elephant. The amulet also revealed an urus-like animal facing right with the usual cult standard in front of it. The third side showed on the extreme right a horned figure with bracelet-adorned arms standing between two trees whose leaves resemble those of the pipal. On the left is what appears like a sacred goat decorated with garlands. Further was a kneeling horned deity, considered a goddess on the basis of its long pigtail, stretching out both hands as if in entreaty. On the extreme left is what looks like a little offering table with something like a bird on it (Mackay 1938, I: 351).

Another amulet, well-preserved and made of terracotta, was retrieved from the upper levels of DK Area. Of the two sides of the tablet, the one which interests us has an object on the left quite similar in shape to a type of pottery vessel common in Mohenjodaro. This type of a vessel with minor differences was also found in beaten silver as well as in copper leading to the assumption that it had some place in ritual. Next to this vessel was the scene of a man on a tree with a tiger walking away looking backwards as if taking a last look at his prey. Since the stripes are missing in this case, but are clearly visible on other similar representations, it can be safely inferred that the animal is a tiger, and not a lion. It has been argued that taken in connection with the obviously religious object next to it, the episode seems allegorical (Mackay 1938, I: 354–5).

Further right were two human figures which were interpreted as either tearing a tree apart to release a tree-spirit, or planting two trees for the tree deity whose apparently leafy and extended arms seem to be blessing the act. Notably, three tablets with the scenes described above were found, and all came from a particular part of the DK Area (Mackay 1938, I: 355).

The animal figures again on a faience amulet which has what has been identified as a gharial in the centre with a long tail and four legs. On either side of the creature's head was a bull facing it with head lowered as if about to charge. Each bull had an indistinct sign above it. Below, an elephant faced to the right, and opposite to it was a feline, probably a tiger looking backwards. The reverse of this amulet, though in bad condition, showed three animals, again most probably tigers tied together at their middles, a reminder of our earlier discussion of a seal carrying the same motif (Mackay 1938, I: 357).

The lower levels of DK Area also yielded amulets showing the animal in similar contexts. An amulet in baked clay had on one side a short-horned bull with a manger. In front of it stood a human figure with one arm raised, pointing to an object which frequently appears amongst pictographic signs. The reverse carried a rhinoceros and a tiger in file (Mackay1938, I: 359). Another retrieval in terracotta was a three-sided prism with an animal on each side, namely an urus bull, a short-horned bull, and a tiger, with a row of pictographs in front of each (Mackay 1938, I: 361).

The animal's absence amongst terracotta figures at the site was noted with the caveat that two motifs might have represented the animal. Both terracotta heads were moulded, and were hollow at the back giving the impression that they were once fastened to bodies made of some other material (Mackay 1931a, I: 348–9).

Another model with a broad slightly open mouth, prominent teeth, and a short beard was with some hesitation related to the tiger. The eyes were represented by inserted pellets of clay, and the long and thin body with its small size was out of proportion to the head. Two fractures indicated where the legs had been broken off (Mackay 1938, I: 294, 313). The specimen, however, seems highly suspect.

At Chanhudaro, Mackay (1943, 147) documented the presence of the carnivore on four seals, three of which are badly mutilated. On a fragment of one, three tigers are joined together by their middles, a reminder of a similar seal from Mohenjodaro. The animal on another broken seal (of which we have a fragment) may have been the sole motif or may have formed a part of the well-known scene depicting a man seated on a tree with a tiger below looking up at him. Part of the tree is indicated on the seal (Figure 3.9).

Figure 3.9 This broken seal from Chanhudaro may have been a part of the popular motif representing a man on a tree and a tiger below looking back over its shoulder

Source: After Mackay 1943, plate LI 19. © American Oriental Society, reproduced with permission.

Most remarkable is the fourth of these seal amulets; it is intact and portrays a man on one knee apparently invoking a tiger which seems to be licking his face, perhaps in anticipation of the feast (Figure 3.10). Above is a tree whose purpose was most probably to suggest that the man had descended from it. The figure carrying water-jars suspended from a yoke on his shoulders was thought to have no bearing on the scene below (Mackay 1943, 147). Mackay's suggestion immediately brings to mind the possibility of the narrative being an extension of the one frequently portraying a man perched on a tree with a tiger below walking away while looking backwards. Could the portrayal be seen as an attempt to visualize the fate that awaited the man if he descended from the tree? Or could this have been a part of another narrative where perhaps the kneeling figure is symbolically trying to befriend the carnivore?

Figure 3.10 An interesting depiction from Chanhudaro of a human figure kneeling in front of a tiger
Source: After Mackay 1943, plate LI 18. © American Oriental Society, reproduced with permission.

Additionally, the face of a clay sealing was impressed with a motif of three animals tied together at their middles. Of the three, one was a tiger, while the others could not be identified. This three-animal motif we have noted before, and it seems to have been a popular one (Mackay 1943, 150).

Though the tiger seems to have been rarely portrayed in clay, at Chanhudaro, we have 'a very rough model' of the animal identified by the stripes of black paint on the body together with its exposed teeth. Coated with a thick red slip, two holes served for nostrils, while three round pellets inserted in a slit in the clay represented the teeth. Oval eyes were pitted in the centre with eyelashes incised above and below. Hair was represented by incisions between the partly missing ears, while the wrinkles above the muzzle were shown by rough scratches (Mackay 1943, 158).

Moving to Pakistani Punjab, the tiger is represented on two seals at Harappa. In one, it is alone with a trough beneath its head, while in the second, it is under an acacia tree in which a 'hunter' is seated. Both the portrayals are rough with the stripes on its body being shown by stiff, conventional lines in relief. Three sealings also carried the animal (Vats 1940, I: 323).

Mythical, hybrid creatures (seen on the seals at Mohenjodaro) are also found on three seals at Harappa. Two broken, the one that survives has a creature with a human face, the trunk of an elephant, horns of a bull, forepart of a ram, and the hindquarters of a tiger with tail erect (Vats 1940, I: 324).

The 'contest' motif with a figure holding two felines (most probably tigers) by the neck recurs at Harappa on a twisted terracotta tablet. Rectangular in shape, it is a largely complete specimen with a plano-convex section. The flat side depicted a person spearing a male water buffalo. Above the hunting scene was a gharial, and behind the hunter was a figure wearing bangles and a horned and plumed headdress, seated in a meditating position (Figure 3.11). On the opposite convex face was a figure holding two felines (tiger and/or lion?) by the neck, and standing above an elephant (Figure 3.12). The figure shows what seems like a well-formed female breast with a single Harappan sign above its head (Meadow and Kenoyer 1997, 162).

Meadow and Kenoyer (1997, 162) also point out that part of the aforementioned motif apparently made from the same mould was depicted on a broken terracotta tablet found by Vats east of the 'working platforms' in the north-east corner of Trench IV, Mound F. Conforming with the more complete representation recovered from the 1994–5 excavations, the convex side of the Vats tablet showed the lower half of a figure standing above an elephant, and holding two felines. However, on the flat obverse of this broken tablet was a different scene portraying a tiger looking back over its shoulder at a figure perched on a tree (Figure 3.13).

Then there were roughly modelled figurines of the animal at the site. For instance, we have a rough terracotta representation, where

Figure 3.11 On this plano-convex moulded tablet from Harappa, a water buffalo is being speared by an individual who presses its head down with one foot, while holding the tip of its horn with one hand. Seated in yogic position, a figure wearing a horned headdress looks on

Source: © J.M. Kenoyer, Harappa.com, courtesy of the Department of Archaeology and Museums, Government of Pakistan.

Figure 3.12 Standing above an elephant, a female(?) figure fighting two tigers, Harappa. A single Indus script is depicted above the head of the figure

Source: © J.M. Kenoyer, Harappa.com, courtesy of the Department of Archaeology and Museums, Government of Pakistan.

Figure 3.13 Moulded terracotta tablet from Harappa with a narrative scene of a man on a tree with a tiger below looking back over its shoulder. The reverse of this tablet showed a figure standing above an elephant, grappling with two tigers

Source: © Harappa Archaeological Research Project, Harappa.com, courtesy of the Department of Archaeology and Museums, Government of Pakistan.

the fore and hind legs were in single pieces. The ears were erect, and the tail short and curled up. A grotesque figure recognizable only due to the duff around its neck came from Mound F of the Great Granary Area, which yielded two more similar models. A roughly executed little seated figure was considered a reminder of conventional tigers and lions common among temple carvings of later times (Vats 1940, I: 308).

Another interesting find from Mound F was a well-executed jugate terracotta tiger head springing from a holed base. Each of the tigers had holes on its head postulated to have been there for fixing horns. The triple, interlaced tigers on the seal from Mohenjodaro were paralleled with this chimaera creature (Vats 1940, I: 25).

The tiger also figured amongst the terracotta animal figurines from Period II (c. 2300–1800/1750 BCE) representing the Mature Indus culture at Banawali in Hisar, Haryana. A notable find from the sitting room of a house in the lower town was an inscribed seal depicting a horned tiger. The house which yielded more than one seal, a few weights, and a large number of jars half embedded in the house floors was postulated to have belonged to a rich merchant.

Another observation hinting towards the probable owners of these seals noted that the seals came only from the lower town, and not the citadel (Bisht 1982, 117, 119).

Additionally, at Rakhigarhi (also in Hisar, Haryana), a seal, the upper portion of which was broken, revealed a composite figure of a man, bull, unicorn, elephant, and tiger in a Mature Harappan context (Nath 1997–8, 45).

In a majestic representation of the carnivore from Kalibangan in Rajasthan, though the head is not visible, the beautifully executed wavy lines are evident markers of the stripes characterizing the predator (Figure 3.14). What is striking, however, are the spike-like projections outlining the body of the animal. Whether this was a device employed with the intention of accentuating the portrayal, or an attempt to depict the hair on the animal's body can at best be speculated.

The man on tree and tiger below theme recurs at Kalibangan, but here the human figure is a little differently positioned (Figure 3.15). There is no attempt to hold on to the tree, and the arm held out in the portrayals at Mohenjodaro and Harappa is instead raised as if in supplication for divine intervention.

Further, the impression of a Harappan cylinder seal from the site shows two warriors wearing their hair in a divided bun at the back of the head (Figure 3.16). While they spear each other, they are both held by the hand by a figure seen wearing a headdress with a long pendant, bangles on the arms, and a skirt. Next to the combat scene, where space seems to have prevented the depiction of more details, there are two figures on each side whose bodies merge with that of the tiger, and their headdress is elaborated with animal horns and a tree branch (Parpola 1994, 253, 255).

In Gujarat, as far as the seals at Lothal are concerned, the tiger motif is employed only once on a steatite specimen where the animal can be identified due to its prominent stripes. The claws are exaggerated and open indicating that the beast is about to strike. The tail has a thick bunch of hair. The missing head portion, however, made it difficult to ascertain whether it was an anthropomorphic figure (Rao 1985, II: 324, 485).

Lothal also yielded a terracotta model of the animal produced from a double mould. It had a large head, slit mouth, and incised

Figure 3.14 A tiger seal (head broken), Kalibangan
Source: © Archaeological Survey of India, reproduced with permission.

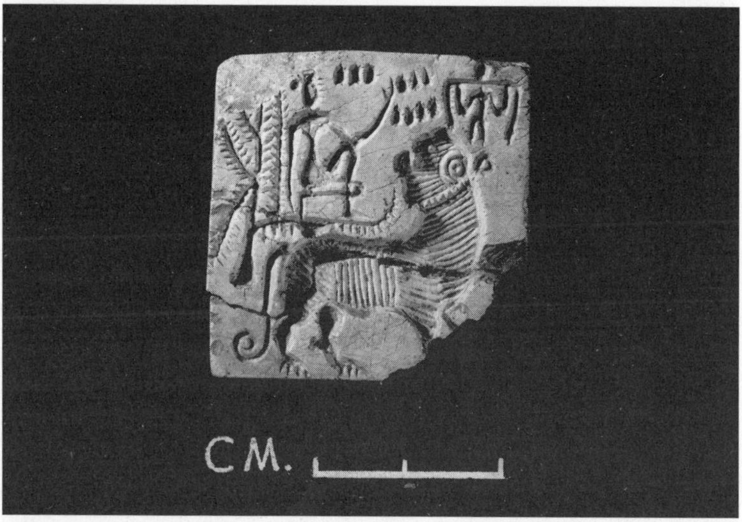

Figure 3.15 Kalibangan yields a variant of the man on tree and tiger below narrative commonly found on Harappan seals
Source: © Archaeological Survey of India, reproduced with permission.

nostrils with the chequered pattern incised on the body of the creature suggesting its stripes (Rao 1985, II: 485).

So, what is it that emerges from the evidence we have in hand? Despite fluid interpretations, there is a certain thread that seems to connect these narratives. The element of contest and conflict is indisputable, and so is the attempt to overpower it. Even though

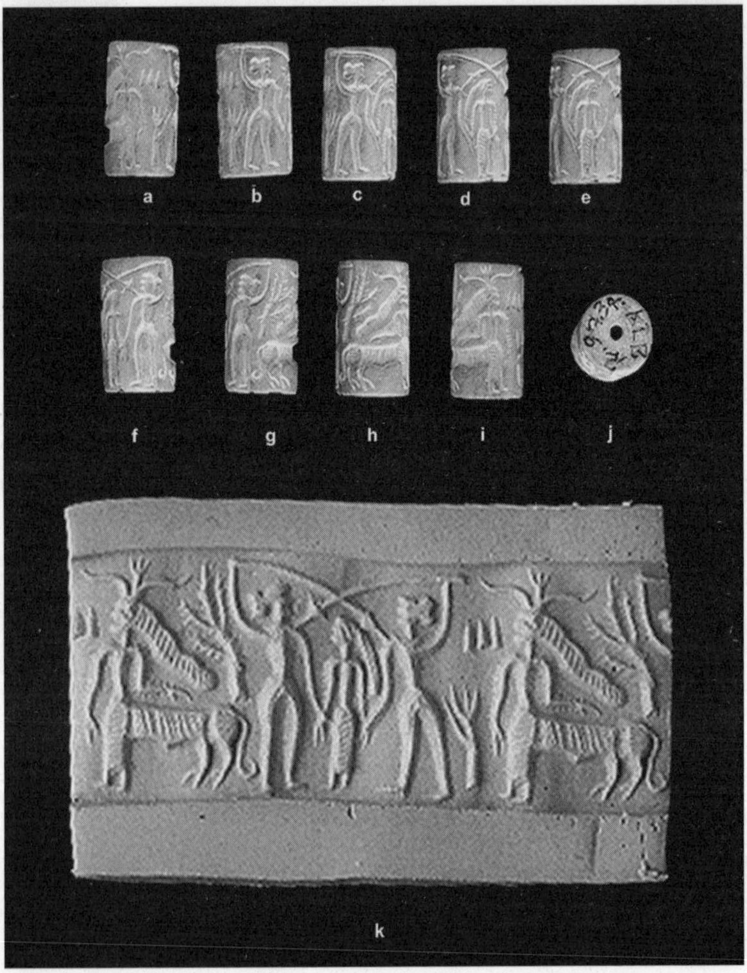

Figure 3.16 Impression of a Harappan cylinder seal, Kalibangan
Source: © Archaeological Survey of India, reproduced with permission.

we cannot determine the precise meanings of these representations, the fact that they are found at different sites suggests that they constitute a vocabulary shared by many communities throughout the Indus region during the period of urbanism.

Keeping in mind the varied representations of the predator, it was argued by Fairservis (1976, 16) that the tiger appears to have had at least three roles in the Harappan world: as one of the wild beasts associated with the horned 'deity'; as an animal to be feared, as in the man on a tree depiction; and as the victim of a hero story. In this sense, according to him, the animal appears to mark three levels of Harappan narrative: myth, epic, and also perhaps legend-folktale.

Until the Indus script is deciphered, such categorization is more in the realm of speculation rather than anything else. What, for instance, has been described as myth, could well be an iconographic representation of a sacred figure. Similarly, it is necessary to keep in mind that what has been read in terms of an epic—with a hero grappling with two tigers—had no direct resonance in the Gilgamesh epic because the animals there are lions, not tigers. Was the Indus representation a consequence of a local modification of the original epic? Or was this a part of some local folktale? Should the oft-repeated man on tree and tiger below representations necessarily be perceived as one fraught with danger for the man who in a bid to save himself from the carnivore perches himself on the tree? Or was the man on the tree symbolically engaging in some sort of a dialogue with the animal? These are questions that cannot be easily answered.

Till the time we have textual support for these portrayals, interpretations as well as ways of perceiving them will remain fluid. What can be said on the basis of the survey is that the sheer diversity of tiger representations in the context of the Harappan civilization undoubtedly suggests a special connection with the animal. The range of narratives woven around the tiger clearly sets it apart amidst the repertoire of animals figuring in the Indus belief system. However, despite the richness of the evidence available, and till the absence of a deciphered script, the elusive Harappan tiger will continue to elude us.

Neolithic Traces

Loebanr III and Aligrama: In a neolithic context, Bruno Compagnoni (1979, 697–700) reported tiger remains from the two settlements of Loebanr III and Aligrama in Swat (north Pakistan). Loebanr III dated back to Period IV of the proto-historic cultures of the Swat Valley, c. 1700–1500 BCE, while Aligrama revealed a sequence extending from Period IV till the Kuṣāṇa period (c. 1700 BCE till fourth century CE). In addition to being settled practically at the same time, the two sites did not lie very far apart, hence their fauna was treated together (Campagnoni 1979, 697). Except for a numerical count of the bones retrieved of each species, no anatomical details were furnished by the report.

The assemblage comprising eighteen forms, nine domestic and nine wild, revealed a mosaic of natural environments. The goral and markhor were high-altitude rock-climbing animals, with the former preferring mountain areas with a dry climate beyond the forest edge and the latter frequenting medium and high altitudes within and beyond the wood line. The deer and the hare, though preferring temperate dry forests, could also occupy different biotopes. Then there were animals such as the porcupine which favoured parklands and open steppes, together with the wild cat inhabiting green bushy steppes, as well as the half-ass which roamed pre-desertic treeless steppes. Together with these was the tiger with its preference for densely vegetated places. With the exception of the tiger and the half-ass, all the wild species were said to be part of the local fauna (Campagnoni 1979, 699). The faunal assemblage, thus, hinted at a mosaic environment at the site, and reiterated the animal's ability to adapt to diverse ecological settings.

Chalcolithic and Early Historic Traces

Atranjikhera: Our next encounter with the actual remains of the animal is at the site of Atranjikhera in Etah district, Uttar Pradesh, where amongst the 927 animal bone fragments that could be identified, one fragment of ulna was attributed to the tiger, and assigned to the Ochre Coloured Pottery (OCP) level placed between 2000 and 1500 BCE. Contributing a miniscule 0.11 per cent of the total faunal assemblage, the animal was postulated to have been a scavenger visiting the site (Shah 1983, 469–70). However, from the find of the bone it was inferred that dense virgin forest could not have been far away (Gaur 1983, 74).

Atranjikhera also yielded the modelled head of a tiger from Phase D (c. 200–50 BCE) of Period IV, which yielded thirty-seven animal figurines from different levels. The solitary representation was treated with a black slip, and had eyes pierced within a pressed triangle. The drooping face was considerably mutilated with traces of a slit mouth. The ears were partially missing, but the mane had prominent grooves between the ears (Gaur 1983, 371).

Madar Dih: Remains of the animal again figure at this site in Jaunpur district, Uttar Pradesh. The faunal sample though small, confirmed the utilization of diverse fauna during the cultural periods identified at the site. An interesting retrieval from Period I representing the NBPW period was a complete maxillary canine of the right side of a tiger. The canine suggested a male between two to three years of age. In view of the tooth not yielding evidence indicating its consumption, it was suggested to have been kept as a personal trophy (Joglekar et al. 2013, 263, 266).

As we can see, the faunal evidence testifying the presence of the tiger at early historic sites is limited. However, we have sites yielding evidence for the tradition of modelling the tiger in terracotta in this period. For instance, Shaikhan Dheri in Pakistan yielded seven specimens belonging to different cultural periods. The city flourished from mid-second century BCE till about the end of the second century CE, and was stratigraphically divided into Phase A or the Kuṣāṇa period comprising Periods I, II, and III, Phase B or the Scytho-Parthian period comprising Periods IVA and IVB, and Phase C or the Greek period comprising Periods VA and VB, and Period VI. The figurines which belonged to Periods II, III (Kuṣāṇa), IV (Scytho-Parthian), and VI (Greek) were mostly damaged, but characteristically represented the carnivore (Dani 1965–6, 102–3).

Similarly, Rairh in Rajasthan yielded one complete model, and several fragments in the form of crude handmade figures. The body was scratched with lines to indicate the stripes, while a strip of clay was added for the tongue and the body. The wide open mouths probably suggested roaring (Puri 1998, 33).

We also have a specimen dated to the second century CE from Patna, where the carnivore is shown with a wide gaping mouth with the tongue sticking out. The exaggerated moustache makes it an attractive specimen. The eyes are protruding and the ears are attached separately. All over the body of the animal are indentation marks.

THE PREDATOR IN TEXTUAL TRADITIONS

The Vedic Corpus

The striped carnivore, it seems, was unknown to the authors of the *Ṛgveda*. At least, that is what it seems from the fact that it does not mention the *vyāghra*, the word for the tiger. Romila Thapar (2004, 157) points out that the *Ṛgveda* was, however, familiar with the *siṃha* or the lion, and this she suggests indicated that the original authors lived in the Indo-Iranian borderlands where lions were common. Thapar, however, does not qualify the grounds on the basis of which she makes the last assertion.

The situation, however, changes when one looks at later Vedic texts where we encounter the tiger with striking frequency. The predator native to the marshy forests of Bengal, and, hence, unknown in the *Ṛgveda*, repeatedly appears, for instance, in the *Atharvaveda* as the mightiest and most dreaded of all the beasts of prey. This can perhaps be explained in terms of a shift to the Gangetic belt which seems to be reflected in the text where geographical and cultural conditions seem to suggest a period later than that reflected in the *Ṛgveda* (Winternitz 1977, I: 123). Given the ritualistic nature of the literature under consideration, it is best to treat the references it offers as mere windows to the larger Brāhmaṇical world view of animals and their symbolism.

The impact the predator had on the ancient human mind is subtly etched out in some hymns. For instance, in a hymn to acquire brilliance, the *Atharvaveda* (VI. 38, 1–2; Whitney 1962, I: 309) includes the animal in a list extolling the brilliance vested in many things, and seeks to draw the same virtue from them. The list includes the lion, the tiger, fire, sun, elephant, leopard, and gold among many others. The night also is said to derive its beauty from the following: 'The eager night has taken to herself the splendor of the lion, of the stag, of the tiger, of the leopard, the horses's bottom, man's [*púruṣa*] roar [? *māyú*]; many forms thou makest for thyself, shining out' (*Atharvaveda* XIX. 49, 4; Whitney 1962, II: 980).

While in the aforementioned contexts, the animal loses out on exclusivity by being one amongst many parameters for comparison, more specific allusions can also be culled. The magnificence of the animal, for instance, is alluded to in the *Śatapatha Brāhmaṇa* (V, 3, 5, 3; Eggeling 1894, III: 81). It is said to have Soma's beauty for when the latter flowed through Indra, he became a tiger.

Though allegorical, the might of the animal also often serves as a point of reference. An amulet encompassing 101 heroisms is likened to a tiger attacking all rivals and rendering them inferior (*Atharvaveda* XIX. 46, 5; Whitney 1962, II: 973). Similarly, in a hymn to plants for the restoration of health, the 'tigerish amulet of plants' is beseeched to save and protect from diseases and demons (*Atharvaveda* VIII. 7, 14; Whitney 1962, II: 500). Even in ritual contexts, the allusion is primarily to the vigour of the animal. In the piling of the fire altar, 'Agni is Rudra; just as a tiger stands in anger, so he also (stands)' (*Taittirīya Saṃhitā* V.5,7,4; Keith 1914). In the building of the sacred fire altar, the shaft is likened to tigers, and the missile to snakes, homage being paid to both (*Śatapatha Brāhmaṇa* VIII, 6, 1, 18; Eggeling 1897, IV: 107).

Its splendid majesty notwithstanding, the animal also exemplifies danger. Its flesh-eating propensity is a recurring theme in the *Atharvaveda* (XII. 2, 43; Whitney 1962, II: 680). In the woods, wild beasts including lions and tigers are said to go about man-eating (*Atharvaveda* XII. 1, 49; Whitney 1962, II: 669). Another hymn enjoins 'I am a vexer [*tápana*] of the *piçācás*, as a tiger of them that have kine; like dogs on seeing a lion, they do not find a hiding-place' (*Atharvaveda* IV. 36, 6; Whitney 1962, I: 210). The *Śatapatha Brāhmaṇa* (XI, 8, 4, 1; Eggeling 1900, V: 131) also preserves a reference to a tiger killing a cow which provided milk. Thus, besides being a threat to human life, the animal is clearly an adversary particularly for the keepers of cattle.

On the occasion of the eruption of the two upper teeth of a child, the teeth are referred to as 'the (two) tigers . . . having grown down . . .' (*Atharvaveda* VI. 140, 1; Whitney 1962, I: 386). The use of the metaphor in this case evidently seems to hint at the ability of the teeth (just like that of a tiger) to bite or dig into all that came in their way.

For a child born on the tiger day, it is prayed that he does not slay his father nor harm the mother who brought him forth (*Atharvaveda* VI. 110, 3; Whitney 1962, I: 361). The *Taittirīya Saṃhitā* (VI. 2, 5, 5; Keith 1914) preserves a reference where it is said that the one who is consecrated is a foetus in the consecration shed which is the womb. If he were to leave the shed, it would amount to the foetus falling from the womb. Hence, to protect himself, he should not leave the shed. Moreover, the fire within a consecration shed is paralleled with a tiger guarding the house. Thus, if the man who is consecrated were to leave the shed, the animal was likely to spring up and slay him. This reference, as we shall subsequently see, has been crucially used by J.C. Heesterman (1957) to establish the connection between the king and the tiger.

Additionally, there is a rather quaint reference to man-tigers along with ogres, thieves, murderers, and robbers in the forests (*Śatapatha Brāhmaṇa* XIII, 2, 4, 2; Eggeling 1900, V: 307–8). Though we are far from certain regarding what is really meant by a man-tiger, it is perhaps a combination of a human and tiger form (could it be a mask donned to intimidate and accost people?), and clearly one which is seen as a threat along with other nefarious elements.

Again despite an apparent awe for the predator, the urge to control or even overpower it is apparent. While a spell to ward off demons and other enemies reflects a general urge to overpower flesh-eating ones seeking to harm others (*Atharvaveda* IV. 36, 3; Whitney 1962, I: 209), a charm against wild beasts and thieves in the text seeks safety from tigers and wolves (*Atharvaveda* IV. 3, 1; Whitney 1962, I: 148). 'Indra-born, soma-born art thou, an Atharvan tiger-crusher (*-jámbhana*)' (*Atharvaveda* IV. 3, 7; Whitney 1962, I: 149), suggests a force able to reckon with and exterminate the dreaded carnivore.

The same hymn evokes greater interest when it remarks, 'Both thy (two) eyes and thy mouth, O tiger, we grind up; then all thy twenty claws' (*Atharvaveda* IV. 3, 3; Whitney 1962, I: 148). 'The tiger first of (creatures) with teeth do we grind up, upon that also the thief, then the snake, the sorcerer, then the wolf' (*Atharvaveda* IV. 3, 4; Whitney 1962, I: 148). What is striking here is the urgency to destroy the tiger before grappling with other potentially harmful elements. Additionally, the allusion to the grinding of its teeth and its claws may suggest the association of the animal with the darker and mysterious realm of magic and superstition. Elsewhere, there seems to be some interest in the hair of the carnivore. As part of the *sautrāmaṇī* sacrifice, in the *Śatapatha Brāhmaṇa*, the hair of a wolf, tiger, and lion are put into cups of spirituous liquor from which libations are made. The three animals are said to have emerged from the body of Indra after he forcibly consumed Soma. Here, tiger hair is said to help secure courage and the mastery over wild beasts, while the hair of the lion helps secure might and the rule of wild beasts (*Śatapatha Brāhmaṇa* XII, 7, 2, 8; Eggeling 1900, V: 218–19).

The mega carnivore held its own even in ritual contexts. As part of the horse sacrifice, where offerings are prescribed for Indra, Varuṇa, Yama, the bull, Soma, and many others, all of whom are referred to as the king, '*Bos Gavaeus*' is offered to the tiger also hailed as the king (*Taittirīya Saṃhitā* V. 5, 11; Keith 1914). The predator itself along with the lion qualifies as an offering to the mighty Indra (*Taittirīya Saṃhitā* V. 5, 21; Keith 1914). On the other hand, once again described as the king, it also figures as an object of worship for the Hotṛ priest (*Vājasaneyī Saṃhitā* XXI. 39; Griffith 1899, 200).

But where did the animal stand vis-a-vis other wild beasts, and particularly the lion? Multiple voices seem to emerge from within the corpus. We could perhaps begin with the legend accounting for the origin of the two big cats. Having slain Tvaṣṭṛ's three-headed and six-eyed son Viśvarūpa, Indra is withheld from drinking the Soma juice, which he eventually ends up consuming by force. As a result, it flowed out in all directions from the openings of his vital airs. A lion sprang from what flowed from the nose, a wolf from what flowed from the ears, and wild beasts with the tiger as their foremost from what flowed from the lower opening (*Śatapatha Brāhmaṇa* V, 5, 4, 10; Eggeling 1894, III: 131). Here, though the lion emerges from the upper part of Indra's body, it is the tiger emanating from the lower part, which is put over and above all wild beasts.

The same legend occurs again in the *Śatapatha Brāhmaṇa*, and here the tiger is said to spring from Indra's entrails. The lion does not trail far behind, emanating from his blood—'from the contents of his intestines his fury flowed, and became the tiger, the king of wild beasts; from his blood his might flowed, and became the lion, the ruler of wild beasts' (*Śatapatha Brāhmaṇa* XII, 7, 1, 8; Eggeling 1900, V: 215). Once again, the tiger emerges from the lower part of

Indra's body, while the lion forms his blood, the river of life. Though figuratively in both cases the lion is put above the tiger, the latter here is referred to as the 'king of wild beasts', while the former is referred to as their 'ruler'. Clearly, both the carnivores were respected and perceived with awe.

Again, we may consider the following verse of a hymn for the success and prosperity of a king which enjoins, 'Of lion-aspect, do thou devour all the clans; of tiger-aspect, do thou beat down the foes; sole chief, having Indra as companion, having conquered, seize thou on the enjoyments of them that play the foe' (*Atharvaveda* IV. 22, 7; Whitney 1962, I: 189). Here, the feat achieved by a king of 'lion-aspect' and of 'tiger-aspect' is equal in impact, the consequence being the destruction of any force contesting his might. Similarly, as mentioned earlier, the hair of the tiger was supposed to impart courage and a sway over beasts, while that of the lion was for might and the rule over wild beasts. 'The tiger is vital vigour, invincible/unassailable the metre. The lion is vital vigour, covering the metre' (*Vājasaneyī Saṃhitā* XIV.9; Griffith 1899, 125; *Śatapatha Brāhmaṇa* VIII, 2, 4, 4–5; Eggeling 1897, IV: 38). Evidently, the endeavour in such instances was to equate the prowess of the two big cats.

Nevertheless, though the might of the lion is manifest as well as acknowledged, it seems to be straggling behind the tiger regarding whose pre-eminence there seems to be little ambiguity if we consider the instances which follow. The *Atharvaveda* (VIII 5, 11 and XIX 39, 4; Whitney 1962, II: 491, 960), for instance, proclaims it as the chief and the highest of the beasts of prey. An amulet against witchcraft is extolled for its efficacy, and the one bearing it is said to become first 'a tiger, likewise a lion, likewise a bull, likewise a lessener of rivals' (*Atharvaveda* VIII 5, 12; Whitney 1962, II: 492).

What is postulated here is reiterated when we move to the royal consecration or *rājasūya* ceremony, a closer look at which has much to convey about the position of the animal. The predator figures prominently here, and the king who has to step on a tiger skin on the day of his anointment is hailed as follows: 'A tiger, upon the tiger's [skin], do thou stride out unto the great quarters; let all the people want thee, the waters of heaven, rich in milk' (*Atharvaveda* IV, 8, 4; Whitney 1962, I: 157). Further, it follows, 'Thus, embracing the tiger, they incite the lion unto great good-fortune; as the well-being ones the ocean that stands, do they rub thoroughly down the leopard amid the waters' (*Atharvaveda* IV, 8, 7; Whitney 1962, I: 158). Though not lucid, the allusion seems to associate the king with the three carnivores: the tiger, the lion, and the leopard.

As is known, the rituals of the *rājasūya* are elaborate, and though not directly relevant here, it would be meaningful to locate the contexts featuring the use of the animal's skin. In the *Śatapatha Brāhmaṇa*, which spells out the details of the *rājasūya*, a tiger skin is spread in front of the Maitrāvaruṇa's hearth with the refrain, 'Thou art Soma's beauty.' The exhortation is rationalized by elucidating that when Soma flowed through Indra, he became a tiger, and, therefore, he is Soma's beauty, and this is why it is also said, 'Thou art Soma's splendour; may my beauty become like unto thine!' The tiger's beauty is, thus, bestowed on him (*Śatapatha Brāhmaṇa* V, 3, 5, 3; Eggeling 1894, III: 81). After the king has symbolically mounted the various quarters (*digvyāsthāpanam*) suggesting his universal rule, on the hind part of the tiger's skin a piece of lead is laid down, which the king kicks off with his foot. The lead symbolizes a demon named Namuci, knocked down by Indra, and trodden upon with his foot. Like Indra vanquished the fiends, so does the king prevail over them (*Śatapatha Brāhmaṇa* V, 4, 1, 9–10; Eggeling 1894, III: 91–2). He is then made to step upon the tiger's skin with the same refrain extolling Soma's beauty and splendour, and linking it with the tiger (*Śatapatha Brāhmaṇa* V, 4, 1, 11; Eggeling 1894, III: 92). Further, the king is made to take the three 'Viṣṇu-steps' within the extent of the tiger's skin with the following words, 'Viṣṇu's outstepping thou art! Viṣṇu's outstep thou art! Viṣṇu's step thou art!' Viṣṇu's outstepping (*vikramaṇa*), his outstep (*vikrānta*), and his step (*krānta*) are said to be the three worlds, having ascended which the king rises above everything (*Śatapatha Brāhmaṇa* V, 4, 2, 6; Eggeling 1894, III: 96). It is equally significant that even when a 'throne-seat' is brought for the king, it is placed on the tiger's skin in front of the Maitrāvaruṇa's hearth. Rendered auspicious, the king is made to sit down on it, and the throne is declared to be the very womb of knighthood (*Śatapatha Brāhmaṇa* V, 4, 4, 1–3; Eggeling 1894, III: 105–6).

The *Aitareya Brāhmaṇa* spells out the details of the *punarabhiṣeka* (repetition of the inauguration ceremony), where the king mounts a throne covered with tiger skin (*Aitareya Brāhmaṇa* 8, 2, 5; Haug

1863, II: 502). Regarding how the king must ascend his throne, it is said that he must spread the tiger skin on the throne in such a way that the hair comes outside, and that part which covered the neck, is turned eastward. Most notably, the tiger is said to be the *kṣattra* (royal power) of the beasts in the forest. The *kṣattra* is the royal prince, and by means of this *kṣattra*, the king makes his *kṣattra* prosper (*Aitareya Brāhmaṇa* 8, 2, 6; Haug 1863, II: 503). In no uncertain terms, then, the king derives his strength and supremacy from the tiger, which by virtue of its association with the summit of political authority emerges as paramount over other beasts.

Heesterman (1957, 108–9) rightly observed a special relation between the king and the tiger: in the *rājābhiṣeka* (royal consecration), he was identified with the tiger, and when the moment for the unction arrived, he was to seat himself on a tiger skin. Drawing on the *Taittirīya Brāhmaṇa* (2, 7, 15, 1), Heesterman picks up another instance from the *rājābhiṣeka*, where in one of the formulas accompanying libations preceding the unction, it is said, 'This tiger wanders in the fire having entered it, the son of the seers.' The reference here, he elaborates, is to the king being poured into the fire in the form of ghee. Similarly, he notes a reference from the *Taittirīya Saṃhitā* (VI. 2, 5, 5), where the womb from which the *dīkṣita* is ritually to be born is also represented by the sacrificial fires in the fire-hut, where he is to spend the night, and if he were to do so outside the hut, the fire would pursue him in the form of a tiger and kill him. Having pointed out earlier (Heesterman 1957, 3) that for the Vedic ritualist, the king was nothing more than a common sacrificer, *yajamāna*, Heesterman reasons that if one was to connect the idea of the *dīkṣita*'s fire womb represented as a tiger with the king's seating himself on a tiger skin in view of his unction birth, the surmise is that the king is to be born as a tiger from a tiger womb (in the form of the tiger skin). The seating on the tiger skin, therefore, he argues, is considered to have a reinvigorating, and reintegrating effect (Heesterman 1957, 108–9). The animal, thus, emerges as the uncontested exemplar of royal authority.

However, this is not to say that there are no aberrations. We consider here the *sautrāmaṇī*, a sacrifice aimed at offsetting the ill-effects of excessive Soma drinking. It is prescribed for the redemption of Indra from the suffering caused by his over-indulgence in the drink.

Significant is an extract where a remedial sacrifice is embarked upon to cure Indra of his sickness and recreate his body. The hair of the wolf is put on his waist and body, tiger hair constitutes the beard on his face, while the lion's hair forms his locks, 'for fame and beauty, worn on his head, his crest and sheen and vigour' (*Vājasaneyī Saṃhitā* XIX. 92; Griffith 1899, 185). A similar but slightly varying hierarchy is suggested in the context of a sacrifice from which a man is born. The hair of the wolf is the hair on his abdomen and below, the hair of the tiger is the hair on his chest and his arm pits, while that of the lion is the hair of his head and his beard (*Śatapatha Brāhmaṇa* XII, 9,1, 6; Eggeling 1900, V: 261). In both contexts, it is the lion, and not the tiger that is deemed worthy of constituting the crown.

What also deserves attention is that while the disintegration of Indra (discussed earlier) as a consequence of the forceful consumption of Soma accounts for the origin of the tiger along with other wild beasts, the carnivore also features in his reintegration, suggesting in a way a circle that has done a full round. Heesterman (1957, 109) expounds that the *sautrāmaṇī* is prescribed for Indra, who disintegrated upon receiving the unction in the *rājasūya*. This reintegration theme, he argues, fits in well with the idea of the king being an embryo in the tiger womb: 'during his stay in the womb he gathers his forces, ripens for his birth, then, after his birth (the unction), he will disintegrate and be gathered up again during another stay in the womb and so forth in an ever recurrent cyclical process'.

But the picture that emerges is far from neat. The tiger serves as a recurring point of comparison—its beauty, strength, and predatory potential being emphatically underlined. As has been reasoned, the hair and skin of the carnivore qualified for use in magic performances and was believed to procure similar strength and sway over men which the tiger exercised over animals. At the same time, what is clearly discernible is that the lion is a very close and an almost equal contender. What is it then that narrowly prevents it from emerging as the animal par excellence? What could account for this cultural choice that invested the elusive tiger with such symbolism?

We could perhaps ponder over the geographical terrain of the later Saṃhitās that mention the tiger. For instance, the *Śatapatha Brāhmaṇa* (I, 4, 1, 14–17; Eggeling 1882, I: 105–6) recounts the well-known legend of Videgha Māthava migrating from the region around the

Sarasvati to the country beyond the Sadānīrā in the east. The land to the east of the river is said to have been uncultivated, very marshy, and quite imaginably for us, the habitat of tigers. The predominantly tiger-face of the land is bound to have impressed people encountering the striped predator for the first time. For people familiar with the lion so far, the carnivore offered a new symbol of strength and might, equal in prowess with the lion. Since the aim was to colonize the land of the tiger, appropriating the animal itself as a symbol of royal authority was a step towards adapting to and assimilating not just the new landscape, but also all that it stood for. The king stepping on tiger skin during the *rājasūya* was, therefore, not just symbolic of his imbibing the prowess of the animal, but perhaps also suggested his subjugation of the animal which represented the land.

Glimpses from the Pali *Tipiṭaka* and Beyond

Though early Buddhist literature is replete with allusions to animals, specific references to the tiger mostly come across as incidental in this early literary corpus. The animal (unlike the lion, whose magnificence and majesty inevitably serve as a standard for comparison) seldom finds mention separately, and is almost always grouped together with other carnivores. Collating the evidence available serves the purpose of providing insights regarding the early Buddhist world view of animals in general, as also taboos and hierarchies embedded within the tradition.

Much is known about the first precept of non-violence or *ahiṃsā* espoused by Buddhism, based on the principle of not causing injury to living things (*prāṇatipātādviratiḥ*). Within this, a major area of emphasis has been the treatment of animals. The *Vinaya* texts, for instance, echo concerns regarding actions entailing injury or destruction of animals by the monastic order. This is evident when a monk wishing to build himself a hut is instructed to exercise caution on the choice of the site. One of the prerequisites is that doing so should not involve an act of destruction. It is further qualified that damaging the abode of creatures such as ants, termites, rats, snakes, scorpions, centipedes, elephants, horses, lions, tigers, leopards, bears, and hyenas is an act involving destruction (*Vinaya Piṭaka, Suttavibhaṅga*; Horner 1949, I: 256). What is noteworthy here is the dispassionate application of the rules of non-injury even to dangerous and destructive species of noxious insects and reptiles as well as deadly predators. It is equally interesting to note (as shall be seen subsequently in this section) that it is also an offence to eat the flesh of some of these animals, for example, the elephant, horse, lion, tiger, bear, and hyena.

Similarly, the use of animal skins was discouraged. While prescribing foot-clothing, seats, and vehicles to be used by the *bhikkhus*, we are told how once when the latter were in the habit of wearing sandals adorned with lion skins, tiger skins, panther skins, black antelope skins, otter skins, cat skins, squirrel skins, and owl skins, they earned the disdain of the people who condemned them for behaving like those who still enjoyed the pleasures of the world. The matter was reported to the Buddha, who accordingly, directed the *bhikkhus* to not do so, qualifying that anybody who transgressed would be guilty of a *dukkata* offence (*Vinaya Piṭaka, Mahāvagga*; Horner 1962, IV: 247). Similarly, the *bhikkhus* were prohibited from using fine skins such as lion, tiger, and panther skins to recline upon. These skins are said to have been cut to fit couches and chairs and spread either on their inside or outside (*Vinaya Piṭaka, Mahāvagga*; Horner 1962, IV: 257). On a slightly different tangent, the *Brahma Gāla Sutta* (*Dīgha Nikāya*; Davids 1899, I: 12) voices disapproval even of coverlets embroidered with figures of lions, tigers, and the like.

One may ask if these proscriptions essentially emanate from the principle of *ahiṃsā* so zealously espoused by the Buddha, or is it possible to offer an alternative rationalization. Such injunctions should perhaps also be juxtaposed with evidence clearly associating the use of these animal skins with status and affluence. Let us take, for instance, a reference in the *Mahā Sudassana Suttanta* where the Buddha while expounding on the ephemeral nature of things, recounts his life as the 'Great King of Glory'. In his elaborate description of the opulence of his bygone life, and his abode in the chief chamber in the Palace of Righteousness in the royal city of Kusāvatī, the riches and fineries include among other things divans of gold, silver, ivory spread with long-haired rugs and magnificent antelope skins as well as chariots with coverings made of the skins of lions, tigers, and panthers (*Dīgha Nikāya*; Davids and Davids 1966, II: 220). Therefore, the embargo on the use of the same by the monastic order seems to stem also from popular derision regarding monks enjoying worldly pleasures considered incongruous with the austerity and frugality of monastic life.

The early Buddhist approach to animals though indisputably compassionate, was at the same time an ambivalent one. This is, for instance, discernible in their stance on meat and fish eating. I.B. Horner (1945) illustrates with examples how the broad principle was that monks aroused no criticism if they ate meat keeping in mind some prerequisites. One of these was that the meat consumed had to be 'the meat of those (aṇimals) whose meat is allowable' (Horner 1945, 450).

Let us elaborate a little on the aforementioned injunction. The *Mahāvagga* alludes to a famine when the people ate elephant meat, and gave the same to the *bhikkhus* in alms. The same was done with horse, dog, and serpent meat, inviting reproach from the people, and the subsequent banning of the consumption of these meats by the Buddha. Again hunters are said to have killed a lion, eaten its flesh, and offered it as alms to the *bhikkhus*, who having eaten it ventured into the forest, and were attacked by lions attracted by the smell of lion's flesh. The same fate befell them when they consumed tiger, panther, bear, and hyena meat offered by the hunters. The list of prohibited animal meats, thus, includes those from elephants, horses, dogs, serpents, lions, tigers, panthers, bears, and hyenas. The raison d'etre behind forbidding the consumption of these animals differ though. While elephants and horses are symbols of royalty, dogs are considered 'disgusting and loathsome', and serpents and wild animals are harbingers of peril for the monks (*Vinaya Piṭaka, Mahāvagga*; Horner 1962, IV: 298–300). In this context, we are reminded of a mention in the *Aṅguttara Nikāya* (Hare 1961, III: 81) of a 'forest-gone' monk's fear of daunting creatures such as the lion, tiger, leopard, bear, or hyena taking his life or causing his death.

What also deserves some reflection is the observation that while elephant, horse, dog, and serpent meat are said to have been eaten by the 'people', the consumption of the flesh of wild animals like the lion, tiger, panther, bear, hyena is attributed to 'hunters'. Clearly, neither seems to have been common practice since the context is that of a famine despite which the use of these animals as food is proscribed. Could it nevertheless possibly be a pointer to groups on the periphery of the social pale that may have hunted and used these animals for food?

It is important to note that the eating of tiger meat was not unknown to ancient India when we consider, for instance, a text like the *Suśruta Saṃhitā* (dated to the early centuries of the Common Era) which classifies various kinds of meats (including those of carnivores as the lion and tiger), and renders them fit for consumption. However, as has been argued, the text being a much later composition, cannot be an indicator of practices in earlier times, nor can it suggest when such practices actually started (Divyabhanusinh 2005, 70). Nevertheless, it does give us a sense of practices prevalent in early India in view of which the consumption of lion and tiger meat (at least by a certain section) in the period under consideration cannot be precluded as a possibility. Moreover, tribal people even today are known to consume the meat of carnivores (Divyabhanusinh 2005, 70).

Returning to the Pali canon, though the meat of the aforementioned animals was forbidden by the Buddha, the *Suttavibhaṅga* tells us that monks were allowed to partake the remains of the kills of some of them. At one time, a company of monks descending from the slopes of the Vulture's Peak, seeing the remains of a lion's kill, tiger's kill, panther's kill, hyena's kill, and wolf's kill had it cooked, and ate it, but were remorseful thereafter. They are then shown the way by being told that 'there is no offence in taking what belongs to animals' (*Vinaya Piṭaka, Suttavibhaṅga*; Horner 1949, I: 98). Hence, the ban on the consumption of the flesh of the eaters of flesh evidently came in view of their potential to harm.

It is also possible to extract subtle pointers regarding notions about hierarchies operating in the wild. That the lion is the king of beasts is a recurring refrain (*Saṃyutta Nikāya*; Woodward 1954, III: 70, 71). It is possible to cite several passages suggesting the same. Let us consider the following ones. 'Like roaring lion, mighty lord of beasts' (*Sutta Nipāta*; Hare 1947, 102). 'Like the beasts before the lion, fall a-trembling, are afraid' (*Saṃyutta Nikāya*; Woodward 1954, III: 72). What is unequivocally conveyed through these insinuations is the predator's sway over lesser animals. On a similar note, the *Pāṭika Suttanta* metaphorically trifles an old jackal's attempts to equate itself with the maned predator. Having flourished on the remains of the latter's food, the scavenger reflects about himself, 'The lion I! I am the king of beasts! And so he roared—a puny jackal's whine.' An aside follows, 'For what is there in common 'twixt the twain—The scurvy jackal and the lion's roar?' It is further remarked for the jackal,

> Roaming the pleasant woods, seeing himself
> Grown fat on scraps, until he sees himself no more,
> A tiger I! the jackal deems himself.
> But lo! he roars—a puny jackal's howl.
> For what is there in common 'twixt the twain:
> The scurvy jackal and the lion's roar?
>
> Feeding on frogs, on barnfloor mice, and on
> The corpses laid apart in charnel-field,
> In the great forest, in the lonely wood
> The jackal throve and fancied vain conceits:
> The lion, King of all the beasts am I!
> But when he roared—a puny jackal's whine.
> For what is there in common 'twixt the twain:
> The scurvy jackal and the lion's roar?
>
> (*Dīgha Nikāya*; Davids and Davids 1965, III: 22–3)

Pitted against the lion, there is an apparent condescension for the jackal. What evokes curiosity, however, is the observation regarding the latter rather abruptly reckoning himself as a tiger, though what follows immediately only serves to reiterate the lowliness of the former vis-à-vis the lion, and not the tiger. Why then does the latter, as if almost jostling for space, suddenly appear as an interloper in a context which fundamentally revolves around the jackal and the lion? Is it possibly because it is a context which engages with the issue of might, and one which subtly suggests that though the tiger is a worthy and close contender, it is the lion which rules the day? No matter how we approach these passages, what seems indisputable is what is implied—the absolute supremacy of the lion over all other beasts.

The sheer number and frequency of similes woven around the lion (a survey of which would be beyond my focus) can also be drawn upon to elucidate the point. Given the relative silence regarding the tiger, portrayals of the lion easily help us ascertain its position in the cultural construction of nature's pyramid. These examples suffice to show that the tiger as an individual entity seldom figures in early Buddhist canonical literature. Though it is to be feared and stayed away from, when it comes to contexts describing might and splendour, it is the lion which is chosen as a point of reference, and irrefutably stands tall over all brute creatures.

The Jātaka Tales

A fascinating treasure trove of narratives, the Jātaka Tales transport us to locales where humans and tigers are very often juxtaposed in ways and contexts which implicitly or explicitly suggest conflict as well as coexistence. In the *Mahā Sutasoma Jātaka* (537), for instance, when people go missing, the suspicion is that some lion, tiger, or demon has devoured them (Francis 1969, V: 248). Similarly, the *Sañjīva Jātaka* (150) revolves around Sañjīva, one of the 500 brāhmaṇ pupils of the Bodhisatta, who was born into the family of a wealthy brāhmaṇ and went on to become a teacher of world renown. From the master Sañjīva learnt the spell for bringing the dead back to life. But though the young pupil had learnt this, he had not been taught the counter-charm. Revelling in the glory of his newly acquired knowledge, he ventured out with his fellow pupils to gather wood from the forest, and chanced upon a dead tiger. To demonstrate the efficacy of the charm he had learnt, he urged his companions (who, anticipating danger, had perched themselves on a tree) to watch what ensued. To allay their apprehensions, Sañjīva repeated the charm, and struck the dead tiger with a potsherd. Calamity struck as the dreaded animal came alive and sprang at Sañjīva, biting him on the neck, and killing him right away. Both fell dead, and having witnessed the tragic chain of events, the group returned with their wood to narrate the same to their master who cautioned, 'Befriend a villain, aid him in his need, And, like that tiger which Sañjīva raised to life, he straight devours you for your pains' (Chalmers 1969, I: 321).

Embedded in this narrative is the counsel that helping an unworthy and undesirable element can be a thankless and even suicidal enterprise. At the same time, the use of the tiger as a motif to drive home the point by portraying it as the 'villain' is meaningful, and suggests an inherent suspicion regarding the predator and its potential to harm fatally.

Once again in the *Vaḍḍhaki Sūkara Jātaka* (283), the mega carnivore appears as an adversary deserving and needing to be exterminated. The narrative recounts that once when the Bodhisatta was born as a tree-spirit, there were some carpenters living in a village near Benares. A visit to the forest for wood is what got one of them to chance upon a young boar fallen in a pit. The carpenter brought it home and nurtured it till the time it grew into a mighty animal with curved tusks. Known as the Carpenter's Boar, the animal would dutifully help him with his chores. However, the man fearing that somebody might eat up the animal, released it in the forest. Looking for a haven, the boar comes upon a cave in a mountainside with plenty of bulbs, roots, and fruits. He is soon joined by hundreds of other boars, and just as he voices his intention of making the place an abode for himself along with the others, words of caution follow. The place, he is told, was dangerous because of a tiger dwelling on a hill which came there in the morning, and seized and carried away anybody who came its way. A few queries ensued after which action was determined as the Carpenter's Boar prepared the others for war against the predator. The crack of dawn saw the boars lined up for battle in rows.

The tiger awoke, and trotted up for his usual trip. But the sight of the boar army made him stop still. Undeterred by being glared at by the dreaded animal, the boars glared back at him, even aping his actions. The tiger's soliloquy tellingly conveys the fear it had previously evoked, an evident lack of which at the moment perplexed the animal. 'They used to take to their heels as soon as they saw me—indeed, they were too much frightened even to run. Now so far from running, they actually stand up against me! Whatever I do, they mimic ... Well I don't see how to get the better of them.' Saying so, he turned away empty-handed to return to his lair, and as he did so, a sham hermit who always partook of the prey the tiger brought home enquired, 'The best, the best you always brought before When you went hunting after the wild boar. Now empty-handed you consume with grief, To-day where is the strength you had of yore?' To this the tiger rejoins, 'Once they would hurry-scurry all about to find their holes, a panic-stricken rout. But now they grunt in serried ranks compact: Invincible, they stand and face me out.'

The hermit exhorts the tiger to not give up, urging that a roar and a leap was all it would take to create panic amongst the boars. The animal submits, and returns to the site of action. With a thundering roar he leaps upon the Carpenter's Boar, which had placed itself between two pits, and dodged as soon as the tiger charged. Consequently, the carnivore tumbled over and fell into the other pit, and that was the end of him as the boar tore him apart with his fangs and tossed him out of the pit for the others to take over what was left of their enemy. The early birds got to eat the animal, but the ones who came in later went about sniffing the others' mouths and asking what tiger flesh tasted like. The next concern was the wicked ascetic who also meets his end as a result of a collective attack by the boar army. A sprite that dwelt in the forest, having witnessed the sequence of events celebrated the triumph thus, 'Honour to all the tribes assembled be! A wondrous union I myself did see! How tuskers once a tiger overcame By federal strength and tusked unity!' (Rouse 1969a, II: 276–9).

Apart from underlining the role of the tiger as the perpetrator of fatality, this narrative also craftily weaves in the element of human greed, a fear of which not only makes the carpenter set the boar free, but is also in the form of the wicked hermit, a reason for the tiger to return to its nefarious activities. What perhaps also deserves mention is that when the same anecdote forms the theme of the *Taccha Sūkara Jātaka* (492), the boars while telling the Carpenter's Boar about the tiger refer to him as follows: 'A king of beasts! striped up and down he is, with teeth to bite ... a beast of might!' (Rouse 1969b, IV: 217).

The carnivore is also a villain in the *Bhīmasena Jātaka* (80) set in Benares. Bhīmasena, a weaver by profession, is recruited by the king after he poses himself as a mighty archer. The first task assigned to him is the capturing of a tiger known to be living in a forest in Kāsi which blocked a frequented road. The animal had devoured many victims. The Bodhisatta, who is the real force behind Bhīmasena, sets the plan and cautions him not to approach the animal's lair alone. He is instead told to muster a strong group of people and march to the site with 1,000 or 2,000 bows. Once he knew that the tiger was awake, he was to dash into the thicket and lie flat on his face. The group would then beat the predator to death, and as soon as he was finished, Bhīmasena was to emerge and move towards the dead animal with a creeper in his hand. He was then to show shock at the sight of the dead brute, and demand to know who was behind the deed while asserting that his own intention had been to lead the animal like an ox by the creeper. The people out of fear would request him to not report them to the king, and he would consequently be credited with slaying the dreaded carnivore. The forest made safe for travellers, Bhīmasena is rewarded for the labour which was never his. It is eventually the wrath of the people which brings the animal to its end (Chalmers 1969, I: 204–5).

Thus, getting the better of the animal was clearly an arduous task, and would often require collective animal (*Vaḍḍhaki Sūkara Jātaka*) or human (*Bhīmasena Jātaka*) action. It is also interesting to note that though the Pali canon emphasizes on *ahiṃsā*, there seems to be no proscription on the killing of the tiger in the Jātakas.

However, despite being synonymous with terror in most situations in the Jātakas, the animal is not evil in every context. In fact, its power to annihilate can also be employed to mete out justice to the aggrieved as in the *Kuntani Jātaka* (343), where a heron serving as envoy to the king avenges the loss of her young ones by seizing and throwing the perpetrators at the feet of a fierce and savage tiger kept in the palace. Scores are settled and justice is delivered as the carnivore crunches them up (Francis and Neil 1969, III: 89–90).

Far more evocative is the *Tittira Jātaka* (438), where a wicked ascetic kills and consumes a learned partridge who happens to be a dear friend of a lion and a tiger inhabiting the same forest. Though the narrative is a longer one, and involves many more characters, I focus here only on the segment where the two mega carnivores stage their entry in search of their avian companion. Realizing one day that some time had lapsed since they had met their friend, the lion sends the tiger to check on him.

The latter arrives at the place to find the wicked ascetic sleeping. In his matted locks, he sees some feathers of the partridge. Sensing something amiss, he wakes up the man with a kick and confronts him. Terrified, the ascetic confesses to all his cruelty except the one involving the killing of the partridge. Far from convinced, the tiger tells him that he deserves to be produced before the lion, the king of beasts, for verdict. The ascetic finally confesses to his crime, and the lion on hearing him speak the truth wants to let him go. The tiger wanting justice for his friend, however, differs, exclaiming, 'The villain deserves to die', and rips him apart with his teeth (Francis and Neil 1969, III: 321–3).

It is interesting to observe how the tiger, usually a slayer itself, transforms into an avenger of the destruction of life in the tales above. Far more telling is the animal's submission to the lion's superiority (in the *Tittira Jātaka*) whether in the case of complying with his desire to go and fetch tidings of their friend or its own perception of the authority of the maned predator to deliver the final judgement by virtue of being the 'king of beasts'.

In fact, a thread that connects tales featuring the lion and the tiger together almost inevitably portrays them as friends. Yet, there is a subtle element of contest if one may say so, either in the form of a dispute between the two predators as in the *Malūta Jātaka*, or simply an attempt to create dissensions between the two (*Vaṇṇāroha Jātaka*), or even a situation where the pre-eminence of the lion is put across as an established fact as observed in the *Tittira Jātaka*.

We encounter the lion–tiger duo once again in the *Malūta Jātaka* (17). The tale recounts their living together as friends in the same cave at the foot of a certain mountain that also served as the abode of the Bodhisatta, a hermit. One day a dispute arose between the two regarding whether it was cold in the dark half of the month (as argued by the tiger), or the light half (as maintained by the lion). The matter was taken to the Bodhisatta who restored peace between the friends by asserting, 'In light or dark half, whensoe'er the wind Doth blow, 'tis cold For cold is caused by wind. And, therefore, I decide you both are right' (Chalmers 1969, I: 50–1).

Let us also consider the *Vaṇṇāroha Jātaka* (361), where a lion and tiger peacefully coexisted in a mountain cave in the forest. With them was a jackal who survived on the remnants of the meats consumed by the two. One day, however, it occurred to him that he had never tasted the flesh of a lion or a tiger, and he engineered a plan to set the two apart in a way that they would fight each other to death. With this thought in mind, he approached the lion enquiring if there was any friction between him and the tiger. On being asked the reason behind his query, the jackal responded, 'Your Reverence, he ever speaks in your dispraise and says, "When I am gone, this lion will never attain to the sixteenth part of my personal beauty, nor of any stature and girth, nor of my natural strength and power."'

The lion refusing to believe the jackal chases him away, and the cunning animal now proceeds to try his luck with the tiger who immediately rushes to check the veracity of what he has heard. After being reprimanded by the lion for having lent ear to such vile talk, the tiger readily submits and begs pardon admitting that it was his fault (Francis and Neil 1969, III: 126–7). The air cleared, the two animals continue to live together happily.

Then there are ecological undertones that subtly run across some tales. The *Vyaggha Jātaka* (272) recounts the birth of the Bodhisatta as a tree-spirit living in a wood inhabited by another tree-spirit. The same forest was also home to a lion and a tiger, and both were so intensely feared that no one dared till the soil or cut a tree there. The ferocious creatures were known for killing and consuming all kinds of creatures, and for leaving the remnants to decay and emit a foul stench. Now this annoyed the other spirit who out of exasperation suggested to the Bodhisatta that both these creatures should be driven away. The Bodhisatta urged him not to do so, asserting the importance of these animals in keeping the forest intact, and cautioning that if driven away, men would cut down all the forest and render it desolate. This did not prevent the foolish spirit from assuming a daunting form and driving away the two carnivores. What ensued was just what the Bodhisatta had forewarned. People inferred from the absence of lion and tiger footprints that they had gone away to another wood, and started cutting down the forest. Distressed by the state of affairs, the spirit turned to the Bodhisatta who sent him off to get the creatures back, an attempt that proved futile since the animals refused to return despite his pleas, 'Come back, O Tigers! to the wood again, And let it not be levelled with the plain; For, without you, the axe will lay it low; You, without it, forever homeless go.' Within days thereafter, the entire wood was cut down, made into fields, and brought under cultivation (Rouse 1969a, II: 244–6).

Multiple themes, thus, seem to run across when we try and bring together images of the tiger as reflected in the Jātakas. Cars or chariots yoked with banners flying free with tiger skin and panther hide make for a gorgeous sight (*Sonaka Jātaka* 529; Francis 1969, V: 132; *Vessantara Jātaka* 547, Cowell and Rouse 1969, VI: 261), while robes of tiger skins are donned by knights and warriors (*Mahājanaka Jātaka* 539, Cowell and Rouse 1969, VI: 30). The predator's grace serves as a figure of speech in the *Kiṁchanda Jātaka* (511, Francis 1969, V: 3), while it is interesting to see how the animal is also allegorically referred to in a scathing attack on the Brāhmaṇical system in the *Bhūridatta Jātaka* (543) when it is said, 'The Vedas have no hidden power to save . . . Brahmins made the Vedas to their cost When others gained the knowledge which they lost . . . Brahmins are not like violent beasts of prey, No tigers, lions of the woods are they; They are to cows and oxen near akin, Differing outside they are as dull within' (Cowell and Rouse 1969, VI: 109–12).

The animal, thus, is dreaded yet admired. Its might is recognized, and more significantly is its crucial role as the protector of the forest. Yet, somewhere beneath these layers is an undercurrent where notwithstanding its strength and grandeur the striped predator voices acceptance of the supremacy of the lion. Keeping in mind the nature of the tales, this comes as a crucial reminder of popular perceptions regarding the animal, and where it stood in the hierarchy of the wild.

Milindapañha or *The Questions of King Milinda*

As can be anticipated in a work of this character, not much can be gathered about animals in general and the tiger in particular. However, fleetingly, it is only the predatory character of the animal along with other big cats that is alluded to in a reference to the females of lions, tigers, and panthers eating hard bits of bone and flesh (Davids 1890, I: 105). Nāgasena expounds that the fear of a black snake, an elephant, lion, tiger, leopard, bear, hyena, wild buffalo, gayal, or of fire, water, or of thorns, spikes or arrows is actually the fear of death, and whosoever is afraid of these is actually in dread of death (Davids 1890, I: 211).

The Jaina World

From a perusal of the texts under study, it seems that the carnivore has a nearly negligible presence in the philosophical discourses of the Jaina canon. Perhaps the only meaningful allusion is encountered in the prohibition on the use by monks and nuns of plaids made of tiger's fur (*Ācārāṅga Sūtra*; Jacobi 1884, I: 158). Also, monks and nuns on begging tours are told to avoid noxious creatures including the tiger (*Ācārāṅga Sūtra*; Jacobi 1884, I: 100).

The Dharmaśāstras

The animal figures in the law code prescribed by Baudhāyana (3.3.6), who in his classification of forest hermits into those who cook and those who do not, mentions the *Retovasiktas*, who used animals produced from semen. The latter collected the flesh of animals killed by tigers, wolves, hawks, or other predators and cooked it, offered the daily fire sacrifice with it morning and evening, and gave portions of it to ascetics, guests, and students, and ate what remained (Olivelle 2003, 309).

Elsewhere, spelling out the details of the Kūṣmāṇḍa rite for the expiation of a man who considered himself impure in any way, Baudhāyana (3.7.12) listed among other things offerings of ghee reciting, 'The strength in the lion, the tiger, the panther . . . ' (Olivelle 2003, 317).

Unfortunately, not much can be gleaned from these references since the contexts though ritual, do not in any way espouse the use of the animal in them. Rather, once again, there is an allusion only to its strength and predatory character.

Ecology, Therapeutics, and Tiger Meat: The Legacies of Caraka and Suśruta

We return once again to the animal classifications embodied in these two medical treatises. The *Caraka Saṃhitā* (I.27.35; Sharma 1994, I: 197), for instance, classifies predators such as the tiger and the lion in the *prasaha* (who take their food by snatching) category of animals. Further, the treatise recommends the serving of the meat of these carnivorous animals to patients suffering from emaciation (*Caraka Saṃhitā* VI.8.153; Sharma 1994, II: 156).

Similarly, we turn to Suśruta's exhaustive catalogue of meats to look for the striped carnivore. Here, the *jāṅgala* category, including the flesh of animals living on dry lands, encompasses within it a sub-division called *guhāśaya* (those having a lair). The *guhāśaya* animals include the lion and the tiger among others (*Suśruta Saṃhitā* I.46.36; Bhishagratna 1963, I: 484). The flesh of the animals of this family is said to be sweet, heavy, demulcent (soothing), and strength imparting. It generates heat, suppresses deranged *vāyu*, and is useful for treating diseases afflicting the eyes and the anus (*Suśruta Saṃhitā* I.46.37; Bhishagratna 1963, I: 484).

Significantly, the *guhāśaya* constitute the 'essential core of the category of *pañcanakha*', the eating of the flesh of which (with a few exceptions) we have seen, was prohibited. As Zimmermann pertinently observes, 'By prescribing broths of the meat of the lion or the tiger, for example, as a remedy for consumption, chronic diarrhea, or hemorrhoids, the Ayurvedic doctrine runs counter to brahminic prohibitions' (1987, 174).

Tiger Encounters in the Epics

As far as the burly predator is concerned, what is common to both the epics is the overwhelming use of the animal in compounds suggestive of supremacy and excellence. An overview of the figures of speech employed makes it evident that the tiger frequently occurs as a motif in contexts meant to convey pre-eminence. This, however, is not to say that the predator holds a singular position in this aspect. When it comes to prowess, the bull, elephant, and the lion (in varying frequencies) intermittently share the slot.

The Rāmāyaṇa

The carnivore is woven into the narrative on occasions which place the creature in its natural surroundings as well as in the form of similes and compounds like *muniśārdula*—'tiger among sages' (*Bālakāṇḍa* 1.30.4; Goldman 1984, 182), *hariśārdula*—'tiger among monkeys' (*Kiṣkindhākāṇḍa* 4.66.31; Lefeber 1994, 193), *ikṣvākuśārdula*—'tiger of the Ikṣvākus' (*Kiṣkindhākāṇḍa* 4.30.16; Lefeber 1994, 122).

Anxious about not having begotten a son to extend his lineage, King Daśaratha on the counsel of sage Rśyaśṛnga embarks on performing the 'son-producing' sacrifice. As he does so, there emerges from the fire a great being with the hair of his body, head, and beard as glossy as that of a yellow-eyed lion, and his gait like that of a haughty tiger (*Bālakāṇḍa* 1.15.10–11; Goldman 1984, 156). The vessel filled with celestial porridge he appears with is then given to Daśaratha for his wives who after partaking it would bear him his much desired sons.

As is well known, the sons born are Rāma to Kausalyā, Bharata to Kaikeyī, and Lakṣmaṇa and Śatrughna to Sumitrā. Virtuous Rāma, the favourite of his father, the old king, is chosen as the prince regent. However, at the behest of the wicked and conniving hunchback Mantharā, Kaikeyī intervenes and demands the consecration of her son Bharata, and the expulsion of Rāma to the forest for fourteen years. Meanwhile, oblivious to these palace intrigues, Ayodhyā revels in the news of the consecration of its beloved prince. The morning of the much-awaited day dawns, and brāhmaṇs make preparations for the ceremony. Among other things, they set out golden ewers, a richly ornamented throne, and a chariot draped with a resplendent tiger skin (*Ayodhyākāṇḍa* 2.13.4–7; Pollock 1986, 109).

As he hears Kaikeyī's ruthless wishes, the old king is a broken man and unnerved like a stag at the sight of a tigress (*Ayodhyākāṇḍa* 2.10.30; Pollock 1986, 104). Rāma departs to the forest with Lakṣmaṇa and Sītā. Grieving Kausalyā is assured after their departure about the well-being of the trio with particular emphasis on the safety of Sītā. The queen is told that in the forest when she would spot an elephant, lion, or tiger, Sītā would slip into Rāma's arms and be safe from fear (*Ayodhyākāṇḍa* 2.54.17; Pollock 1986, 200).

Nevertheless, the inconsolably despairing queen knowing no other way, rebukes her husband for having brought this fate upon their tenderly raised son. Even if her son returned in the fifteenth year, Kausalyā was certain he would spurn both the kingship as well as the treasury since Bharata will have possessed them. 'A tiger', she contends, 'will not eat the food another beast has fed upon. In the same way the tiger among men will scorn what another has tasted.' (*Ayodhyākāṇḍa* 2.55.12; Pollock 1986, 201). In a similar vein, she argues, 'Rāghava will not suffer an insult of this sort, any more than a powerful tiger suffers having its tail pulled' (*Ayodhyākāṇḍa* 2.55.15; Pollock 1986, 201).

Meanwhile Rāma, Lakṣmaṇa, and Sītā embark on their fourteen-year exile in the forest fraught with its perilous wild beasts and demons. As they move into the wilderness of Daṇḍaka, a world starkly different from the one they have left behind awaits them. Rāma beholds the heart of the forest, the haunt of wolves and tigers, and crowded with herds of various other animals. A tangle of trees, vines, and shrubs with marshy pools, different sorts of birds fallen silent, and swarms of chirping crickets surround them (*Araṇyakāṇḍa* 3.2.2–3; Pollock 1991, 88–9). The forest is an abode not only for wild beasts, but also for fierce *rākṣasas* or demons, encounters with whom form an intrinsic part of the exile period. The dreaded *rākṣasas* are huge, their strength great, and their pride like that of a tiger (*Araṇyakāṇḍa* 3.21.8–10; Pollock 1991, 132). Virādha, for instance, forms a monstrous and ghastly sight clad in a tiger skin dripping with grease, and splattered with blood. Roaring deafeningly, he encounters the exiled trio with an iron pike on which were impaled three lions, four tigers, two wolves, ten dappled antelopes, and the massive head of an elephant with its tusks intact and smeared with gore (*Araṇyakāṇḍa* 3.2.5–8ab; Pollock 1991, 89).

As they settle down to their life in Pañcavaṭī, trouble strikes in the form of the dreadful *rākṣasi* Śūrpaṇakhā, who is consumed with desire on seeing Rāma. Spurned and mutilated, she returns to her brother Khara seeking justice, who then sets out to avenge the mutilation of his sister at the hands of Lakṣmaṇa. Khara is himself, however, seized with terror witnessing the reverses the *rākṣasa* ranks face, fleeing like deer frightened by a tiger (*Araṇyakāṇḍa* 3.26.19; Pollock 1991, 143).

Eventually Khara is vanquished, and the task left unfinished by him is taken up by Rāvaṇa. Revengefully, Sītā is abducted, and distraught Rāma begins to look for her frantically in the vicinity of the leaf hut. Battling the pain of separation, he questions all and sundry comprising trees and animals like the deer, elephant, and tiger. 'Or you there, deer, do you know anything of fawn-eyed Maithilī? . . . Elephant, I am sure you would know if you had seen her. Her thighs are just as smooth as your trunk . . . Tiger, if you have seen Maithilī, my moon-faced love, speak out in full confidence; you have nothing to fear' (*Araṇyakāṇḍa* 3.58.20–2; Pollock 1991, 214). It has been pointed out that unlike the previous two verses which compare Sītā with the deer and elephant, the last one reflects no characteristic shared by Sītā and the tiger whereby the animal might recognize her. Though commentaries suggest that it is the similarity in their gait that is meant, Sheldon Pollock (1991, 336) contends that when Rāma says 'you have nothing to fear' (*na te bhayam*), he only means to assure the animal that he is not out hunting, and it need not flee from him.

The search for Sītā gains strength after Rāma allies with Sugrīva and helps him in ousting his brother Vālin. The defeated and dead monkey king is likened to a hero felled by a hero as a great stag slain by a tiger for its flesh (*Kiṣkindhākāṇḍa* 4.19.21–4: Lefeber 1994, 96). Sugrīva is consecrated elaborately and is brought, among other things, tiger skin and boar skin sandals (*Kiṣkindhākāṇḍa* 4.25.21–5; Lefeber 1994, 109).

Following the consecration of Sugrīva, Rāma proceeds to Mount Prasravaṇa, which resounded with the cries of tigers and wild beasts and the roars of lions. Covered with all kinds of bushes and vines, and thick with trees, the mountain always abounded in pure water (*Kiṣkindhākāṇḍa* 4.26.1–3; Lefeber 1994, 110). Sugrīva instructs his monkey troops to go in search of Sītā, and to look for her also amongst the Kirātas known as tiger-men (all commentators agree that these are literally half-tiger, half-men where the upper part is that of the animal, and the lower part that of a man) who lived on islands (*Kiṣkindhākāṇḍa* 4.39.23cd–27; Lefeber 1994, 142). Urged by Jāmbavān, Hanumān sets afoot Mount Mahendra to embark on his legendary leap across the ocean towards Laṅkā in search of Sītā. It was canopied with all sorts of trees, and its grassy plots were frequented by deer. Abounding in waterfalls, it was inhabited by lions, tigers, and frequented by rutting elephants (*Kiṣkindhākāṇḍa* 4.66.34–6; Lefeber 1994, 193).

Particularly noteworthy are contexts which juxtapose the two big cats. Accompanied by his two ministers, Śuka and Sāraṇa, Rāvaṇa stands atop his snow-white palace, gazing around to see a land completely covered with leaping monkeys. Sāraṇa, familiar with the leaders of the forest-dwelling monkeys, sets out to describe them to his lord. The monkeys are likened to lions with their four great fangs, as unassailable as tigers (*Yuddhakāṇḍa* 6.18.37; Goldman et al. 2009, 164). Similarly, the *rākṣasa* Akampana proceeding towards the battleground, is said to have shoulders as massive as a lion's, and a stride like that of a tiger (*Yuddhakāṇḍa* 6.43.9; Goldman et al. 2009, 237). In the bloody battle, Nīla and Prahasta slashing each other with their razor-sharp fangs are like a lion and a tiger, and are said to even look like them (*Yuddhakāṇḍa* 6.46.38; Goldman et al. 2009, 248). Despite the juxtaposition, the verdict is delivered in favour of the lion when tidings arrive regarding Sītā having been found. Sugrīva is eager to see the monkeys led by Hanumān, who had accomplished this mission and whose pride was like that of the lion, the king of beasts (*Sundarakāṇḍa* 5.61.27; Goldman and Goldman 1996, 285).

The Mahābhārata

Moving to the *Mahābhārata*, references to the tiger (*vyāghra/śārdūla*) abound mostly in the form of figures of speech. Though such expressions primarily convey the admiration the animal commanded by virtue of its might and grandeur as also the obvious fear that it evoked, it is imperative to delve into some of them in order to have a sense of the ways in which the animal was perceived, as also for glimpses of where these cultural perceptions placed the animal within hierarchies governing the wilderness.

Predominantly, the animal occurs as the final element in compounds denoting valour and supremacy. For instance, strong cultural resonances regarding the predator are reflected in compounds such as *puruṣavyāghra*—'tiger among men' (*Sabhāparvan* 2.22.34; Buitenen 1975, 2, 74), *rājaśārdūla*—'tiger among kings' (*Āraṇyakaparvan* 3.106.9; Buitenen 1975, 2, 426), *bhṛguśārdūla*—'tiger of the Bhṛgus' (*Ādiparvan* 1.13.45; Buitenen 1973, 1, 71), *yaduśārdūla*—'tiger of the Yadus' (*Sabhāparvan* 2.22.2; Buitenen 1975, 2, 73), *bharataśārdūla*—'tiger of the Bharatas' (*Āraṇyakaparvan* 3.106.6; Buitenen 1975, 2, 426), *kuruśārdūla*—'tiger of the Kurus' (*Āraṇyakaparvan* 3.83.97; Buitenen 1975, 2, 397), *vṛṣṇiśārdūla*—'tiger of the Vṛṣṇi clan' (*Strīparvan* 11.16.42; Crosby 2009, 271), which are scattered across the epic, and encountered all too frequently.

When it comes to might, the animal serves as a frequently employed simile. The Pāṇḍavas, that is Yudhiṣṭhira, Bhīma, Arjuna, and the twins Nakula and Sahadeva, are tigers among men, proud of their strength like tigers (*Āraṇyakaparvan* 3.255.3; Buitenen 1975, 2, 719). Similarly, the battle between Bhīma and the *rākṣasa* Kirmīra is described as dreadful like that between proud tigers armed with claws and fangs (*Āraṇyakaparvan* 3.12.54; Buitenen 1975, 2, 242). Bhīṣma acquaints the blind old Dhṛtarāṣṭra with the might of their adversaries, the Pāṇḍavas, and cautions, 'While modest, all these tigerlike men are proud of their strength like tigers, and superhuman in speed, striking power, and combativeness . . . When they attack your army, proud of their strength like tigers, they will trounce it in battle' (*Udyogaparvan* 5.166.22, 25: Buitenen 1978, 3, 489).

Associated with its might is the predatory temperament of the carnivore which runs across illustrating situations fraught with contest and conflict. The aforementioned combat between Bhīma and Kirmīra, for instance, proves fatal for the *rākṣasa* when brawny Bhīma strangles him like a beast at a sacrifice, as a tiger kills small game (*Āraṇyakaparvan* 3.11.24; Buitenen 1975, 2, 239). The wrestling contest in the court of King Virāṭa amplifies the same when Bhīma is summoned to take on the mightiest wrestler. The tiger-like Bhīma enters the ring with the loose step of a tiger (*Virāṭaparvan* 4.12.18; Buitenen 1978, 3, 43), and roaring, pulls the wrestler with his arms as a tiger attacks an elephant (*Virāṭaparvan* 4.12.21; Buitenen 1978, 3, 43). Even while killing Kīcaka who had lusted after Draupadī, he is compared with a tiger that is hungry for meat, and has caught a large deer (*Virāṭaparvan* 4.21.58; Buitenen 1978, 3, 60). Rescued from peril, Draupadī is a reminder of a young but spirited doe that had been terrified by a tiger (*Virāṭaparvan* 4.23.12; Buitenen 1978, 3, 62).

Jubilant after their triumph in the fateful game of dice, Duryodhana issues orders for Draupadī to be presented in the assembly hall. Visualizing the impending doom the act spelt, Vidura cautions, 'The incredible happens through people like you, You don't know it, nitwit, you are tied in a noose! You hang over a chasm and do not grasp it, You dumb deer to anger tigers!' (*Sabhāparvan* 2.59.2; Buitenen 1975, 2, 138).

Subsequently, as events build up to war, Karṇa is certain that his infallible arrows will scatter the Pāṇḍavas and Pāñcālas to the ten horizons like bullocks before a tiger (*Udyogaparvan* 5.165.21; Buitenen 1978, 3, 488). Dhṛtarāṣṭra, however, is conscious of the folly his son Duryodhana is about to commit by going to war with the mighty Pāṇḍavas, and cautions that like tigers among herds of antelopes, the latter would kill all their leaders (*Udyogaparvan* 5.57.20; Buitenen 1978, 3, 325). Elsewhere, we find the old king expressing his fear of Bhīma, likening it to the fear a sturdy antelope has of a tiger (*Udyogaparvan* 5.50.2; Buitenen 1978, 3, 312).

The sleeping tiger serves as a point of allusion both in the context of approaching danger as well as bravery. 'Do not awaken a lying tiger, Or a venomous snake that licks its mouth: If you kick its head with your foot, you will Not escape unbitten, be sure of that!' (*Āraṇyakaparvan* 3.134.3; Buitenen 1975, 2, 478).

While such counsel cautions against the dire consequences of provoking a formidable adversary, a warrior slain in battle is also likened to a sleeping tiger (*Strīparvan* 11.22.5; Crosby 2009, 299).

Overpowering the animal is an act of bravery emanating from extraordinary prowess. The infant son of Duhṣanta and Śakuntalā is one such outstanding being, fettering tigers along with lions, boars, buffaloes, and elephants to the trees around Kaṇva's hermitage, and running about playing, riding, and taming them (*Ādiparvan* 1.68.5; Buitenen 1973, 1, 165). Similarly, Bhīma asserts his might when he claims to have faced and killed many tigers along with lions, buffaloes, and elephants in battle (*Āraṇyakaparvan* 3.176.4; Buitenen 1975, 2, 561).

Moving beyond these figures of speech which serve as windows regarding the animal, the search for more meaningful perceptions about the mega carnivore itself often captures the animal in its natural habitat. The setting of Dadhīca's hermitage on the other bank of the Sarasvati, for instance, is a reader's delight. Wooded with all kinds of trees and creepers, buffaloes graze everywhere together with swine, marsh deer, and yaks bereft of the fear of tigers. Yet the latter could not have been far away since the hermitage, we are told, resounded with the mighty roars of lions and tigers, and of other creatures living in caves and hollows (*Āraṇyakaparvan* 3.98.14, 16; Buitenen 1975, 2, 417).

The forest is abode not only for the terror-evoking predator (*Āraṇyakaparvan* 3.61.2, 25, 83; Buitenen 1975, 2, 336, 337, 340), but also teems with leopards, buffaloes, lions, wild boars, bears, *ruru* deer, and elephants (*Āraṇyakaparvan* 3.61.2, 123; Buitenen 1975, 2, 336, 341). The animal is also met with on 'Mount Himālaya, ornamented with many-shaped and mineral rich peaks . . . Rivers and arbors and swells adorned it, it abounded in water, and lions and tigers inhabited it, lying in caverns and caves' (*Āraṇyakaparvan* 3.107.4–6; Buitenen 1975, 2, 428). The unprepared are said to encounter them along with flies, gnats, mosquitoes, lions, and snakes on Mount Gandhamādana, but those prepared do not see them (*Āraṇyakaparvan* 3.142.27; Buitenen 1975, 2, 493–4).

Most tellingly, our search for the carnivore leads us to the *Udyogaparvan*, where the narrative craftily integrates the big predator to convey a profound ecological message. Vāsudeva impresses upon Saṃjaya, the charioteer and confidant of Dhṛtarāṣtra, the injustice brought upon the Pāṇḍavas, acknowledging the futility of war, yet validating its need for the sake of securing what justly belonged to them.

Dhṛtarāṣtra and his sons, he asserts, were the forest, while the Pārthas were the tigers. What ensues in this dialogue is crucial in view of its ecological implications. Vāsudeva cautions Saṃjaya,

> Do not cut down the forest with its tigers, and don't banish the tigers from the forest. The tiger perishes without the forest, and the forest without its tigers is cut down. Therefore the tiger should stand guard over the forest and the forest protect its tiger. The Dhārtarāṣṭras are of the nature of creepers, Saṃjaya, and the Pāṇḍavas are *sāla* trees; a creeper never grows without clinging to a big tree (*Udyogaparvan* 5.29.47–8: Buitenen 1978, 3, 245)

The same wisdom is imparted by Vidura to Dhṛtarāṣṭra when the former reiterates more clearly that there would be no forest but for its tigers, and no tigers but for their forest; for the forest is protected by its tigers, and the forest protects its tigers (*Udyogaparvan* 5.37.41–2; Buitenen 1978, 3, 276). What is even more remarkable here is the centrality the carnivore shares with the lion (which though not an animal of the forest, is, nevertheless, referred to as one) in the ecological pyramid when Vidura furthers likens the old king and his sons to the forest, and the Pāṇḍavas to forest lions, and goes on to emphasize, 'A forest is doomed when empty of lions, And without the forest the lions die' (*Udyogaparvan* 5.37.60; Buitenen 1978, 3, 277).

Elsewhere, the allusion to the tiger creating a lot of carrion for the sake of his own stomach, and other animals of lesser strength living upon that conveys a sense of predator–scavenger relationships operating in the wild (*Śāntiparvan* 12.17.8; Fitzgerald 2004, 7, 200).

A consciousness of such symbiotic equations within the natural world is not new for ancient India. One is reminded of the narrative in the *Vyaggha Jātaka* that also unambiguously conveys the same idea.

Further, amidst the heart-rending scene of the devoted Damayantī despairing for Nala in a desolate and dangerous forest, the animal is unambiguously declared to be the king of the forest. Wailing bitterly, she calls out to her lost companion,

> Are you sleeping in this dreadful forest teeming with lions and tigers . . . Here comes the illustrious king of the forest, a four-tusked, broad-jowled tiger, to meet me, I am not afraid, I ask him, 'Sir, sovereign of animals, you are the master of this jungle ... comfort me, king of beasts, have you seen Nala? Of if you will not speak of Nala, king of the forest, then eat me, great beast and deliver this wretch from her misery!' The king of beasts hears my lament in the wilderness and goes off on his own to the clear waters of the river that flows to the ocean. (*Āraṇyakaparvan* 3.61.25, 30–4; Buitenen 1975, 2, 337)

At the same time, the predator seems to meet its match in the lion (*siṃha/hari/kesarin*). With blade, tooth, and nail, Jayadratha and Abhimanyu clash and fiercely strike each other like tiger and lion (*Droṇaparvan* 7.14.68; Pilikian 2006, I, 131). Moreover, if the Pāṇḍavas are likened to tigers, they are also compared to lions (*Udyogaparvan* 5.166.21–2, 25: Buitenen 1978, 3, 489). Similarly, if the prowess of Bhīma reminds Dhṛtarāṣṭra of the fear an antelope has of a tiger (mentioned earlier), it is also a reminder of the fear evoked in a weak animal by a lion (*Udyogaparvan* 5.50.3; Buitenen 1978, 3, 312). Elsewhere, it is as perilous to kick a sleeping lion (*Āraṇyakaparvan* 3. 252.7; Buitenen 1975, 2, 711) as it is to kick a sleeping tiger (referred to earlier).

However, though the striped predator seems to dominate the cultural landscape, and is proclaimed the king of beasts, the maned carnivore subtly gains ascendancy when the two are pitted against each other. Lamenting over the lifeless body of her slain son Duryodhana, Gāndhārī likens him to a tiger struck down by the lion Bhīma (*Strīparvan* 11.17.19; Crosby 2009, 279). Similarly, grieving for Karṇa, she refers to him as a tiger slain by the lion Arjuna (*Strīparvan* 11. 21. 5; Crosby 2009, 297).

The skin of the animal also holds its own in the epic saga. There are allusions to it being donned as well as to its use in other ways. Karṇa is to be consecrated seated on the tiger skin (*vyāghracarma*) (*Udyogaparvan* 5.138.16; Buitenen 1978, 3, 444). We also catch Duryodhana arraying his armies: 'With spare axle trees, quivers, bumper bars, javelins, arrow holders, spears, arrow cases ... covers of tiger-hide, leopard skins... with all this the handsome, colorful armies blazed like fires' (*Udyogaparvan* 5.152.6; Buitenen 1978, 3, 468).

The tigress (*vyāghravadhū/śārdūlī*), however, is not encountered very often except as a mode of expression where the latter is juxtaposed with a jackal (*Āraṇyakaparvan* 3.248.17; Buitenen 1975, 2, 707; *Āraṇyakaparvan* 3.262.28; Buitenen 1975, 2, 735).

Along with the cultural canvas, it is equally or perhaps more crucial to consider the ecological landscape that emerges from the two epics. It may, therefore, be worthwhile to first have a sense of the geographical horizons being considered. As far as the *Mahābhārata* is concerned, it has been argued that the setting of the great battle on the field of Kurukṣetra, and the location of the capital at Hastināpura indicate very clearly the geographical focus of the main narrative, which is set in the area of modern Haryana and the upper Doab. This, however, is different from that of the *Rāmāyaṇa*, which is set in the Middle Ganga Valley with a fair knowledge also of the Indus region (Brockington 1998, 198, 420).

The diversity of fauna reflected in both the texts suggests a mosaic topography comprising forests, grasslands, and marshy patches. What also gets reiterated is ancient India's close familiarity with the natural world. That alone can perhaps explain the exhaustive scenic descriptions embellished with a wide range of flora and fauna with attempts to capture the temperaments and habitats of the latter. For instance, though tigers are not tied to a particular habitat type or temperature regime, the emphasis on the presence of water bodies where the animals are is a clear example of a sense of what transpired in the wilderness, since the predator needs constant access to water. However, certain aspects can also be attributed to the poet's imagination which, for instance, turns the dreaded tiger into a passive audience to Damayantī's woes.

The *Arthaśāstra*

The *Arthaśāstra* has already figured in the narrative when the manner in which this political treatise engages with our first protagonist, the rhinoceros, was examined. While the rhinoceros finds mention in the context of the utility of its body parts in fashioning equipments of war, the preoccupation with the elephant permeating the text (as shall be seen in Chapter 5) is perfectly intelligible given the nature and purpose of its composition. It is, however, with a certain amount of apprehension that one sets out in search of the carnivore in a text that primarily addresses concerns central to the state.

Let us begin with contexts which seem to indirectly suggest the presence of the tiger in the text. In general, apprehensions regarding wild animals loom large in the *Arthaśāstra*. We are told, for instance, that the country is to be protected against eight great natural calamities: fire, floods, disease, famine, rats, wild animals (*vyālāḥ*), serpents, and evil spirits (*Arthaśāstra* 4.3.1, Kangle 1963, II, 303). I pause here to reflect on the term *vyālāḥ* which has been variously interpreted in translations. While Kangle (1963) understood it as 'wild animals', readings have also construed the term as specifically denoting the tiger (Shamasastry 1967, 236; Gairola 1977, 434). In such a situation, it would be worth examining the dictionary meaning of the term. In his epic lexicon, Monier-Williams (1963, 1038) defines it as a vicious elephant, a beast of prey, a snake, a lion, a tiger, a hunting leopard, and so on. Hence, the word as can be seen, can broadly be applied to wild animals of various kinds with the potential to harm, and evidently human encounters with such species were frequent enough to elicit repeated attention in the text.

In a departure from the affairs of the state, the text also devotes some thought to the leisure of the king. The disposal of non-agricultural land, therefore, stipulates the creation of a park for the king's recreation. Apart from having trees without thorns and shrubs, and bushes bearing sweet fruits, it was to be stocked with tamed deer and other animals including wild ones with their claws and teeth removed (*Arthaśāstra* 2.2.3, Kangle 1963, II, 67). Implicit herein, seems the conception of a natural enclave with diverse flora and fauna which possibly seems to have included predators like the lion and the tiger. The *Arthaśāstra* further enjoins that on the border of this park or depending on the availability of land, another animal park was to be set up where all animals were to be welcomed as guests, and given full protection (*Arthaśāstra* 2.2.4, Kangle 1963, II, 67). The idea of creating an exclusive protected space for animals of all kinds also deserves particular attention in view of its being an early attempt at conserving the natural world.

A more direct reference occurs while suggesting measures for instigating sedition in the enemy camp. Among many other devices prescribed, it said that the diligent should be stirred up by alluding to the ordinary donkey, and those secretly put to test, by the tiger skin (*vyāghracarmaṇā*), and the death-trap (*Arthaśāstra* 13.1.16, Kangle 1963, II, 552). The allusion though not immediately comprehensible, is explained as suggesting that the king is ferocious like a tiger (Kangle 1963, II, 553).

Additionally, the predator figures in the list of animals whose skin, bones, bile, tendons, eyes, teeth, horns, hooves, and tails are considered as valuable forest produce (*Arthaśāstra* 2.17.13, Kangle 1963, II, 149). Apart from pragmatic reasons for valuing the animal, occult practices prescribed for deceiving the enemy extol the virtues of the bone marrow and semen of the tiger along with that of the lion, the leopard, the crow, and the owl in enabling one to walk a hundred *yojanas* sans exhaustion (*Arthaśāstra* 14.2.43, Kangle 1963, II, 582).

Piecing together the evidence, what we have are scattered references that drive home a picture where wild animals in general are to be guarded against. As for the tiger specifically, its ferocity qualifies it as a parameter for comparison with the king. At the same time, however, when the latter is to spend his moments of leisure in a park packed with tamed deer and wild animals, it is emphatically qualified that they should be sans their teeth and claws, that is, the animals are to be divorced of their ferocity and ability to harm. At yet another level, the teeth and other body parts are held valuable not only as forest produce crucial for the state, but also in the dark and mysterious world of magic for their strength-imparting qualities.

Looking West

The images of the animal retrieved from the corpus of classical texts range from the fabulous to the real, a statement easily explained by the fact that these narratives were often based on hearsay rather than on personal experience. Treading cautiously then, what we first encounter is a bizarre animal called the *martikhora* described by Ktesias. However, as already mentioned, since fragments of the *Indika* of Ktesias have come down to us only through the writings of later classical authors who cited him, I shall assemble glimpses of the *martikhora* as they chronologically occur in these works. Consequently, it may be worthwhile to point out at the very outset that though the most complete account of Ktesias's version of the *martikhora* comes to us through the account of Photios (mentioned in Chapter 2), I draw on the latter much later in my narrative since the aim as far as possible is to follow these writers in chronological sequence.

One can begin with the fragment of the *Indika* that is retrieved from Aristotle (fourth century BCE) who acknowledges Ktesias as his source. Said to be equal in size to the lion, and resembling it in its claws and having shaggy hair, the creature is said to have a triple row of teeth in each of its jaws with a face and ears like those of a human being. Blue eyes and hair of the colour of cinnabar (vermillion) are attributed to the animal whose tail is reported to resemble that of a land scorpion containing the sting, and with a growth of prickles which it could discharge 'like shafts shot from a bow'. We are also told about the ferocity of the animal, and its avidity for human flesh. Its voice is said to be like the sound of the pipe and the trumpet blended together, and it ran very fast, being as nimble as a deer (McCrindle 1882, 38–9).

Pausanius, a Greek traveller and geographer of the second century CE, on the other hand, was convinced that the wild beast described by Ktesias and called the *martikhora* by the Indians and the *androphagos* (man-eater) by the Greeks, was the tiger. He is evidently quoting Ktesias while describing the animal, arguing that the Indians accepted this version due to their intense fear of the ferocious creature. He also maintains that they were mistaken even regarding its colour, for when the tiger was seen by them in the sunlight, it appeared to be all red owing to the speed with which it ran, and even if it was not doing so, it appeared so due to the agility with which it turned its body from one side to the other. This was, contends Pausanius, also because it was not possible to get a closer view of it without putting oneself at risk (McCrindle 1901, 210–11).

Similarly, Aelian's work *On the Peculiarities of Animals*, spells out (quite elaborately) the peculiarities of the *martikhora* as described by Ktesias, adding that anyone who considered the Knidian a competent authority on such subjects should be content with the account furnished by him. He attributes to Ktesias the observation that its favourite food was human flesh, and that to satisfy its lust for the same, it killed many by springing from ambush not upon solitary travellers, but on a band of two or three for which it was more than a match. Far more telling in terms of the probable identity of the so-called *martikhora* is the remark that except the lion, which it could not subdue, all the beasts of the forest yielded to its prowess. Its love for human flesh earned it the name *martikhora* which was an Indian word meaning man-eater (McCrindle 1882, 40–2).

However, the most elaborate portrayal of this enigmatic creature is found in the abridgement of the *Indika* by Photios (ninth century CE). The account is a rather strange one where the creature is said to be as big as a lion, red in colour like cinnabar, and with a face like that of a man. What adds to the macabre apparition are the three rows of teeth, human-like ears, pale blue human-like eyes, and more than a cubit long tail armed with a sting like the land scorpion. Besides stings on each side of its tail like the scorpion, an additional sting on the crown of the head is attributed to the creature. With it, it is said to have stung anyone going near it, the wound in all cases proving mortal. If attacked from a distance, it defended itself both in the front and the rear—in front with its tail, by lifting it and darting out the stings like shafts from a bow, and in the rear by straightening it out. It could strike to a distance of 100 feet, and no creature except the elephant could survive the wound it inflicted. The stings were about a foot in length, and not thicker than the finest thread. Ktesias further elaborates that the name *martikhora* meant in Greek man-eater, and it was so called because it carried off men and devoured them though it preyed upon other animals as well. In fighting, it used not only its stings but also its claws, and fresh stings grew up to replace those lost while fighting. The animals numerous in India, Ktesias informs us, were killed by the natives who hunted them with elephants, from the backs of which they shot darts (McCrindle 1882, 11–12).

Having collated from various classical writers what Ktesias has to say about the *martikhora*, one is now confronted with the task of ascertaining which animal he had in mind while describing it. What cannot be denied is that it is a big predator mighty enough to prey upon humans after lying in ambush for them. That it is not the lion is also evident since it is compared in size with it. The possibility of it being the tiger is raised by the observation regarding its hegemony over all other beasts of the forest except the lion.

But if it was the tiger, how close was Ktesias to describing the real animal? Valentine Ball (1879–88, 310–11) attempts to unveil the *martikhora* for us arguing that though some commentators suggest that it was the tiger, none of them seem to have realized that the statements were based on actual facts. He brings forth one such detail, which though not widely known, is according to him mentioned in some works on zoology, and one which he from his personal knowledge attests to as being familiar to Indian *shikaris*, that is, at the extremity of the tail of the tiger as well as of other *felidae*, there is a horny little dermal structure like a claw or a nail which the natives compared to the sting of the scorpion. Ball additionally points out that the whiskers of the tiger were by many natives considered capable of causing injury. Similarly, the idea of the three rows of teeth, according to him, is likely to have had its origin in the three lobes of the carnivorous molar, which vastly differed from the molars of ruminants and horses. Notwithstanding the distortions in the telling, Ball asserts that the martikhora was the tiger, and that the account of it embodied actual details.

Did Ktesias see the *martikhora* himself, or was he writing on the basis of hearsay? According to Aelian, Ktesias claimed to have seen one of these animals in Persia when it was sent as a gift to the Persian king from India (McCrindle 1882, 42). It can be argued that given that Ktesias served as the royal physician at the Persian court, the possibility of his having seen the animal, if at all one made its way there, cannot be ruled out. What remains open to speculation, however, is the point that if Ktesias had seen the animal himself, why was his account full of such exaggeration?

From the enigmatic *martikhora* I now move towards writings which unambiguously mention the tiger. From Curtius Rufus, for instance, we learn that presents for Alexander brought by Indian ambassadors included tame lions and tigers of extraordinary size (McCrindle 1896, 252). We can also turn to what Nearchus (the commander of Alexander's fleet) has to tell us about the predator. A fragment retrieved from Arrian says that though Nearchus had not seen the animal itself, he had seen its skin. Indians had told him that the tiger is equal in size to the largest horse, but the swiftness and strength of no other animal could be compared with it. When confronted with an elephant, it could leap on the latter's head and strangle it with ease (McCrindle 1877, 217).

Embedded in these classical narratives are also details which often hint at the geographical distribution of the mega carnivore. We have the account of Megasthenes, fragments of which are available through the works of later writers like Strabo, who in his *Geography* attributes to the latter the observation that the largest tigers, almost twice the size of lions, were found in the country of the Prasii (the eastern stretch of the country). They were so strong that a tame tiger led by four men seized a mule by the hind leg, overpowered it, and dragged it to him (McCrindle 1877, 56).

Similarly, *The Periplus of the Erythrean* Sea, a work ascribed to the first century of the Christian Era tells us the following:

> From Barugaza the coast immediately adjoining stretches from the north directly to the south, and the country is therefore called Dakhinabades, because Dakhan in the language of the natives signifies south. Of this country that part which lies inland towards the east comprises a great space of desert country, and large mountains abounding with all kinds of wild animals, leopards, tigers, elephants, huge snakes, hyenas, and baboons of many different sorts, and is inhabited right across to the Ganges by many and extremely populous nations. (McCrindle 1879, 124)

The *Natural History* of Pliny the Elder (23–79 CE) also carries numerous references to India, and is divided into thirty-seven books. The sixth, dealing with the geography of the land draws largely on the *Indika* of Megasthenes. Based on this, Pliny furnishes us with a general account of the basins of the Indus and the Ganges, and goes on to list the 'races' inhabiting north India (McCrindle 1877, 129–54).

> The hill-tribes between the Indus and the Iomanes are the Cesi; the Cetriboni, who live in the woods; then the Megallae, whose king is master of five hundred elephants and an army of horse and foot of unknown

> strength; the Chrysei, the Parasangae, and the Asange, where tigers abound, noted for their ferocity . . . These are shut in by the Indus, and are surrounded by a circle of mountains and deserts . . . Below the deserts are the Dari, the Surae, then deserts again for 187 miles, these deserts encircling the fertile tracts just as the sea encircles islands. (McCrindle 1877, 142–3)

The names, McCrindle (1877, 143) remarks, are obscure, while suggesting that the locality occupied by the Chrysei, the Parasangae, and the Asange must have been to the north of the Ran between the lower Indus and the chain of the Aravali mountains.

Keeping in mind the habitat of the tiger (dense forests), it can be argued that while there may be an element of truth in the aforementioned references to the distribution of the mega carnivore, they mostly come from writers who were writing on the basis of hearsay (with the exception of Megasthenes). Errors, therefore, were not unlikely, an example of which we see in the sixth book of Pliny where he describes the island of Taprobane (Ceylon or Sri Lanka), and observes how their festive occasions were spent in hunting—their favourite game being the tiger and the elephant (McCrindle 1901, 106). This clearly is an inaccurate piece of information since we know that though we have geological evidence for the presence of the lion and the tiger during the Late Quaternary of Sri Lanka, both species have since long been extinct in the region (Manamendra-Arachchi et al. 2005, 423–34).

Beyond geography, Pliny also devotes some attention to the animals of India, and in the eighth book of his *Natural History*, he attributes to India, the tiger, an animal tremendously swift according to him, a characteristic particularly evident when the female was deprived of all her whelps, which were always numerous (McCrindle 1901, 115). Additionally, Pliny also refers to a breed of Indian dogs, 'large, powerful, and of untamable ferocity', attributing their savage disposition to the 'tiger blood' that ran in their veins. According to him, Indians asserted that these dogs were begotten from tigers, for which the bitches when in heat were tied up in the woods (McCrindle 1896, 363). What sense do we make of such an account? A plausible explanation may be that these classical writers were impressed with the might of the predator which consequently served as a point of reference for contexts engaging with the demonstration of power.

Then there are more accounts of animals being gifted to kings. Aelian tells us that the Indians brought to their kings tigers made tame, domesticated panthers, and oryxes with four horns (McCrindle 1901, 144). Similarly, Dion Cassius (grandson of the famous Greek rhetorician and philosopher Dion Chrysostom), writing his *History of Rome* in the second century CE, mentions embassies coming to Augustus, and the Indians formerly having entered into a treaty of alliance, concluding it with gifts, along with other things, of tigers—animals, which according to him the Romans, and possibly also the Greeks, saw for the first time (McCrindle 1901, 212).

Though tigers and lions are known to have been tamed in history, these are rare events. Divyabhanusinh (personal communication) suggests the possibility of these so-called tamed lions and tigers being under the effect of opium, and, therefore, seeming compliant. We, however, have no means of ascertaining this from the classical accounts we have in hand. The images which emerge from them, thus, need to be dealt with judiciously, and should best be seen as windows to Western perceptions (often warped) of a distant land, and all that it encompassed.

Silently shall I endure abuse as the elephant in battle endures the arrow sent from the bow: for the world is ill-natured.

They lead a tamed elephant to battle, the king mounts a tamed elephant; the tamed is the best among men, he who silently endures abuse . . .

Be not thoughtless, watch your thoughts! Draw yourself out of the evil way, like an elephant sunk in mud.

—*Dhammapada*
(Müller and Fausböll 1881, 77–8)

4

TRUNK CALLS IN ANTIQUITY

THE ELEPHANT IN ARCHAEOLOGY AND ART

SILHOUETTED AGAINST A DARKENING SKYLINE, the magnificent elephant stands tall over all that surrounds it. The gait makes for poetic stuff, while the dexterous trunk is a gentle reminder of the wonders of nature. The massiveness overwhelms, yet the animal is real.

Denizens of a range of habitat types, Asian elephants are generalists inhabiting grasslands, tropical evergreen, semi-evergreen, moist deciduous, dry deciduous, and dry thorn forests. Once widespread in the subcontinent, habitat loss, degradation, and fragmentation driven largely by an expanding human population are pushing the pachyderm to the verge of extinction.

Dramatic illustrations of the magnitude of the human-induced attrition of Asian elephants come from historical references to former vast stocks reflecting wild populations of a size far greater than those of today. A startling reminder of the journey of this mega mammal across millennia comes in the form of Figure 4.1, which when compared with Figures 6.1, 6.2, 6.3, and the *gaja vanas* listed in the *Arthaśāstra* (discussed in Chapter 5), makes it amply clear that the population of wild elephants has greatly diminished since ancient times.

As an animal that has amazed as much as it has terrified, elephants have pervaded the human past as well as the present through art, legend,

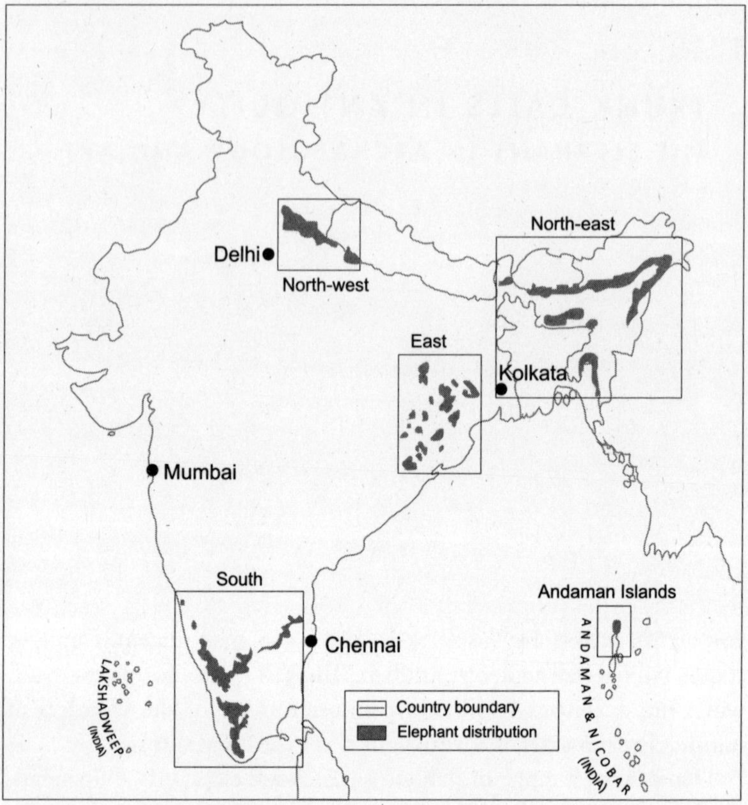

Figure 4.1 Regional distribution of the Asian elephant in India (for representative purposes only)
Source: Based on Baskaran et al. 2011, 48.

literature, and religion. The fact that India is home to more than 50 per cent of Asia's elephants is testimony to the long-standing relationship of its culture with the pachyderm, which it has considered sacred yet ruthlessly exploited and persecuted. It is the story of this complex interaction that I attempt to reconstruct by weaving together glimpses from art, archaeology, and, subsequently, literature.

I begin my narrative by taking cognizance of the fact that much has already been written about the evolution of elephants. This probably explains why 'before riding the elephant wave', Sukumar (2003, 22) insists on attending a sideshow of the cast of characters in 'the great proboscidean drama'. Proboscidea is the group of mammals which includes elephants and their extinct relatives. Hence, what Sukumar refers to is the complex evolutionary history of elephants with numerous players that marched across continents and millennia battling for survival.

Though fascinating, this evolutionary saga is beyond the purview of my quest for the extant Asian elephant. Therefore, I straightaway embark on unearthing the earliest remains of the pachyderm within the geographical confines of the study area. A quick glance at some of the Pleistocene fossil evidence yielded by the subcontinent can serve as a point of entry.

The Pleistocene witnessed the presence of some highly evolved genera of elephants such as *Stegodon*, *Hypselephas*, *Archidiskodon*, and *Elephas*. However, since it is only the genus *Elephas* which concerns us here, it would be meaningful to mention the extinct *Elephas hysudricus*, and the extant *Elephas maximus*. There were many more, some of which never evolved further, while others became extinct (Badam personal communication).

Moving on to the presence of *Elephas* in the fossiliferous horizons within the study area, Helmut De Terra and Thomas Thomson Paterson (1939, 82, 93, 124, 214) mentioned its occurrence along with other mammals in a lower Pleistocene context from localities such as Sombur, Magam, and Ningle Valley in the Karewas formation of Kashmir. Subsequently, Badam (cited in Badam 1985, 124) reported *Elephas hysudricus* in the same context from Baramula, Zangam, and Gandarbal localities of the region. Shopian and Wapzan in the same region also yielded fossil traces of *hysudricus* (cited in Badam 1985, 124). In an upper Pleistocene context, a complete skull of *Elephas antiquus* was retrieved from the older alluvium near Gurgaon, Haryana (cited in Badam 1985, 126). The Belan and Seoti Rivers in Allahabad district, Uttar Pradesh, have also preserved faunal remains including those of *Elephas* sp. of the Quaternary period (Badam 1985, 126).

Fossil and sub-fossil species of the surviving *Elephas maximus* have been reported from the Malay Archipelago and the Siwaliks of north India, as also from the younger alluvium of the Narmada and Deccan river valleys (Badam 1979, 220).

Mention must also be made of the presence of *Elephas maximus* in a late Pleistocene context in the faunal assemblage from Kalpi, district, Jalaun, Uttar Pradesh. The retrievals included a tusk broken into seventeen pieces. Additionally, two fragments of tusks were collected from the surface of Event II at the site. One of these had a cut mark, while its tip showed marks of polishing. This was perceived as an attempt to make a larger tool. Another conical tusk fragment displayed polishing on all sides. However, what remained unclear was whether this was an artefact, or the result of elephants rubbing their tusks on tree barks (Joglekar 2012–13, 170).

Significant finds in terrains that are no longer hospitable to the animal include fossilized bones of an elephant from a gypsum bed of the Quaternary age at Bhadawasi village in Nagaur district of western Rajasthan. The report emphasized the significance of the finds, being the first remains of a large mammal in the Quaternary sediments of the great Thar Desert of western Rajasthan, and also for 'understanding the environmental conditions and the rapid climatic changes that the Thar Desert has witnessed and how a green fertile land along the mighty river system (Vedic Sarasvati?) got desertified'. The study presumed a possible late Pleistocene age of the elephant fossil–bearing gypsum horizon. It also did not completely rule out a middle or early Holocene age in view of the evidence for arid conditions in the region around 4800 ^{14}C yr BP and > 10,000 ^{14}C yr BP (Paliwal 2003, 1188–91).

For more tangible evidence of the long and intimate association humans have had with the pachyderm, I now turn to the Holocene remains of this mega herbivore retrieved from different cultural levels at archaeological sites in the Indian subcontinent. In the tabulation that follows, I shall not only synthesize the available archaeofaunal data, but also take a look at representations in rock paintings and terracotta in order to comprehend the frequency of depictions, and its implications for the fortunes of the animal.

THE MESOLITHIC

Collating the evidence gathered from published faunal reports, the elephant is reported in mesolithic contexts at Sarai Nahar Rai, Mahadaha, and Damdama in Pratapgarh district, Uttar Pradesh.

Sarai Nahar Rai: Alur (1980, 207–10) documented the presence of the elephant (*Elephas indicus*) at the site in association with species like cattle including bison (*Bos indicus* and *Bos gaurus*), sheep/goat (*Ovis vignei blyth-Capra hircus aegagrus*), hippopotamus (*Hippopotamus palaeindicus*), deer, stag (*Cervus duvauceli-Cervus axis*), and tortoise (*Chelonia* sp.). He argued that the find of a large number of bones of wild animals, and the evidence of roasting, chopping, and extraction of marrow from them indicated their use for purposes of food. However, Alur's assertion regarding the presence of the elephant at Sarai Nahar Rai was not confirmed by later reviews of faunal material at the same site (Thomas and Joglekar 1994; Chattopadhyaya 2002).

Damdama: As mentioned in Chapter 2, the presence of the elephant along with other big mammals such as the rhinoceros, gaur, and wild buffalo in the faunal assemblage was confirmed by Thomas et al. (1995, 29–36; 1996, 255–66). According to their investigations, in the assemblage comprising a total of 4,054 identified bone specimens, 26 were elephant bones accounting for 0.64 per cent of the collection. Like the rhinoceros and other big mammals, the elephant was not considered in the list classifying species according to their live weight in order to look for a pattern of resource exploitation. Rather, the carcasses or isolated bones of these animals were suggested to have been collected and utilized for making bone tools (Thomas et al. 1995, 31; 1996, 258–9).

Mahadaha: The analysis of the assemblage by Joglekar and his team revealed five bone fragments of the elephant in the Eastern Area (butchering complex), and one each in the Western Area (cemetery-cum-habitation area), and the Lake Area. The one from the Western Area was 'an almost square piece of ivory' whose length, width, and thickness were 29, 28, and 27 millimeters respectively. The one from the Lake Area was a long ivory piece that displayed no marks of working or any attempt at modification. The Eastern Area remains included three rib fragments which were charred, and one of them showed a cut mark. A charred fragment of the right scapula of an adult elephant was also found. An almost complete vertebra was attributed to a young elephant since its central plates were unfused (Joglekar et al. 2003, 75). Overall, the faunal assemblage at Mahadaha and Damdama were found to be similar in species composition. The broad-spectrum resource base based on the consumption of diverse mammalian species was underlined, while the carcasses of big mammals were suggested to have been scavenged for obtaining meat for consumption or bones for making tools (Joglekar et al. 2003, 116).

The Issue of Subsistence: Some Possibilities

The evidence though scattered certainly hints at the recognition of the economic worth of the animal since early times. An important line of enquiry relates to the issue of the inclusion of such big mammals in the diet. Notwithstanding apprehensions regarding the same, a line of reasoning which seems rational and scarcely deniable is that big mammals like the elephant would have provided large packets of meat at one go which would have been difficult to abstain from. In arguing thus, one finds support in Sukumar's intervention in this regard.

He emphasizes that where available, elephant meat must have been consumed by early humans since it is unlikely that hunting-gathering bands would have easily forgone the 'opportunity provided by a mountain of meat' (Sukumar 2011, 21). Apart from meat, what the animal provided would have been 'a bonanza for humans, the most obvious being skin, bone and ivory for shelter, clothing, tool-making or creative expression'. The pressing question for him is not whether elephant meat was a part of the diet of early humans but the manner in which this meat was retrieved, or the extent to which elephant meat was procured through scavenging as against active hunting, and the stage of human evolution when elephant flesh became a part of the diet of humans.

Sukumar (2011, 21) further points out that most anthropologists now believe that the advanced abilities not only in terms of weaponry but also in hominid brain development needed to successfully hunt big mammals like elephants became possible with the evolution of the modern human *Homo sapiens* sometime between 200,000 and 100,000 years ago. Though direct evidence of elephant exploitation by early humans is wanting in the case of the Indian subcontinent, Sukumar (2011, 24) is emphatic in asserting that this did occur. In the absence of sufficient fossil evidence hinting at the cultural association between modern Asian elephants and *Homo sapiens*, he proposes to direct his enquiry to the interface between early humans and elephant forms that preceded *Elephas maximus*. He persuasively argues that if in Africa naturally dead elephants were scavenged for their meat, marrow, and brain by *Homo erectus* armed with heavy tools such as choppers, hand-axes, and cleavers, there is no reason to believe otherwise for their Asian counterparts with similar toolkits.

While such perspectives certainly open up fresh avenues of thinking, one should perhaps also reflect on the available faunal evidence. Contrary to Sukumar's (2011, 23) assertion that there is nothing to directly prove that early humans ever hunted or butchered elephants, I would argue that the concentration of the bones of large mammals in the eastern part of Damdama along with cut marks on most of them as well as the retrieval of five fragments of elephant remains from the butchering complex or the Eastern Area at Mahadaha needs to be underlined. While there seems to be little doubt regarding the exploitation of this mega herbivore for its ivory, and for purposes of fashioning tools of its bones, direct evidence for its meat being consumed is still awaited. Nevertheless, it is only reasonable to speculate that if the bone and ivory yielded by the animal were put to use, its flesh, when and where available, would also have been used for dietary purposes by early humans.

As far as the animal being hunted is concerned, pursuing it in its trail to the domain of art may perhaps offer some clues. Varma (1996, 330) mentions the elephant as the subject matter of the earliest representations in the rock paintings of the Vindhyan region, but is emphatic in qualifying that despite its presence, it does not seem to have been a very popular animal.

In a preliminary study of a newly discovered rock art site in Dadikar and Hajipur villages of Dehara village panchayat in Alwar district of Rajasthan, a numerical count of the motifs revealed only one elephant representation attributed to the mesolithic period (Sharma and Meena 2011, 92) (Figure 4.2).

Elephant figures were, however, quite common in the rock paintings of Mirzapur in Uttar Pradesh. Rakesh Tewari (1990, 13) recorded about fifty-five figures of the pachyderm in the rock shelters of Kauva-khoh, Ghormangar, Hathvani, Kandakot, Jhariya, Likhaniya, Mukhadari, and others. At Ghormangar, the animal is shown lifting grass with his trunk while at Kauva-khoh and Kerwa the representations are those of pregnant elephants. Tewari (1990, 20) records about seventy hunting scenes in various parts of the district. Interestingly, however, he observes that elephant-hunting scenes are few in number, being recorded from Dhandhraul and Mukhadari. At Dhandhraul, a hunter with an attractive headdress is shown attacking an elephant with a long multi-barbed spear, the point of which is embedded in his body.

Figure 4.2 Mesolithic elephant and bird, Harsora Hill, Rajasthan. Note the animal is without tusks

Source: © Indira Gandhi National Center for the Arts (IGNCA), New Delhi. Reproduced with permission.

At Mukhadari, on the other hand, as the illustration shows, hunters are shown confronting two elephants (Figure 4.3). The foreleg of one has been pierced by an arrow, and both have raised their trunks in anger. One of them has thrown a hunter on the ground, and put his leg over him (Tewari 1990, 25–6). Neumayer (2013, 140) takes a closer look at this painting and highlights the presence of anthropomorphic figures along with square-shaped figures possibly delineating women (a detail he verifies in view of a painting group in the vicinity). A finely-contoured cross is seen in front of these elephants. Such geometric symbols, he contends, are common in the mesolithic rock art of the Kaimur range.

Neumayer (2013, 140) also reinforces the rarity of elephant depictions, but asserts that right from the beginning, the pachyderm is shown as game. What is also significant is his observation regarding most of the early elephant depictions being found in the central Vindhyan region, while the pachyderm appears rarely in the Kaimur Hills. Though he does not spell out the implications of this difference, Neumayer goes on to observe that quite a few elephant depictions occur in 'irrational' compositions suggesting that the elephant was not just another big animal in the forest, 'but more so a spiritual being central in the mythological world of the hunter of the late Stone Age' (2013, 140).

Figure 4.3 Elephant hunt scene from Mukhadari, Mirzapur, Uttar Pradesh
Source: After Tewari 1985.

What seems apparent from this overview is that given the availability of softer targets like various species of deer and the paucity of representations depicting elephant hunts, the latter were evidently not resorted to frequently. Plausibly there is merit in the contention that big mammals such as the rhinoceros and the elephant were hunted under particular circumstances or in self-defence. However, what also needs to be kept in mind is the observation emerging from an evaluation of elephant-hunt scenes which at times portray the act as an orchestrated one where the animals are taken on by groups rather than individuals. Conscious of the multiple ways of perceiving such depictions, I would nevertheless argue that though few, at least some of them can possibly be seen as expressions of experience in actively hunting this big mammal. What remains to be asked is what became of the animal after it was hunted? Reason acquired from such visual depictions permits us to infer that once overpowered, the animal would be put to use in its entirety.

It may also be underlined that elephant hunts, though undoubtedly perilous, were not unachievable. One would do well to remember that apart from the direct method of hunting, ancient humans may have devised other methods of procuring game which did not lead to direct and fatal confrontations with the pachyderm. Paintings in the Vindhyan region, for instance, points out Varma (1996, 332), show that traps were laid to catch animals. In one method of trapping, concentric circles have been shown just before the animal. These circles, according to him, possibly represent deep pits, and suggest a method of catching animals practised till today by the tribals of the region. According to this method, circular pits are dug on paths generally followed by animals. The pits are carefully covered with grass, and tree branches, spikes, and thorns are sometimes placed in them. The hunters watch from a distance, and as soon as an animal falls into the pit, they beat the animal to death and slaughter it.

THE NEOLITHIC

Mehrgarh: Situated at the foot of the Bolan Pass in the Kachi district of Baluchistan, Mehrgarh provides a continuous record of the beginnings of agriculture and animal husbandry, and its subsequent development. What may also be pointed out at the outset is that in the same time period, the elephant is present (as discussed earlier in this chapter) in a very different cultural setting—that of the mesolithic of the Central Ganga Plains. Of the seven cultural periods represented at Mehrgarh, Periods I and II correspond to the neolithic period. The faunal assemblage retrieved from aceramic sub-period IA (seventh millennium BCE) was largely dominated by wild species, amongst which Meadow (1989, 66) suggested the possible presence of the elephant together with gazelle (*Gazella dorcas*), wild sheep (*Ovis*? *orientalis*), wild goat (*Capra*? *aegagrus*), barasingha (*Cervus duvauceli*), nilgai (*Boselaphus tragocamelus)*, wild cattle (*Bos*? *namadicus*), water buffalo (*Bubalus bubalis*), chital (*Axis axis*), half-ass/onager (*Equus hemionus*), blackbuck (*Antilope cervicapra*), and pig (*Sus scrofa*), albeit with the caveat that these forms were represented by very few identifiable specimens. Meadow also underlined the preferred habitats of these animals—foothills, plains, and riverine environments.

A spectacular faunal find from a compartmented building of Period IIA at the site was a part of an elephant tusk, grooved longitudinally on three sides in preparation for local manufacture of artefacts

(Meadow 1984, 35). Nerissa Russell's (1995) comprehensive study of the bone tool industry at Mehrgarh and Sibri, the two sites located 5 km apart in Kachi district, Baluchistan, Pakistan, noted the rare but sporadic occurrence of elephant remains (ivory) through the sequence starting in Period IIA. However, the period did not yield any finished ivory artefacts (Russell 1995, 589). It was argued that despite the identification of one fragment of elephant long bone in the Mehrgarh fauna from Period IA, it was certainly not hunted regularly at the site. In the tabulation of the taxa of animals that served as raw material for different types of bone tools at Mehgarh and Sibri, *Elephas* remains were used for fashioning points, button-seals, and also figured as waste, constituting a meagre 0.4 per cent of the collection (Russell 1995, 596–7).

Chirand: Elephant remains reported by Nath and Biswas (1980, 122) from the neolithic period at the site consisted of a fragment of the upper molar tooth, and the unciform bone of manus. These remains were found to closely resemble those of modern specimens in the Zoological Survey of India collection, and were used to infer a swampy wooded area near the habitational site.

Lala Aditya Narain's (1972, 1–24) study of the techniques of neolithic bone-tool making at Chirand noted 150 artefacts. Ivory along with tortoise bone was reportedly used for making bangles.

THE HARAPPAN EVIDENCE

Having documented the presence of the animal at mesolithic and neolithic sites, I now move on to examine the much larger volume of Harappan evidence in the form of bones and representations. The attempt will also be to examine certain facets of Harappan interactions with the pachyderm.

Faunal Remains

Within the Harappan cultural zone, elephant remains have been found at Mohenjodaro and Chanhudaro in Sindh, Pakistan, and Harappa in Punjab, Pakistan, Rupar and Bara in east Punjab, Kalibangan in Rajasthan, and Lothal, Surkotada, Rojdi, Kanmer, and Khirsara in Gujarat.

Mohenjodaro: The remains reported included the upper articular surface or caput of the femur and tips of small tusks. Although these were the only fragments obtained of the pachyderm, Sewell and Guha (1931, 653) emphasized the thorough acquaintance of the inhabitants with the animal. The actual finding of a part of the skeleton was considered as sufficient evidence for the presence of the animal in the region (Sewell and Guha 1931, 653).

The report emphatically clarified that it made no attempt to correlate the various zoological finds with the different periods of the civilization at Mohenjodaro, and randomly grouped them according to the depth at which they occurred. Correspondingly, the later presence of *Elephas maximus* along with *Bos bubalis* and *Ovis* sp. in the upper level, and their apparent absence from earlier strata was interpreted as indicating a modification of the state of civilization, where more animals were introduced as domesticated herds. It is significant that in the classification of the faunal assemblage retrieved from the site, the elephant features in the list of animals argued to have been maintained 'in a state of domestication' (Sewell and Guha, 667–8). I shall ponder over this issue subsequently.

Pieces of ivory were common, and the site yielded at least one large fragment of a tusk (Sewell and Guha 1931, 653). After John Marshall's (1931) excavations at Mohenjodaro, the contention was that ivory objects (mostly casting sticks, handles) were few (Mackay 1931c, II: 562). However, further excavations at the site established beyond doubt the substantial use of ivory for carving utilitarian items. In fact, ivory was so commonly used as a raw material at Mohenjodaro that bone seems to have taken a subordinate position, clearly indicating the popularity of the medium with the Harappan people (Mackay 1938, I: 579).

It may be relevant to look at what these early reports have to say about how ivory was procured. Discarded tusks of elephants were initially perceived as the only possible source of ivory. This was Mackay's (1931c, II: 563) contention on the grounds that if the animal was considered sacred, as it appeared from its frequent portrayal on seals with or without the cult object before it, it might not have been considered lawful to hunt the animal either for its meat or ivory. There were also some concerns regarding whether these supplies were imported into Sindh from other parts of India or were available locally (Mackay

1938, I: 579). However, in view of the skeletal remains of the pachyderm at the site as well as its prolific occurrence on seals, there seems to be no doubt that ivory was abundantly available at Mohenjodaro.

Additionally, Mackay's (1938, I: 117) account of the find of two elephant tusks amidst nine human skeletons at Mohenjodaro is noteworthy. While on one tusk lay skulls 3 and 5, as also some of the bones of body 6 (it seemed certain that all were buried at the same time), the other tusk lay a little to the south with the arm-bones of burial 7 just below its broader end. The elephant tusks like the human remains showed signs of great antiquity. In view of the proximity of these bodies to an exit from the city, Mackay's sense was that these were the remains of a family that had tried to escape with their belongings at the time of a raid, but were stopped and slaughtered by the raiders. He further conjectured that one or more of the family may have been ivory workers, and only the tusks for which the raiders had no use were left behind. Though the hypothesis is clearly speculative, what can perhaps be safely inferred is the close association of the Harappans with the pachyderm.

Chanhudaro: Though Glover M. Allen's identification of animal remains at the site did not report the presence of the elephant, a tusk of the animal was found beneath the middle buttress supporting the south-western wall of a room at the site. This was intriguing since there were no indications of the room having been used for ivory working. Also, since the room was devoid of objects, and in view of the evident care that had been taken to conceal the tusk, it was suggested to have been stolen property (Mackay 1943, 14). Dwivedi (1976, 39) is, however, hesitant in accepting this view, but contends that it certainly indicated valuable possession, which probably explains why the owner tried to conceal it beneath the buttress.

Ivory objects retrieved from the site were few, suggesting that the working of this material was not extensively practised. The finds included dies, dice of ivory, an ivory pin, a comb, a crossbar, and an ivory peg (Mackay 1943, 10, 58,171,194,196, 233).

Harappa: Prashad's (1936) report of the faunal remains at Harappa did not document the presence of the animal at the site. However, more than two decades later, Nath (1959, 1–14) suggested otherwise. Despite efforts, the report could not be procured even from the Zoological Survey of India. Hence, the title of the paper, 'Remains of the horse and the Indian elephant from the prehistoric site of Harappa (West Pakistan)' is all we have as evidence testifying the presence of the pachyderm at Harappa.

The ivory finds at the site included an inscribed pointed ivory rod, combs, a cup, handles, kohl sticks, pins, awls, a gamesman, a spatula, and some small balusters (Vats, 1940, I: 459–61).

Rupar: From the collections of animal remains retrieved from excavations at Rupar in Ambala district of east Punjab, Nath (1968, 69, 78) reported the remains of *Elephas maximus* Linn. in the form of one proximal fragment of the left scapula with glenoid cavity, neck and without blade from cultural Period I (c. 2000–1400 BCE) representing a late phase of the Mature Harappan culture.

Bara: The entire faunal assemblage at Bara in the same district belonged to the Harappan cultural phase, but on the basis of slight variation in potteries it was divided into Phases A and B. The latter yielded a well-preserved, and quite intact first phalanx of the left second digit of the forefoot belonging to an adult specimen of *Elephas maximus*. Overall, the remains closely resembling modern specimens in the Indian Museum collection were few, but proved beyond doubt the acquaintance of the inhabitants with the animal (Nath 1968, 78).

Kalibangan: Elephant remains included a broken phalanx of the left forelimb, and a fragment of rib from a late level of Period II, and proximal fragments of tibia retrieved from the middle level of the same period representing the Mature Harappan phase commencing around 2600 BCE (Banerjee et al. 2003, 269–70).

The ivory objects reported from Early Harappan levels at the site were few and included a small ivory wheel with a hub and an ivory spatula, both from an early level of Period I (Bala 2003, 234).

Lothal: The faunal remains initially studied by Nath and Rao (1985, II: 641) included three fragments of the right femurii, the first phalanx of the fourth digit of the right hindlimb, and one fragment of the radius, all of which closely resembled their counterparts of the modern specimen of the elephant in the Zoological Survey of India collection. The occurrence of the leg bones of the pachyderm near the dock at Lothal was treated as evidence for the presence of elephants in Kathiawar in Harappan times. As pointed out in Chapter 2, the faunal report does not mention the cultural period to which the remains belonged though we know that Period A, Phase III, revealed an ivory tusk partially cut for further working from the ivory worker's shop in the acropolis (Rao 1985, II: 631).

Placing the Harappan culture of Lothal between 2200–1700 BCE, a subsequent report listed among the remains of the proboscidean the broken right horizontal ramus of mandible, head of right humerus, broken right humerus, distal part, broken shaft of right humerus without epiphysis, head of left humerus, broken distal epiphysis of left humerus, broken distal part of left ulna, broken carpal bone on left forelimb, second phalanx of forelimb, broken left pelvis with a portion of acetabulum (sawed), broken lateral condyle of left femur, broken left tibia, distal part and another one (sawed), broken proximal part of left tibia (sawed), and broken right patella (Saha et al. 2004, 159).

Ivory finds included a solitary seal, an awl, a stopper, a kohl rod, ceremonial ivory knives, a lunate-shaped ivory object probably used as an engraver, ivory antimony rods, and gamesmen. The most interesting object of ivory was a scale. A significant observation was that the phase-wise distribution of ivory objects suggested that Lothal enjoyed greater prosperity in Phases II and III than in Phases IV and V (Rao 1985, II: 626–31).

Rojdi: A massive rib retrieved from this site in Rajkot district of Saurashtra was possibly that of an Indian elephant (Kane 1989, 183).

Surkotada: The presence of the elephant in a Late Harappan context was confirmed by the retrieval of parts of tusks from the uppermost levels of Period IC (1790 BCE to 1660 BCE). A.K. Sharma (1990, 380), who studied the faunal remains at the site, argued that they were probably imported from neighbouring areas for making ornaments. He further maintained, 'In Indian cultural context it is normal not to find skeletal remains of elephants in the sites as their meat is never eaten and after death they are buried away from the habitation' (Sharma 1990, 380). From the evidence we have so far, it seems difficult to accept this contention, and we have also already explored the possibility of elephant meat being consumed. Also, Sharma's assertion that the case was similar at Mohenjodaro, Lothal, and Kalibangan, from where only tusks of *Elephas maximus* Linn. were recovered, warrants caution since all the three sites have yielded skeletal remains even though fragmentary.

It may be added that in the same report, C. Margabandhu (1990, 339) mentioned the find of a huge elephant tusk from the period (Figure 4.4). On the basis of the variety of ivory objects found at the site (from the earliest to the late levels), he also postulated that the ivory industry was a flourishing one. The objects included carefully executed kohl sticks, stylus, comb, pendant, and polishers. Though similar shapes have been reported from Mohenjodaro, Kalibangan, Chanhudaro, and other sites, the polishers at Surkotada were observed to be a class apart as no published report mentions similar objects (Margabandhu 1990, 339).

Shikarpur: Fresh excavations from 2007–8 onwards brought to light a three-fold sequence in the Harappan occupation at the site. Phase I yielded classical Harappan ceramics along with Anarta pottery. Phase II was distinguished by the preponderance of Sorath Harappan pottery, while Phase III, representing the last phase of

Figure 4.4 Elephant tusk from Surkotada, Gujarat

Source: After Joshi 1990, plate LXXI B. © Archaeological Survey of India. Reproduced with permission.

occupation at the site, was marked by Late Harappan material (Joglekar and Goyal 2011, 15). The elephant was represented in the faunal assemblage of Phase II by two fragments of ivory (Joglekar and Goyal 2011, 22).

Kanmer: A muticultural site in the Kachchh district of Gujarat, Kanmer revealed a fivefold cultural sequence where Periods I and II corresponded to the Early and Mature Harappan phases respectively. Analysis of the faunal remains brought forth several animal taxa including mammals, birds, fish, reptiles, and molluscan species. The elephant was represented in the Mature Harappan (KMR II) faunal material by the presence of two fragments of ivory (Goyal and Joglekar 2012, 784).

Khirsara: Excavations at Khirsara, a small village in western Kachchh, revealed a fortified Harappan settlement of the urban phase. Amongst the species identified at the site, the elephant was represented by an ivory fragment retrieved from layer 2. The fragment had been modified in order to fashion a craft object or utilitarian item out of it. The object, however, seemed incomplete (Joglekar et al. 2013, 4).

Seals, Terracottas, and More

Additional information regarding the Harappan interface with the animal comes from iconography including models in terracotta and copper as well as engravings on seals and copper tablets. In the absence of a deciphered script, I shall now turn to these in order to try and situate the animal in the Harappan scheme of affairs.

I begin with a survey of the modelled figurines of the animal, which are few. A very well-preserved baked clay model awaits us at Mohenjodaro, which also yielded model elephants in a poor state of preservation. In the well-preserved piece where the whole body is found intact and faithfully modelled, the rotund body of the animal makes it clearly distinguishable (Mackay 1938, I, 290). Mention may also be made of a well-executed copper statuette cast in the round with the feet having been lost (Sahni 1931, 210). Though there was some uncertainty amongst the excavators regarding the species, the long snout, a portion of which was perhaps broken off, seemed to suggest an elephant (Mackay 1931b, II: 506).

At Harappa, there is an elephant in terracotta hollow from inside (Figure 4.5), while other models impress us with eyes consisting of pierced round pellets and a well-modelled upraised trunk with traces of paint (Vats 1940, I: 308).

A statistical evaluation of the animal figurines retrieved from Mohenjodaro and Harappa by Marcia Fentress (1976, 237) clearly showed that the elephant was infrequently modelled at these two sites, an inference reinforced by evidence from other sites as well.

A terracotta elephant on a paved floor, which probably served as a toy, is a valuable find at Chanhudaro (Figure 4.6). The elephant has

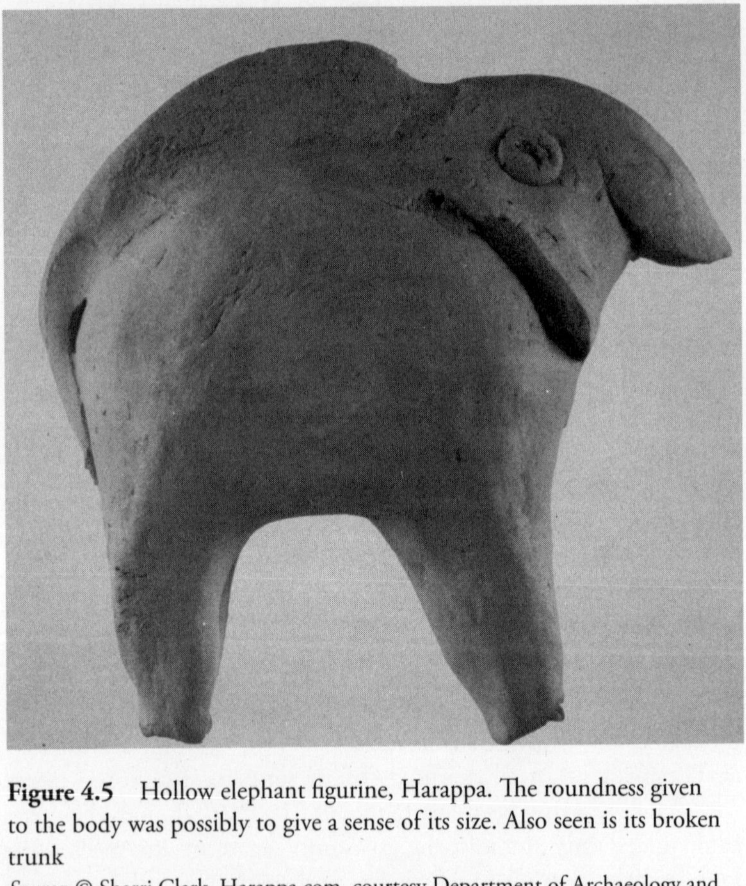

Figure 4.5 Hollow elephant figurine, Harappa. The roundness given to the body was possibly to give a sense of its size. Also seen is its broken trunk

Source: © Sharri Clark, Harappa.com, courtesy Department of Archaeology and Museums, Government of Pakistan. Reproduced with permission.

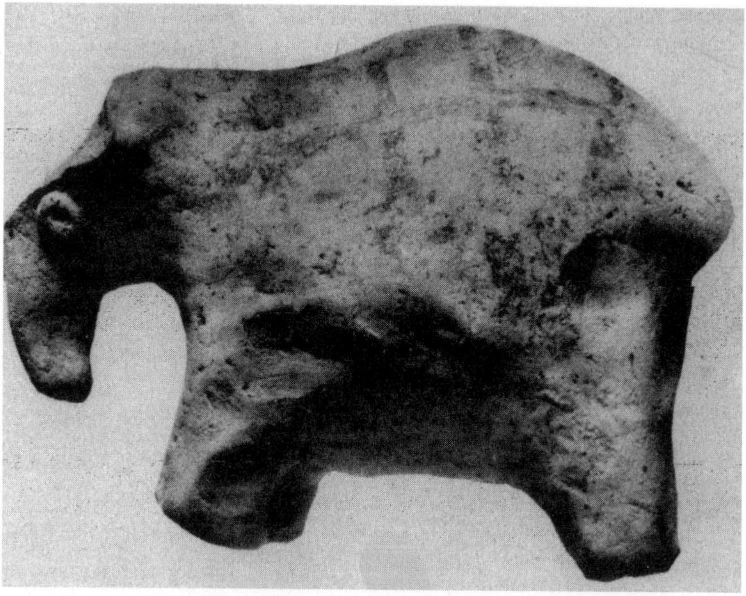

Figure 4.6 Though a crude model, this terracotta elephant from Chanhudaro is a significant find in view of the trapping marks on its body
Source: From Mackay 1943, plate LVI, 9. ©American Oriental Society. Reproduced with permission.

short legs, spinal ridge, and sloping hindquarters. The trunk, however, is too short. The animal carried vertical and horizontal lines all over its body, probably representing trappings. A ring of red round the end of the trunk and a painted line down its front were reminders of those seen on elephants dressed for state and festive occasions (Mackay 1943, 159).

A crudely executed specimen with stumpy legs, and a comparatively small trunk is also reported from Kalibangan (Sant 1997, 80).

At Lothal, the terracotta head of an elephant with its long trunk is an equally welcome find, and a specimen better finished than the one from Chanhudaro. Though damaged, the trunk was easily distinguishable. The eye sockets were marked by incisions, and the retina by a pellet. The appliqué technique was employed to plant a short tusk on one side of the trunk, which though had fallen off from the other. A nail-punch mark indicated the mouth (Rao 1985, II: 484).

Notwithstanding limited representations in clay or metal, the proboscidean figures prolifically on Harappan seals and amulets. In terms of the frequency of their use as field symbols on seals, unicorns take the lead (1159) followed by short-horned bulls (95) and elephants, which occur on 55 stamp seals, sometimes with a trough in front (Mahadevan 1977, 793). It is crucial to mention that no seal depicts the animal in a wild disposition, permitting the inference that the taming of the animal was an established fact in Harappan times (Sinha 1955, 519). In fact, in some seals the elephants depicted are decorated with paint and some are draped in cloth or other material suggesting a close relationship of the animal with the Indus people (Possehl 1999, 192). A closer look at representations on some of the seals and copper tablets would not be out of context as it can yield vital clues regarding the familiarity of the seal-cutters with the animal.

Marshall's (1931) excavations at Mohenjodaro revealed fifteen seals with the pachyderm carefully portrayed on most of them, in some cases even down to the wrinkles on its back. The animal was said to rank next in order of popularity to the bull. On four seals, the sloping back recalled the African species though the comparatively small ears negated any such connection. Three seals showed folds on the back of the animal. It, however, remained uncertain whether these were trappings or the vertical folds of skin at the junction of limbs (Mackay 1931b, 388).

This problem was subsequently addressed by Zeuner while tracing the earliest evidence for the use of elephants. Referring to the fifteen seals retrieved from Mohenjodaro, he observed that six showed a line extending down from the back of the elephant behind the foreleg. He, however, dismissed attempts to explain it as a part of its natural skinfolds arguing that if a living Indian elephant is observed, it will be found that there are no such large skinfolds. Hence, what seemed plausible to him was that the line represented a covering cloth or wrapping on the back, indicating the 'domesticated' status of the animal by the time of the Indus civilization (Zeuner 1963b, 286).

Another remarkable insight regarding the elephant seals from Mohenjodaro pertained to the presence on five seals of two processes on the trunk (Mackay 1931b, II: 388). By processes, I mean finger-like projections at the tip of the trunk. It is worth mentioning that while the Asian elephant is known to have a single finger-like feature at the end of its trunk, the African elephant has two.

Notwitstanding the two processes, it was emphasized that in view of the unmistakable smallness of ears, flatness of foreheads, and profiles of backs typical to the Indian species, the animal portrayed on the seals represented the same (Mackay 1931b, II: 388). The contention that the animals on these seals might have represented a species now extinct in India seems far-fetched since Holocene faunal remains from archaeological sites attest to the presence of only *Elephas maximus* in the subcontinent. Nevertheless, what appeals is Mackay's argument that similar portrayals on five seals could scarcely have been a mistake on the part of the seal-cutter (though he does not dismiss that mistakes were made as in the case of wrongly placed eyes on two seals, and a strange ear on one). What is questionable is his contention that the elephant was possibly not so well known to the inhabitants of Mohenjodaro, and that it may never have been wild in Sindh, and was possibly used by a few people only for purposes of state (Mackay 1931b, II: 388). A survey of the repertoire of elephant seals negates this position, and clearly suggests that with a few exceptions, the fidelity of the depictions cannot be accounted for if the Harappans were not adequately familiar with the animal.

Returning to the curious observation regarding the presence of two processes at the end of the pachyderm's trunk on some seals, notwithstanding analyses like those by Mackay (1938, I: 670–1), it may be worthwhile to consider the possibility of this being a manifestation of the Harappan familiarity with the African counterpart of the Asian elephant.

With the exception of one, seventeen more seals from the site showed the Indian elephant with a nearly straight back and comparatively small ears. In the best-cut figures (on five seals), careful attention was paid to the rugous folds of the trunk which again terminated in two distinct processes. The cautiously cut forelegs (on three seals) were a departure from the wooden appearance usual in representations of the animal. The considerable attention paid to the representation of bristles along the outline of the body and the head (on four seals) was perhaps to suggest a younger animal. No traces of tusks were found on three seals suggesting the possibility of these having been female elephants (Mackay 1938, I: 329). Moreover, a close inspection of some of the seals also showed that in some cases one of the hindlegs of the animal is raised, indicating an animal in motion (Mackay 1938, I: 671).

Mohenjodaro also yielded eighty copper tablets, rectangular or square in shape, primarily attributed to the later periods. The majority (forty-seven) of the tablets carried an animal or human figure on one side, and an inscription on the other. Amongst the animals represented on these tablets, the elephant took the lead. The figures though were roughly executed in contrast to the usually careful portrayals on seals. These tablets were suggested to have been used as amulets, wrapped in some material, and worn around the neck or wrist or sewn into clothing. The animals on the tablets were postulated to have been symbols of certain deities, the wearers having desired the patronage of the particular deity symbolized thereon (Mackay 1931b, II: 398–401).

At Chanhudaro, the mega herbivore figured on one seal where it was carefully engraved. The row of bristles along the back suggesting a young animal was remarkable in terms of the emphasis given to realism (Mackay 1943, 147).

At Allahdino, Pakistan, all the seals were in steatite with the exception of a single example in clay, considered a local imitation of the Harappan seals, and one of the most interesting of the group. Both the script and the pictorial portion of the seal were slashed into the clay. While the script signs dominated, the elephant motif was tucked into the lower left behind the 'manger' device. The inscription and iconography was not hollowed out to permit a clear impression nor was the boss on the back pierced. In fact, the emphasis on the script and the linear treatment of the pictorial portion suggested that the possession of the seal was perhaps more important than its use. It was also argued that such a copy being made in clay indicated the status implications of the genuine seal (Fairservis 1976, 7, 9).

Harappa yielded only six seals with an elephant motif, in which the animal appeared with small ears and other features characteristic of the Indian elephant (Figure 4.7). On two, however, we once

Figure 4.7 Faience tablet depicting an elephant on one side and script on the reverse, Harappa

Source: © Harappa Archaeological Research Project, Harappa.com, courtesy Department of Archaeology and Museums, Government of Pakistan. Reproduced with permission.

again encounter a double finger-like process at the tip of the trunk similar to that seen on the Mohenjodaro seals (Vats 1940, I: 322). A point worth noticing is that animals such as the bull, buffalo, bison, elephant, and tiger, which were generally depicted with remarkable fidelity on seals, do not figure at all on painted pottery (Vats 1940, I: 290).

An interesting terracotta cylindrical sealing was also reported from Rakhigarhi in Hisar, Haryana. It depicted a scene in four parts, with side A showing a unicorn facing right with a double cross symbol in front. The other file on this side showed the hind portion of an animal, which cannot be identified due to its incomplete profile, and eleven Indus signs. However, it is side B which interests us because here the lower file depicts a humped bull, elephant, and a goat (?), all moving towards the right. The upper file depicts two animals, possibly bulls, also moving right. In the absence of a deciphered script, the significance of these representations (like those on seals and sealings found at other Indus sites) remains uncertain, yet the importance of the sealing lies in its being the first of its kind discovered from Haryana (Bhardwaj 1997, 153–5).

It may be meaningful to pause and reflect on the significance of the elephant motif on Indus seals. There have been less credible explanations arguing that the elephant on Indus seals probably represented one of the lunar mansions in the Harappan calendrical scheme since it is depicted with horns to signify the crescent of the moon appearing in that constellation (Ashfaque 1974–86, 152–3).

A more tangible line of enquiry, however, would be regarding the ritual significance of the animal, if any. Though the absence of a deciphered script necessitates the exercise of caution while attributing such significance to most of the animals represented, it may be worthwhile to take note of evidence hinting at the same. Mohenjodaro, for instance, yielded a seal with a low manger or cult object placed before the animal (Mackay 1931b, II: 388). As pointed out earlier, it seems reasonable to concur with Marshall's explanation that rather than having been a sign of its 'domestication', it suggested instead, the offering of food to the animal as an object of worship. Interestingly, Kalibangan yielded lumps of clay with seal impressions of the image of a horned elephant (Figure 4.8). Visibly, the animal seems caparisoned with what seems to be a cloth flung across its back. Additionally, there are twenty Harappan seals showing a fabulous

Figure 4.8 Seal with the impression of a horned elephant, Kalibangan
Source: ©Archaeological Survey of India. Reproduced with permission.

animal with the body of a ram, horns of a bull, trunk of an elephant, hindlegs of a tiger, and an upraised serpent-like tail (Mahadevan 1977, 794). In fact, the elephant forms a vital component of a large number of such composite figures.

No seal with an elephant motif was found at Lothal. The site, nevertheless, yielded ten sealings bearing impressions taken from a single seal showing a running elephant in great detail (Rao 1985, II: 324). Though the elephant is now extinct in Kutch and Kathiawar, we have sufficient archaeological evidence in the form of bones, tusks, and representations to suggest its existence in western India in the third and second millennia BCE.

Some Aspects of the Interface

Having strung together the available evidence, it is also important to mention the two positions regarding the presence of the elephant in the Harappan context. If we consider Raikes and Dyson's (1961, 276) contention that the climatic conditions of the Indus region were not materially different from those of today, then there is little to deny that present-day conditions are not conducive for elephant habitats. Given the extent of the Indus civilization, these scholars considered the importation of these animals from its periphery a perfectly reasonable possibility without, however, suggesting any particular region from where this importation could have taken place. The other view is of scholars such as Zeuner (1963b) who not only asserted that the elephant was found within the Harappan cultural zone but also argued that the Harappan civilization provides the earliest evidence of elephants in captivity or tame elephants.

Though the 'domestication' of the elephant by the Harappans has been strongly advocated, it would not be out of context to look at some arguments regarding the suitability of the animal for the purpose, the possible uses it was put to, and how the animals were acquired. R.C.D. Olivier's (1984, 187–8) contention, for instance, is that it is debatable whether the Asian elephant can be considered a domesticated animal since until relatively recently there has been practically no breeding under control, and of the little there has been, it has not entailed any selection away from the wild type. To be viable as a full domesticant, he argues, an animal must breed and reach full productivity quickly, and elephants, he informs, do not reach breeding maturity until they are over seven years old. Additionally, the animals conceive only once in two to four years, carry their young for nearly two years, and the usual single calf is dependent on the mother for at least five years, thus, hampering her own work for almost seven years. The calf itself cannot actively contribute to work for over fifteen years, entailing loss for the owner. Also, much of the energy they derive from the vast amounts of food they consume is used in body maintenance, so that they can labour for only limited hours at a stretch. All these considerations put together, inevitably restrict the economic use of elephants.

Then why at all would there be attempts to tame and train an animal that has so much to its disadvantage? The elephant is evidently more effective as a beast of burden, and can by virtue of its manipulative trunk and tusks do work that no other animal can do. Even in battle in later times, it was realized that elephants as super-heavy cavalry could be formidable for the enemy (though they could be so for their own side as well), each animal carrying several warriors instead of only one (Olivier 1984, 188).

How it all began can only be speculated with one view suggesting that stray calves taken as pets may have sparked the idea, thus, paving the way for a steady cultural relationship with the pachyderm (Olivier 1984, 188). Once started, it was obviously easier to find and capture elephants because of their then large numbers. Since keeping elephants is an expensive proposition, it is cheaper to catch wild elephants at an age and size when they can be easily tamed and trained for work. Once fully trained (within two to three years), the animal can be controlled by verbal commands and tactile signals from the mahout. Tame elephant stocks are thereby built up and maintained (Olivier 1984, 188).

Apart from the use of its ivory, what purposes the animal was put to by the Harappans can only be left to speculation in the absence of substantial evidence. Sinha (1955, 51) suggests the possibility of the animal having served as a mount for the 'ruling aristocracy'. What he further contends is interesting though clearly conjectural. According to him, the timber used in the architecture of the Indus civilization may have been procured from the Himalayan regions of the Punjab, and here elephants must have been crucial in having served as beasts of burden and means of carriage and transport. Despite apprehensions regarding the taming of the animal for transport or haulage (Mackay 1938, I: 329), it has been reasonably argued that if tamed by the Harappans, the animal would have been excellent for heavy lifting and pulling (Possehl 1999, 194).

POST-HARAPPAN CHALCOLITHIC AND EARLY IRON AGE TRACES

The Harappan cultures were succeeded by a number of non-urban and farming cultures, which either devolved from the Harappan tradition or were independent regional phenomena characterized by the use of copper and stone tools. The economy was a combination of plant and animal husbandry along with hunting, where the latter continued to supplement subsistence activities crucially (Thomas and Joglekar 1994, 188–90).

Though a large number of chalcolithic sites have been excavated in the subcontinent, our knowledge of the animals associated with them is patchy and limited to only a handful of them. Even a quick look at the tabulation of the fauna identified at these sites is enough to indicate that the distribution of large mammals such as the elephant and rhinoceros is not uniform (Thomas and Joglekar 1994, 189).

In the absence of faunal remains, I turn to iconography. Given that visual representations are important social documents of the milieus that produce them, it would be worthwhile to explore the fortunes of the pachyderm in terms of the frequency with which it was modelled, and to see if patterns indicating its popularity can be discerned.

It is only fair in this context to acknowledge the debt owed to studies documenting and interpreting early Indian terracotta art (Banerji 1994; Sant 1997). For instance, a survey of the early terracotta art of northern and western India reveals that while terracotta bull figurines are ubiquitous in all cultures, it is only during the early NBPW phase that the elephant rises to a position of preeminence in the terracotta repertoire along with the bull (Banerji 1994). The discussion that follows draws from such reviews, but where available, the evidence on which the analyses were premised has been cross-checked, and one shall see that the tradition of modelling the elephant in terracotta continued in the intervening period as well.

I begin with the copper-using sites known from different parts of Rajasthan, dating between circa third and second millennium BCE. These have been broadly divided geographically and culturally into three distinct groups, all having distinct cultural characteristics. The chalcolithic in southeastern Rajasthan is represented by several sites classified as belonging to the 'Ahar culture', which is distinct and distinguishable from the Pre-Harappan, Harappan, and Post-Harappan cultures known from northern and northwestern Rajasthan as well as from the copper-age Ganeshwar–Jodhpura 'culture' sites of central and northeastern Rajasthan (Hooja and Kumar 1997, 324).

Though no evidence regarding the pachyderm comes forth in the form of faunal remains, excavations at the sites of Ahar and Gilund have clues to offer in the form of terracotta figurines of the animal. At proto-historic Ahar in Udaipur, Rajasthan, the bull took the lead, followed by the elephant and horse (Sankalia et al. 1969, 181–2). A fragment of a voluted trunk of an elephant well fired in black clay came from chalcolithic Ahar Ib (c. 1725±110 BCE), while Ic (ranging between c. 1270 and 1550 BCE) yielded another example in the form of a pointed trunk probably of an elephant with the body broken (Sankalia et al. 1969, 182). A bust of an elephant with stumpy legs and pinched up head with signs of being over-burnt also survived from chalcolithic period IA at Gilund, 72 km to the north-east of Udaipur (Sant 1997, 104).

In the context of the early iron-age cultures in the study area, we have some doubtful evidence from the north-west, where iron occurs for the first time in Period III (ninth and sixth centuries BCE) of the Gandhara grave culture (Dani 1967, 48). The evidence is in the form of the terracotta remains of an indeterminate animal resembling an elephant or wild boar with a suggested mane and broken tail at the grave culture at Zarif Karuna (Banerji 1994, 152).

More evidence is at hand in the pre-NBPW deposits in the Central Ganga Valley. At Rajghat, for instance, the pre-NBPW iron-bearing deposit represented by sub-period IA (800 BCE to 600 BCE) yielded four coarsely made elephant figurines (Prakash 1985, 69). Unfortunately, most of the elephant forms were damaged, but the trunk portion survived to demonstrate the detailed and naturalistic treatment of the animal. A fragmentary trunk of the pachyderm in black colour with a highly polished surface and cream painting was unearthed at the site. Another representation from the same site showed only the forepart of the animal where the trunk was realistically modelled with a deep cut at its end, and the forehead slightly raised. The overall treatment suggested a close observation of the animal by the artist (Banerji 1994, 169). With the exception of these, the near absence of terracotta representations of the animal in the pre-600 BCE levels both in the Upper and Middle Ganga Valley was specifically underlined (Prakash 1985, 63).

At Buxar in Shahabad, Bihar, the finds from Period I represented by red, black, and grey wares included four animal figurines representing the elephant, ram, and horse, which carried three horizontal strokes painted in yellow on the leg, back, and tail. The finds were unique since the occurrence of such painted terracotta figurines in a horizon earlier than the NBPW was observed for the first time (*IAR 1963–64*, 8).

EARLY HISTORICAL AND HISTORICAL TRACES

Faunal Remains

At a general level, urbanism and the presence of the elephant need to be emphasized and historicized. However, within the geographical limits of the study, evidence testifying to the presence of this mega mammal at iron-age and early historical sites is sparse. Overall, in the faunal collections of the period, there seems to be a gradual decrease in the proportion of wild animals (Thomas and Joglekar 1994, 197).

Hastinapur: *Elephas maximus* remains were reported from the early levels of Period III (early sixth to early third century BCE), and Period IV (early second century BCE to the end of the third century CE) at this early historic site in Meerut, Uttar Pradesh. The remains included a tip of the small tusk of a young elephant from an early level of Period III, another tip of the small tusk of a young one from an early level of Period IV, and a pisiform bone of the left forefoot from the same level. Though few, the vestiges left no doubt regarding the familiarity of the inhabitants with the animal. It was argued that the find of a part of the skeleton along with a few pieces of ivory adequately proved that the animal lived in this region (Nath 1954–5, 110).

Autha: At this site in Gurgaon, Haryana, the fragmentary skeleton of an elephant was exposed as a result of erosion and diggings by villagers (Figure 4.9). A small-scale excavation undertaken by K.M. Srivastava established that the earliest occupation at the site had been due to the Painted Grey Ware (PGW) people. It was, however, not possible to assign an exact date to the elephant skeleton since it had been lying on the slopes of the mound, on the top levels of the deep pit cut into the PGW deposits, and sealed by eroded material of Period II and later periods. It was, therefore, roughly placed in the last stages of the NBPW period around second century BCE (*IAR 1964–65*, 33–4).

Balathal: Though the mound of Balathal in district Udaipur, Rajasthan, yielded habitation deposits of the chalcolithic and the early historic periods, the faunal material mostly belonged to the early historic period and testified to the presence of the elephant through the retrieval of a rib of the animal from layer 4. Since this was found to be sturdier than the bones of other animals, it was postulated to have been brought to the site with the aim of fashioning an object or tool (Thomas and Joglekar 1996, 92).

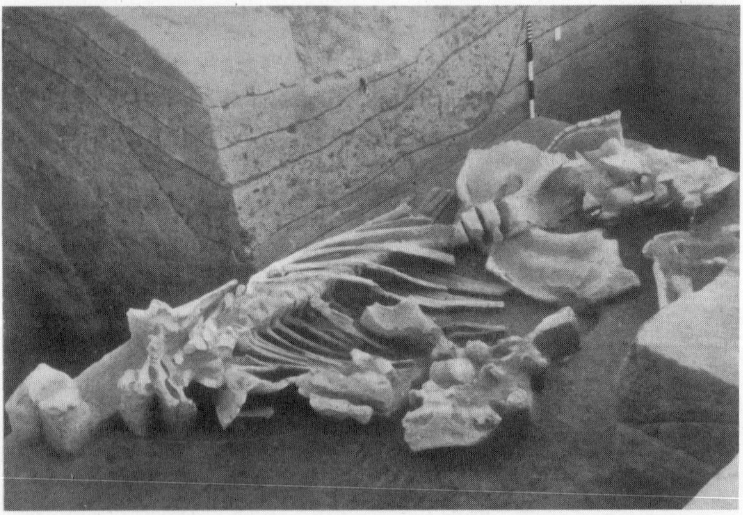

Figure 4.9 Skeletal remains of an elephant, Autha, Gurgaon, Haryana
Source: After *IAR 1964–65*, Plate XXIV. © Archaeological Survey of India. Reproduced with permission.

Art and Iconography

In view of published faunal profiles being limited, we can certainly derive vital glimpses from visual representations. I begin with rock paintings, which are often thematically and stylistically different from the earlier paintings. According to Neumayer (1983, 35), the most striking new theme is war, scenes of which pervade the shelter walls. Since absolute chronologies for rock art still elude us, in the discussion which ensues, I take into consideration examples that have been broadly put within the historical period.

In his survey of the rock paintings of Mirzapur, Manoranjan Ghosh (1932, 16) elaborately described a painting in the Likhunia shelter (Figure 4.10). The subject he recounted was what seemed like the capture of a wild elephant by horsemen assisted by a tame elephant. The central object of the scene is a driver on a big elephant, carrying in one hand a long pointed spear-like weapon, and in the other, a round object, probably a shield. He is pursuing a tusked elephant. Among his retinue are shown in the upper field two archers, two men with spears and shields, and a man without any weapon. Below him are a man on foot with a spear and a shield, and three horses of which one has lost its rider. The two riders are armed with spear and shield.

Below this elaborate representation was another depicting the catching of birds, to the left of which was an elephant and a horse.

Figure 4.10 A scene from Likhunia, Mirzapur, Uttar Pradesh
Source: From Ghosh 1932, Plate VI a. © Archaeological Survey of India. Reproduced with permission.

Though Ghosh (1932, 20) placed the paintings of the region between the fourth and tenth centuries CE, Wakankar and Brooks (1976, 83) chose to broadly attribute it to the early historic period.

Stylistically, Varma (2012, 65) points out that in paintings of the historical period, elephants are painted with mounted figures forming part of some procession or war scene, and such depictions are always painted away from the prehistoric ones. In a study of the rock art of the north Vindhyan region, he leads us to Hanumana shelter, where the painting is in red ochre and the mount is holding a sword in his hand.

In Rajasthan also, where both petroglyphs and paintings are encountered in the rock art repertoire, we come across an elaborate battle scene at Kunjota in Jaipur (Figure 4.11). The illustration shows an engraving on a rock surface of a battle scene with animal riders and foot soldiers. All the warriors including the elephant rider to the left and the camel rider to the right are armed with swords and shields. The figures, both human and animal, are executed in silhouetted style (Biswas 2012, 93).

From these, I move up northwards towards the rock carvings of the Karakorum and the Upper Indus Valley. In 1978, explorations by

Figure 4.11 Battle scene (historical period) with elephant and camel riders, Kunjota, Jaipur, Rajasthan

Source: © Indira Gandhi National Center for the Arts (IGNCA), New Delhi. Reproduced with permission.

Karl Jettmar and Ahmad Hasan Dani brought to light thousands of petroglyphs and inscriptions in the region, concentrated mainly in the area east and west of the village of Chilas under the shadow of the famous Nanga Parvat, in the Dyamar district of north-eastern Pakistan.

At Chilas II, Jettmar (1982, 13–14) pointed to a rock forming a platform. This, he contended, must have been a place suitable for religious ceremonies. The main theme in the decoration of this evidently Buddhist structure was the veneration of stupas. While a discussion of their structural details would be out of context, what elicits attention is Jettmar's mention of worshippers who are warriors, riding or leading their horses or approaching on foot. Equally important is his observation regarding elephants, which are realistically rendered but do not seem to be religious symbols. There were other animals like zebu and goat. Based on the design of the stupas, the attire of the warriors, and Achaemenoid reflections, a pre-Kuṣāṇa or Kuṣāṇa date of the drawings was inferred.

Dani's subsequent (1983) account of the rock carvings of Chilas II, lying close to the Indus about 1.5 km from the municipal checkpost of Chilas on the Karakorum Highway, tells us that the site consists of a rough rock pile sloping down towards the river. There are two rock formations with only the western rock carrying the carvings. Since the river face of the rock was first used for engraving, the earliest figures and inscriptions are found on the vertical face of this side, and can chronologically be attributed to the Scythian period (first century BCE) on the basis of Kharoshthi inscriptions, although some Kharoshthi inscriptions of the Kuṣāṇa period have also been recorded. Based on chronology, the engravings at Chilas II were, thus, placed in three main groups: Group A falling between first century BCE and second century CE; Group B, not associated with any writing, was placed between the second and fifth centuries CE; and Group C belonging to a much later period between the sixth and seventh centuries CE (Dani 1983, 91– 2).

Described in order of recesses as they occurred from east to west, Recess No. 3 in Group A is the biggest recess on the site. Figures, inscriptions, and scenes are depicted at different levels. On the left top is a stupa, to the right of which is a man (whose head has disappeared) in a running pose, driving an elephant to the right. His left hand is raised, while the right is on his waist. The elephant with a raised trunk and front leg is shown with a body filled like that of the man. Animals below the stupa, and technically related to it include a humped bull and a peacock with filled-in bodies on the right, and on the left respectively. Other animals such as horses, ibexes, and markhors add variety to the scene (Dani 1983, 98, 100). Considered a late scene, the same recess also carries a depiction of an elephant carrying a log of wood.

While it may be difficult to ascertain the exact meaning of such portrayals, what is evident is the appropriation of the animal in varied contexts. While on the one hand, we have it at the centre of the elaborate battle scenes of Mirzapur and Rajasthan, representations from the north-west present a different world view where the pachyderm even without religious manifestations occurs in indisputably Buddhist contexts.

Beyond depictions in rock art, a closer look at some representations in terracotta available through published reports can furnish insights regarding the eminence accorded to the animal in this period. What is evidently perceptible is an unmistakable preference for elephants in the terracotta repertoire from c. 600 BCE onwards. This seems evident from the significant number of elephant figurines found at sites such as Hastinapur, Atranjikhera, Mathura, Sonkh, Rajghat, Prahladpur, Kausambi, Bhita, Sravasti, Vaisali, Buxar, and many others.

Before I embark on a broad survey of the figurines that occur in profusion within the study area, it is important to qualify that my work makes no claims of undertaking an all-inclusive survey of representations of the animal in terracotta. What is attempted instead is to examine the geographical spread of these models and to locate important markers like frequency of depictions, variety, and decorations employed while delineating the pachyderm. I have arranged my data in geographical clusters within which I try to follow a broad chronological sequence so that patterns that emerge can be easily discerned. Put together, the evidence is likely to give us some sense of the familiarity with the animal and cultural representations of it during the period being considered—c. 600 BCE–300 CE. The aim of these details is also to underline that terracottas are an indispensable window for understanding the deep and intimate familiarity of ancient north India with the anatomy of the elephant.

I begin with a survey of the relevant sites in Uttar Pradesh, which not only gives us the bulk of our earliest evidence for the historical period, but also facilitates an overview spanning centuries at a stretch. The primary focus will be on sites that have yielded long cultural sequences.

Excavations at Hastinapur yielded 105 hand-modelled animal figurines (Figure 4.12). With the exception of four specimens, generally well-modelled elephants dominated the terracotta animal repertoire of Period III (early sixth to early third century BCE). A popular way of decorating the animals in this period was to cover them with circlets which were punched, stamped, pierced, or notched, but far more artistic were the elegant *cakras* and leaves stamped on the bodies and trunks of the animals. The attention to detail is noteworthy in most cases. For instance, in one fragment from an early level of Period III, the tusks were shown coming out of their root-sheaths, which were indicated by a thin coating of clay applied around the tusk at the joints. Similarly, a middle level of the same period offered

Figure 4.12 Terracotta elephant, Hastinapur
Source: From Lal 1954–5, Plate XLIII, 4. ©Archaeological Survey of India. Reproduced with permission.

the figure of a stylized elephant with grotesque features consisting of longish ears, applied tusks, and notched circlets on the forehead (Lal 1954–5, 85–6).

At Atranjikhera in Etah district, Uttar Pradesh, one encounters representations of the animal in terracotta in Period IV (c. 600–50 BCE) representing the NBPW tradition. Amidst a total of thirty-seven figurines that were recovered from the four phases of Period IV, the bull took the lead, followed by the horse, the elephant, and the ram (Gaur 1983, 366).

Of the four elephant figurines, the one from Phase B (c. 500–350 BCE) was presumably that of an elephant with bold and heavy hind legs with the front part and tail missing. The hind part of the model, treated with a red slip, had an appliqué clay-band decorated with a row of punched circlets. The elephant figurine from Phase C (c. 350–200 BCE) of Period IV was a model in grey colour with traces of black slip. Though the legs, trunk, ears, and tail were missing, it had a robust body. The tusks emerging from their root-sheaths were indicated by a thin coating of clay at the joints. The eyes were formed by incised circlets within incised lozenges. The back, forehead, and the sheath were decorated with incised circlets while an appliqué incised clay band passed over the head running down the ears. Of the two representations from Phase D (c. 200–50 BCE), one was an exquisitely modelled head of an elephant treated with a slip. The forehead was decorated with small peckings, and the tusks near the root by a ringed band with tiny dots at regular intervals. The eyes were pierced within incised circlets, the ears were fan-shaped, but the trunk and tusks were partially missing. The second was a fragment of a modelled animal, presumably an elephant treated with a bright red slip. The heavy body was decorated with a stamped leafy pattern and *cakras*, and there were traces of an appliqué band on the back probably meant for a rider (Gaur 1983, 369–70).

A significant find was a moulded terracotta plaque of Gajalakṣmī retrieved as part of the remains of an apsidal temple belonging to the earlier levels of Phase D of Period IV. On the brick platform found within the four-walled enclosure were found pottery pieces of the late NBPW phase along with antiquities such as a corroded copper coin, a copper bangle, an antimony rod, a few terracotta balls, and the aforementioned plaque (Gaur 1983, 257). Treated with a red slip, only the upper part of the plaque was extant. It showed the deity with round *kuṇḍalas* being sprinkled over the head by two elephants on either side with their trunks twisted above (Gaur 1983, 366).

Relevant evidence also comes from Mathura, Uttar Pradesh, where Period II (placed between the closing decades of the fourth century and c. 200 BCE) marked an expansion of settlement evincing a transition from ruralism to urbanism. A study of the terracotta animal figurines by M.C. Joshi and C. Margabandhu (1976–7, 22–3) revealed that the elephant was the most popular amongst the animal figurines of this period. Mostly handmade, the figurines were both of the solid and hollow variety (Figure 4.13). The trunks and body were richly decorated with appliqué, punched, and incised designs.

Figure 4.13 Black terracotta elephant, Mathura, Uttar Pradesh, third century BCE, 5.25 × 2.5 × 6 in (13.34 × 6.35 × 15.24 cm)
Source: Courtesy of Los Angeles County Museum of Art, Gift of Paul F. Walter (AC1994.233.2).

Rhombic eyes were stamped on either side of the head, which had a decorative strip suspended downwards with the end bearing a floral knob. In some cases, the head also carried a fillet-like strip. Parts of elephants, which are usually painted even now, were embellished with incised lines and punched circlets. The hind part of the animal also carried decorative stripes. The frontal treatment was realistic enough to stand comparison with the Dhauli elephant on the Aśokan pillar.

Interestingly, mention was also made of some male figurines with a concave bottom clearly indicating that they were meant to be fixed on some other object, most probably on elephants, some of which bore such impressions on their back. One of the male figurines had demonic facial features, while others had a pleasing youthful countenance. It was postulated that some of these figurines were used as elephant riders representing popular deities including Indra (Joshi and Margabandhu 1976–7, 22). Though this seems conjectural, the attention given to decoration certainly seems to suggest that these elephants had some kind of religious or ritualistic uses.

Bautze (1995, 23) contends that the elephant-and-rider motif was already in vogue at Mathura during the Mauryan period, and draws attention to a specimen in black-slipped terracotta chronologically placed between third century BCE and second century BCE (Figure 4.14). It showed riders whose feet were not indicated by the artist. The riders were portrayed with moulded heads and stumps rather than fully-developed arms. The animal with impressively big tusks had its head adorned by lotus pistils, while a garland or *cāmara* hung from each ear. The trunk was hand-modelled, and decorated with appliqué rosettes belonging to a wreathed garland on top of the head between the large rhombic eyes.

The elephant emerged as the favourite in Period II at Sonkh in Mathura, Uttar Pradesh. The pachyderm accounted for twelve out of twenty-nine figurines tabulated for the period, which coincided with the Maurya and Śuṅga cultural phase. It is also important to mention that of the 306 figurines and fragments that the site yielded, elephants accounted for 46, being third in line after horses, which accounted for 75 figurines, while humped bulls accounted for 91. Since the fragments and figurines were arranged according to stratification, changes in style and the decoration of animals like elephants and horses could be studied by comparing their appearance in successive periods (Härtel 1993, 163). An elaborate specimen, for instance, showed tusks

Figure 4.14 Elephant with riders, Mathura, Uttar Pradesh, third–second century BCE. Terracotta, 11.25 × 4.125 × 11.5 in (28.58 × 10.48 × 29.21 cm)
Source: Courtesy of Los Angeles County Museum of Art, Gift of Mr and Mrs Subhash Kapoor (M.85.72.1).

emanating from their root-sheaths, but broken short. The preserved right ear protruded sidewards showing an incised line dividing the ear in two halves. The eyes were incised as lozenges within which the pupil was pricked. The forehead was decorated with punctured holes, while a band hung over the ears and ended in tassles. Another band decorated with incised strokes encircled the body of the animal in its back part just before the hind legs (Härtel 1993, 164).

Elephant figurines in an NBPW context were also retrieved from Period II at Batesvara in Agra, Uttar Pradesh. The radiocarbon date for this period (2520 ± 160) placed it in the middle of the first millennium BCE (*IAR 1975–76*, 43).

A rich collection of animal figurines was retrieved from Stratum VIII (300 BCE–200 BCE) at Ahichchhatra in Bareilly, Uttar Pradesh. Elephants accounted for forty-one models decorated with different types of punched, impressed, and incised strokes, circlets, and plant designs. The figurines, eleven in red, and the rest in grey with a black slip could be further divided into undecorated moulded figurines, and solid ones with punched decorations (Prakash 1985, 94).

A spectacular find from the site was a round plaque depicting different elephant poses on one side, and birds of various kinds on the other. Full of life and movement, the elephants were both male and female, and young and old. The animals were organized in three bands with a front-facing full-grown tusker in the centre, a row of eight elephants in the second band, and twelve in the third. Considered by V.S. Agrawala (1947–8, 162) as 'a unique specimen of its kind in Indian art', the specimen was not dated. However, Margabandhu (1991) brackets it with early historic circular plaques. What one sees in Figure 4.15 is a pencil sketch of the plaque (which went missing) by Sivaramamurti.

Terracotta elephant figurines occur mostly in a post-NBPW context at Sravasti, 160 km north-east of Lucknow, Uttar Pradesh. Excavations by K.K. Sinha yielded a total of seventy animal figurines, all modelled by hand. Such figurines were the largest in number in Period II, and in many cases the body was embellished. The period was divided into three phases: early phase c. 275–200 BCE; middle phase c. 200–125 BCE, and the late phase c. 125–50 BCE. From the evidence examined so far, it seems plausible to concur with Sinha's contention that in view of the wide distribution of elephant figurines at early historical sites of north India, chiefly from Mauryan levels, and the presence of decorative and symbolic motifs on the bodies of most, it may be assumed that the animal in this period was regarded as one of some importance and as an object of respect (Sinha 1967, 56).

Of the four specimens illustrated in the report, the elaborate decorations on two deserve particular mention. Both came from a lower level of Period II, and were elephant heads. Realistically modelled, one carried fan-like ears and lozenge eyes. The forehead and body were decorated with a stamped circle filled up with raised dots in the quadrants, and three vertical rows of sigmas adorned the trunk portion. The other had a short forehead decorated with stamped rosettes and a pointed trunk with a tree-like design (Sinha 1967, 57).

Figure 4.15 An exceptional early historical terracotta plaque from Ahichchhatra, Bareilly, Uttar Pradesh
Source: After Agrawala 1947–8, 162. © Archaeological Survey of India. Reproduced with permission.

Hand-modelled representations of the animal also come from the site of Sringaverapura in Allahabad, Uttar Pradesh. Excavations by B.B. Lal (1993, 144–5) revealed that all the elephant figurines came from deposits of Tank C, which was a part of the large brick tank that was built between the second half of the first century BCE and the end of the first century CE during Period IV at the site. This was a post-NBPW deposit characterized by red ware. It is crucial to mention the occurrence of a large number of terracotta figurines of gods and goddesses in Tank C, some in the lower deposits, but most in the debris indicating the existence of some religious structures by its side. The possibility of some of the animal and bird figurines having had some religious significance was also suggested. What was far more definitely asserted was the use of some of them as toys in view of the devices for propulsion in them. The latter view can, however, be reconsidered if we remind ourselves of the Daimabad bronzes, where the animals carrying devices of propulsion were suggested to have been meant for religious processions (Dhavalikar 1988, 21).

The specimens selected for discussion revealed interesting facets. Most remarkable were representations of the animal mounted on platforms. In one such illustration, two elephants stood side by side on a common, low platform. The models, though damaged, had their hind parts conjoined with a broad vertical strip which had two holes, one at the back of each elephant. A relatively narrow strip bearing notched decoration ran from the back of one elephant to the other. The front legs of the elephants were separated by means of a large hole without any apparent function. On another broken platform, there was only one elephant, though the possibility of the presence of another one on it was raised. The treatment meted out to the animal was similar to the mounted specimen described before (Lal 1993, 145).

Apart from the mounted specimens, there was a large-sized figure of the animal, evident from the portrayal of the ears, of which the left one was nearly intact. A part of the head was also extant. An oval piece with a central depression rested on the back, and possibly represented the howdah. The tail was slightly turned to the left side. The legs as also a major part of the head and trunk were broken (Lal 1993, 144).

A more complete head of a large-sized elephant showed excellent delineation of various parts, and in particular of the eyes and the raised knobs of the head. Though the trunk and tusks were broken, the remnants sufficed to suggest the fidelity of the depictions. In another retrieval, a fairly large-sized incurved trunk survived as also did the part from where the tusks had emanated. The thin folds of the skin on the underside of the bent-in trunk were also very well delineated. Another specimen, though considered slightly doubtful because of its broken face, deserves mention because of the stamped design of leaves on the 'seat' resting on the broad back of the animal and on the hanging that passed over the neck (Lal 1993, 144–5).

At Kannauj in Uttar Pradesh, the broken head of an elephant bore a lustrous black slip generally associated with NBPW, and there were tan-coloured strips on the head of the animal. Two mould-made plaques depicting Gajalakṣmī, a popular deity during the Śuṅga period (second century BCE) were also documented. In one of the examples, the goddess stood erect holding the stalk of lotus buds in both her hands. On each of the buds, an elephant stood pouring water on the head of the goddess. Another specimen was a mould-made bust of Gajalakṣmī that was executed in the early phase of first century BCE. She is shown flanked by elephants pouring water from jars held in their mouths. An elephant-rider figurine in the collection was attributed to the Kuṣāṇa period (Kala 1992–3, 118).

Erich in Jhansi, Uttar Pradesh, also yielded an elephant-and-rider depiction assigned to the Kuṣāṇa period (Srivastava 1991, 21). Further, a terracotta fragmentary elephant trunk with a tusk was reported from Period III dated between first century CE and sixth century CE at the site of Charda, about 200 km north-east of Lucknow, Uttar Pradesh (Tewari 2002, 123).

In eastern Uttar Pradesh, at Rajghat in Varanasi, the period under review was illustrated through sub-period IB (600 BCE–400 BCE), which yielded fifty-one animal figurines, among which twenty-nine were figurines of bulls, fifteen of elephants, five of dogs, and two were *nāga* figures. Of the elephant models, ten were very fine figures in NBP fabric, while the remainder were in red ware with simple incision and appliqué decoration (Prakash 1985, 68–9).

That the animal was carefully modelled can be seen, for instance, from a hand-made solid elephant trunk and head in NBP fabric, with the front portion of the trunk twisted towards the mouth. A big hole on one side indicated the place of the tusk, while the other side showed a broken tusk. A broad forehead with a slightly raised clay portion was seen in the middle just below the eyes made by appliqué clay circles. The big ears and the head portion were delineated by clay projections with the help of finger-pressing (Prakash 1985, 70).

As subjects of representation, the attention the animal received also becomes evident when one notes the diversity of figurines in the collection of Rajghat IB. The first category contained solid and hand-made elephant figurines with shining black surface and creamish strokes. Other features were a grey core and multiple vertical and horizontal painted strokes on the forehead, neck, ears, trunk, and legs. The eyes were represented through appliqué pellets. It was pointed out that comparable specimens came from corresponding levels at Prahladpur (mid-phase of Period I) and Bhita (primitive).

The second category consisted of figurines in red colour bearing incised and appliqué decorations. The five examples in this context were modelled with muscular legs and pellet eyes. One figurine had a U-shaped clay band (with linear incisions) on the forehead. Application of clay bands on the back was also noticed in some cases. Incised lines on the back of some figurines possibly indicated a saddle. The period also yielded an elephant figure with rider, a type also found at Bhita and Taxila (Prakash 1985, 69–70).

Period IC (c. 400 BCE–200 BCE) yielded fifteen elephant figurines of varying sizes, mostly hand-modelled. Trappings were found around the belly in certain cases. As in Period IB, the repertoire once again reflected a diversity of forms including profusely decorated figurines, elephants with riders, as also a realistic representation of curves and bulges in an elephant with uplifted trunk and prominent lower lips (Prakash 1985, 96–7).

Similarly, sub-period IB (c. 500 BCE–163 BCE) at Prahladpur in Varanasi, Uttar Pradesh, yielded a trunk portion of an elephant in grey colour with traces of slip in red and black colours. The realistic execution of the twisted and pointed ends not only suggested skilled artisanship, but also close observation of the animal. Another fragmentary trunk portion in dark grey contained two round appliqué eyes. The occurrence of orange dots on the black lustrous surface also elicited attention (Narain and Roy 1968, 45–6; Banerji 1994, 193).

Characterized by NBPW, sub-period IB at Masaon in Ghazipur, Uttar Pradesh, yielded an elephant with a row of impressed *cakras* over the back, and a row of leaves over the trunk with eyes formed by pierced holes within lozenges (*IAR 1965–66*, 51).

At Kausambi in Allahabad, Uttar Pradesh, the earliest animal figurines recovered from the site were those of elephants. The figurines were mostly modelled by hand, though occasionally stamps were used to shape the head or other limbs. In the case of elephants, the chains and the richly decorated seat were produced by stamps (Sharma 1969, 69). The elephant figurines ranging in date from the early fourth century BCE to the middle of the first century CE have some remarkable depictions (Figure 4.16).

In what was the best preserved of all the elephant figurines, the head and trunk of a specimen from sub-period IB were considerably

Figure 4.16 Lakṣmī lustrated by elephants (Gajalakṣmī), Kausambi(?), Uttar Pradesh, first century BCE; terracotta, 5.75 × 3.5 × 0.625 in (14.61 × 8.89 × 1.59 cm)

Source: Courtesy of Los Angeles County Museum of Art, Indian Art Special Purpose Fund (M.85.62).

larger in proportion to the body. A band was tied around the neck, while another on the forehead was decorated with punched circles enclosing dots. The two ends carried tassels. The eye was a rhomboid punch enclosing a dot. In another figurine from sub-period IB, the decoration of the seat on the back of the elephant was very elaborate. The seat had a covering on which designs and patterns were represented by punched circles, showing dots in each quadrant. The disproportionately large trunk was decorated with punched circles. The neck carried a punched band representing a garland. A mould-made example from sub-period V showed an elephant with a chain in the right hind leg probably binding it to a pole, and the animal in rage trying to release itself from the chain. The strain of the struggle is effectively conveyed. Alternatively, it was suggested that the elephant may have been performing an acrobatic feat, balancing himself on a narrow pedestal for the amusement of the people (Sharma 1969, 71).

S.C. Kala's (1950, 41) study of the terracotta figurines from Kausambi takes note of a number of plaques depicting chained elephants or the animals uprooting trees. His survey illustrated an interesting plaque where a winged lion was shown attacking an elephant. The lion pounced on the elephant's back from above, terrifying the pachyderm so much that it let out balls of stool (Kala 1950, 41). It is difficult to explain the potter's keen and realistic approach towards his subjects unless we assume a close familiarity with them. Another plaque showed a fight between a lion and an elephant (Kala 1950, 50). Whether these were manifestations of scenes witnessed in the wilderness or images that emerged from animal fights organized for recreation is worth considering.

An elephant in full profile with three riders perched on its back appears on a moulded terracotta plaque from Kausambi, from around first century BCE to first century CE. The first rider sits immediately behind the elephant's neck typically like the *mahāvats* or elephant-drivers of India today. He drives the animal with the goad in his right hand. The man behind him seems to be carrying what appears like a flask by the neck, but which could also be a leather bag. The third rider holds an eel-like fish with tail bent downwards. Bautze (1995, 23) considers the possibility of this being the elephant's tail, in which case, the identification of the tail-like extension at the elephant's back below the 'eel' is rendered difficult. There is a rope going through the mouth of the elephant.

At Bhita near Allahabad, John Marshall reported the spectacular finds of a child's rattle in the form of an elephant with rider, and a fragment of an elephant's head that may have been part of a jar from the Primitive group belonging approximately to eighth century BCE. A mutilated miniature elephant, apparently a handle of a vase, came from the group attributed to the Mauryan period. Another mutilated specimen with its head and the pad on its back stamped with floral designs came from the Śuṅga and Āndhra group. Compared with these, the model assigned to the Kuṣāṇa group was crude with its three legs and the trunk broken (Marshall 1915, 71–5).

The town complex of Ganwaria, about a kilometre south-west of the ancient site of Piprahwa in Siddharthanagar, Uttar Pradesh, also adds to our evidence. The elephant figurines here mostly belonged to Period III representing the Śuṅga period beginning in the second century BCE, and ending by the beginning of the Christian Era. A striking specimen assigned to the second century BCE showed the animal with a rider whose head was damaged. He held the head of the elephant with both hands as if holding the reins. The pachyderm was shown with a raised trunk. Similarly, a damaged specimen dated to the beginning of the Christian Era was significant for bearing embellishments all over the body in the form of stamped circles. Another model dated to the same period portrayed the animal drawing its trunk into the mouth. The period yielded other decorated specimens of the animal. Period IV, which was characteristically Kuṣāṇa, and chronologically placed between the beginning of the Christian Era and the close of the third century, yielded the head, tusk, and trunk of a hollow elephant figurine dated to the first century CE (Srivastava 1996, 56, 221–2).

The mega herbivore (represented by 52 excavated specimens of a total of 114 recovered) seems to have been a favourite at the ancient site of Tilaura-kot close to Taulihawa, which is 21 km to the north-east of Shohratgarh, a town in Basti district of Uttar Pradesh. While the beginning of settlement at the site was traced back to about the sixth century BCE, it was suggested to have been occupied till about the third century CE (Mitra 1972, 15). It is interesting to note Mitra's observation regarding the chief inspiration behind terracotta animal figurines being a juvenile interest in toys. What was also suggested was the possibility of these being intended as votive offerings to deities (Mitra 1972, 109).

Some broad observations were made on the basis of the assemblage of elephant figurines retrieved from the site. Most specimens were plain, and only some were decorated with stamped designs such as leaves, solar symbols, an eight- or four-spoked symbol with a pellet in each angle, and diamonds or circles with a central punctured dot. Ornaments were rarely represented, and trappings were conspicuous by their absence. In some models, the excavator noticed a sound sense of form based on a close observation of the animal, while in others this realistic touch was lacking and the animal could be identified only by virtue of its tusks or trunk (Mitra 1972, 109).

Among the numerous specimens described, mention may be made of a nearly complete (but for the missing three legs, tusks, and tail) and lavishly ornamented surface find. The ears were disc-like with a central shallow hole. The neck, waist, and base of the tusks were ornamented with applied bands decorated with stamped circlets, the neckband being distinguished by a bell-shaped pendant and the waistband by three pendants, all with stamped circlets. The forehead was embellished by a similar applied band, crescent-shaped, and with two rows of circlets. On the crown was a deep hole, and the model was treated with red slip (Mitra 1972, 112).

At Pakkakot in Ballia district of eastern Uttar Pradesh, remains of elephant figurines were retrieved in the form of a head fired to red slip, and a tusk belonging to Period IV, representing the Śuṅga–Kuṣāṇa period. The site yielded a total of twenty-three animal figurines with all the specimens being hand modelled. Numerically, birds were the most prolific with the elephant and the horse following closely (Dubey et al. 2012, 180–1).

Narhan in Gorakhpur, Uttar Pradesh, yielded forty-two terracotta animal figurines comprising stylized figures of bulls, elephants, horses, rams, dogs, and camels. In the assemblage spread over the five occupational deposits identified at the site, humped bulls were numerically the most prolific accounting for eleven specimens followed by five of the dog, three of the ram, and the same for the elephant (Singh 1994, 153). Keeping in mind the chronological limits of the study, only the specimen belonging to Period IV (c. 200 BCE–300 CE) is relevant to our enquiry. The Narhan find (mostly damaged) clearly suggests that elephants were more popular at Hastinapur, Kausambi, and Rajghat (Singh 1994, 153).

In Haryana, excavations at Sugh in Ambala revealed a number of crudely hand-modelled terracotta animal figurines. The elephant figurines were embellished with applied or incised bands on the back and motifs like the stamped leaf, *cakra*, and punched circles. They had prominent temples, lozenge-shaped eyes with a punch-marked pupil in the centre, and were mostly assigned to c. 300 BCE. Most figurines were broken, but the head and trunk of a well-modelled specimen belonging to c. 200 BCE deserves particular mention because of its rare character. The decoration consisted of an impressed *cakra* and *nandipada* design in the ears, temple, and trunk. In another model, the figurine was adorned with a chain-like decoration on the back and temple with applied bands designed by oblique incised lines and punched circles. Yet another depiction showed a circular punching decoration on the temple and trunk indicating a cloth (Singh 1990, 7–8).

Similarly, Thanesar in Haryana, about 160 km north of Delhi, yielded a well-fired elephant head in grey colour placed between c. 100 and 200 CE. The model carried an appliqué strip around the neck decorated with incised dot marks. The eyes were applied and incised. The face and tusk were adorned with a beautiful appliqué strip decorated with dot marks. The left ear was made separately, and stuck at the proper place, while the right ear and tusk were partly broken (Singh 1989, 407).

Excavations at Farmana in Rohtak, Haryana, brought to light a terracotta elephant attributed to the Maurya–Śuṅga period. Hand-modelled, the specimen was decorated with appliqué, stamped and incised dots, and grooves (Uesugi 2011, 388).

In Rajasthan, the tradition of modelling the animal in terracotta is attested to by the site of Noh in Bharatpur district. Period IV marked by the emergence of NBPW, yielded terracotta animal figurines, the most important being an elephant with incisions all over the body (*IAR 1966–67*, 30–1).

At Rairh in Sawai Madhopur district of Rajasthan, the elephant with rider representation was second in order of popularity following representations of horses with riders between c. 200 BCE and 300 CE. However, only one model complete with the rider was found. The elephant figurines had hollow and solid bodies, where the latter had disproportionate figures, and were crudely finished and ill-conceived. The hollow specimens on the other hand, were fine examples of the modeller's art, with the ears, tusks, and details of the forehead realistically shown. The head bore faint traces of white linear pattern on the forehead suggesting decorative designs similar to those of today. Traces of red paint on some of the models revealed that these including the rider were coated with a thick coat of red ochre. There was also a third type of representation that seems of interest because of its mythological character. It showed the animal with its trunk uplifted resting on its flat legless abdomen with a tiny hump on the back, and wings behind the ears (Puri 1998, 31–2). Additionally, a few examples of pottery mould also depicted the elephant figure in relief (Sant 1997, 149).

Moving towards Gujarat, an analysis of the terracotta objects at Nagara, a small village in the Cambay taluka in Kaira district revealed four specimens of the animal. The one chronologically relevant to this study belonged to Period III (placed between the beginning of the Christian Era and the eighth and ninth centuries CE). Except for the colour and broken state of the model, no specific details were imparted regarding the delineation (Mehta 1968, 96). The evidence, nevertheless, is of importance to us because of its geographical provenance.

Mention may also be made of the small village of Shamalaji in Bhiloda taluka in Sabarkantha district. Despite the absence of terracotta elephants, I mention this site because of a representation of the animal in relief on a potsherd made of Red Polished Ware, a finely lavigated red pottery found in the layers ascribed to the early centuries of the Christian Era (Mehta and Patel 1967, 40). The find is unique in itself since it is not usual to find elephant representations on pottery.

Moving east, at Vaisali in Muzaffarpur district, Bihar, chronologically relevant evidence for our quest comes from cultural Period II representing the NBPW period between c. 600 and 200 BCE, and cultural Period III representing the Śuṅga period between c. 200 BCE and 200 CE (Sinha and Roy 1969, 7–8). A look at some of the illustrated specimens revealed mostly hand-modelled representations. From a mid-level of Period II came a fragment in red colour with a very fat body. The head and legs were missing. A rope tied around the body for holding the saddle was also partly missing. An early level of Period III offered a terracotta elephant with the rear portion broken. The feet had transverse holes for holding the axles of a toy cart (Sinha and Roy 1969, 166–7).

A fairly large number of animal figurines also come from the site of Chirand in Saran district, Bihar. Some of these were said to constitute toys for children. Relevant evidence comes from Periods III and IV, where Period III dated between c. 800 BCE and 200 BCE marked the emergence of NBPW, and Period IV dated between c. 200 BCE and 300 CE represented the Kuṣāṇa period (Verma 2007, 17).

Of the specimens from Period III, one had holes in the rear legs, possibly for providing axle and wheel. In this specimen, conjectured to have been used as a toy for children, the forehead of the animal was decorated with the incised and notched method. A broken piece of an elephant figurine with a chain encircling the neck of the animal was also retrieved. Another was a complete figure of an elephant toy cart. The trunk though slightly damaged, had a hole in the middle for inserting a thread to pull the cart provided with wheels. A remarkable specimen was a piece of a bowl where the head portion of the animal was depicted as the handle of the pot. Amongst the specimens from Period IV, mention must be made of a complete figure of an elephant with the rider missing. The trunk and the forehead of the animal were embellished using the applied and punctured method. The tail was perfectly depicted. The pachyderm was shown excreting represented by the appliqué dot by the side of its tail (Verma 2007, 184–5).

A roughly made thin elephant figurine with an elongated neck and visible ears came from Period III (c. 100–300 CE) at Kumrahar, situated about 4 km to the east of Patna, Bihar. Most of the animal figurines were not well-baked, which was sought to be explained by their general use as toys (Altekar and Mishra 1959, 119).

Senuwar in Rohtas district, Bihar, yielded twenty-four animal figurines, the highest concentration being observed in Period IV representing the Kuṣāṇa period between the first and third centuries CE. With the exception of a mould-made elephant and a horse belonging to Period IV, all the specimens were hand modelled. Leaving the moulded examples, which were fashioned out of fine-grained clay and treated with slip, the remaining were mostly made of coarse clay, and usually without any surface treatment. There was also no attempt to decorate them. The only decoration on these figurines were pricked dots. Except for being mould made, there was nothing spectacular about the specimen from Period IV, where the trunk, legs, and tail of the animal were broken (Singh 2004, 348, 350).

Welcome evidence also comes from the north-western part of the subcontinent, where excavations at Shaikhan Dheri by Dani brought forth seventeen figurines comprising both undecorated and decorated elephants. Two examples from Period I (Kuṣāṇa) and Period IV (Scytho-Parthian) were specimens that originally had riders, while five had a holed leg evidently for the attachment of wheels. In a crude delineation from Period V (Greek), the legs were paired together, but the remaining elephants had separate thickset legs. The ears where preserved, were of broad fan-shape, though some were saucer-like. The eyes, generally circular, were diamond-shaped in two cases. An example from Period V with modelled eyes clearly showed the eyelid and the pupil. The tusks in most cases were preserved (Dani 1965–6, 88–90).

Terracotta elephants, with and without riders, were found at both the Bhir Mound and Sirkap at Taxila. Marshall's (1951, II: 454) meticulous cataloguing of the same revealed that some were quite plain, others were with trappings, while yet others were adorned with stamped designs. The earlier specimens from the Bhir Mound were found to be more ornamental than the later ones from Sirkap. Most of the Bhir Mound specimens were embellished with a range of small-patterned squares or triangles or circles or lozenges stamped on the surface of the head and trunk. Only one elephant of this kind came from Sirkap, though in this case there was just a single stamped device instead of several, and the workmanship had become noticeably cruder. In the case of Sirkap, one elephant showed some effort towards decoration, but the ornament consisted of only a row of incised circles to indicate the girth, and a group of similar circlets on the head to indicate what could have been the headcloth.

Examples of elephants with riders were few and came exclusively from the Bhir Mound. These were easily recognizable from the way in which the legs and seats of the riders were flattened against the bodies of the elephants. Marshall's contention was that the 'toy-makers' of Taxila had greater difficulty in giving the riders a natural seat on the backs of these great beasts than did the sculptors of Sanchi. He went on to attribute this apparent unease to a lack of familiarity with the animal in this region as compared with that of the craftsmen of central India and Hindustan. What also seemed evident to Marshall was the greater popularity enjoyed by these 'toy elephants' in the pre-Greek period represented on the Bhir Mound than later on (Marshall 1951, II: 454).

Interesting finds also came in the form of elephant goads from stratum II (Maurya) at Bhir Mound, and stratum III at Sirkap (first century BCE to first century CE) (Marshall 1951, II: 551).

Having collated substantial evidence, one should perhaps mention studies that have attempted to chart broad patterns within the tradition. Pratibha Prakash's (1985) study of terracotta animal figurines in the Ganga–Yamuna Valley between c. 600 BCE and 600 CE stands out as one such endeavour. Her observations based on an examination of the terracotta repertoire of the region revealed that the workmanship in specimens dated between c. 600 BCE and 300 BCE was simple, with the figures being solid and generally heavy, having fat legs and eyes indicated by either incised circles or round clay in appliqué. She argued that between c. 300 and 150 BCE, almost all the figurines were treated with elaborate incised and stamped motifs like *cakras*, leaves, wheels, and knobs with floral designs on the forehead, temples, trunks, and backs of the animal. The use of these symbols was possibly associated with the veneration of the animal during this period (Prakash 1985, 84).

Prakash further observed that between c. 150 BCE and 50 CE, the animal's previous popularity as a choice for representation dwindled as there was an unmistakable plunge in the frequency of depictions, coupled with factors including the absence of stamped designs like wheel and leaf, and scant decorations on the bodies of the animals as compared to the profusion of the same in the preceding period. This trend, however, witnessed a reversal between 50 CE and 300 CE, coinciding broadly with the Kuṣāṇa cultural phase. About thirty figurines were reported from Period III alone at Rajghat, and similarly, a substantial number came from Vaisali, which revealed a total absence of terracotta elephants in the late level of Period II. Elephant figurines from Ahichchhatra also lent support to this observation. Noteworthy features of the animal figurines of this period were their large sizes, and a marked decrease in decorations on the bodies of the figurines.

Of far greater significance is Prakash's observation regarding the use of certain motifs being exclusively reserved for some animals. Between c. 600 BCE and 300 BCE, the use of the appliqué motif consisting of a U-shaped design formed by applying a band of clay on the forehead and back was restricted to elephants and bulls only (Prakash 1985, 67). Equally enthralling is the observation that though elephant figures in red fabric continued to occur, only the horse and elephant were selected to be fashioned out of NBP fabric, the premium ware of the period (Prakash 1985, 63). This is an important choice whose significance has thus far not been analysed. It is most likely that this choice must have had to do with the iconic status these animals had come to attain because of their importance as war animals as also symbols of royalty.

While Prakash's study needs to be tested on a wider spatial canvas, what emerges from the picture constructed so far leaves us with implicit certainty regarding the importance of the animal in early historical and historical times. We have seen a range of models from simple to ornate ones, from ones portraying the animal as it was seen, to ones which were constructs of imagination (winged elephant at Rairh). What also calls for attention is the fact that the depictions, mostly of male elephants, suggest an engagement predominantly with the same.

The definite purpose of modelling these figurines in terracotta, however, eludes us. Apart from being used as motifs for designing household items as at Bhita, Arundhati Banerji's (1994, 196) suggestions regarding the use of these figurines as votive offerings, or as possible play items for children, seem plausible.

5

TRUNK CALLS IN ANTIQUITY

THE ELEPHANT IN TEXTUAL TRADITIONS

HAVING RECONSTRUCTED THE STORY ARCHAEOLOGY tells us, I now sift through the reams of ancient Indian literature in search of the pachyderm. Its all-pervasive presence in textual sources not only stands testimony to the close interface with the animal, but also necessitates being selective while trying to retrieve facets of its association with humans.

THE VEDIC CORPUS

It has been argued that the Vedas and their commentaries make merely passing references to the elephant, which was natural of a people more familiar with the horse, who were still in the process of acclimatizing to the culture of the land they had migrated to. Such incidental references then scarcely permit firm conclusions (Sukumar 2011, 35). The argument itself seems to be based on the premise of the non-indigenous origins of the Aryans, which incidentally is also a matter of dispute.

Notwithstanding the limitations posed by the nature of this corpus, I will try and assimilate some glimpses of the animal in Vedic literature. Though this has been attempted before in scholarly writings on the elephant in ancient India (Singh 1965; Sukumar 2011), I undertake the enterprise not only in order to take my narrative forward, but also to link it up with subsequent periods of Indian history, since the concern central to this chapter is the evolving human relationship with the mega herbivore across millennia.

The elephant pervades the Vedic corpus as the *mṛga hastin*, *ibha*, *gaja*, *nāga*, *vāraṇa*, and *śukladant* (Macdonell and Keith 1958, II: 571–2). The multiple epithets used for the animal are indicative of its qualities and habits. Its ability to move freely over all terrains—hill, valley, or water—earned it the soubriquet *nāga*. It is known as *gaja* because of its thunderous roar, and *vāraṇa* because it wards off hostile forces (Iyer 1977, 44).

Chronologically, the obvious starting point is the *Ṛgveda*, which mentions the *mṛga hastin* (I. 64, 7), *mṛga hasti* (IV. 16, 14), *ibha* (I. 84,17; IX. 57, 3), and *vāraṇa* (VIII. 33, 8; X. 40, 4) to denote the elephant. In the text as well as later, *mṛga* is interpreted as the generic term for wild beast (Macdonell and Keith 1958, II: 171). However, it has also been argued that *mṛga* may not necessarily imply 'wild', but may simply mean an 'animal' or may imply a distinction between 'wild' and 'domesticated' elephants (Singh 1965, 75). The possibility of both having been known cannot be ruled out if we keep in mind that the elephant is rarely bred in captivity, and that even the ones that are trained, start out as wild beasts. Therefore, even if the early Vedic people were describing them as wild beasts, it does not imply that they were not tamed or trained (Singh 1965, 77).

Conflicts over semantics notwithstanding, based on the compound name *mṛga hastin/hasti* (animal with a hand), there has been a tendency to emphasize the 'newness of the elephant to the Vedic Indians' (Roth cited in Macdonell and Keith 1958, II: 171–2). The appellation, Sukumar (2011, 35) contends, was the most natural portrayal of the animal to a people hitherto unfamiliar with a creature possessing a proboscis or trunk. He also argued that in contrast to the Indus civilization, where the elephant and not the horse figured prominently on seals, the latter and not the former received greater importance in the *Ṛgveda* as also the three later Vedas. This contrast he again attributes to the non-indigenous origins of the Vedic people. The sense that emerges from these writings seems to suggest that the intense familiarity of the Harappans with the pachyderm gave way to a certain curiosity regarding the animal armed with a trunk.

In this context, it is important to bear in mind the necessity of separating terminology from actual content. As we shall see, references to the elephant in the *Ṛgveda* give an inkling of a considerably close acquaintance with the animal, which would have been scarcely possible if we work with the hypothesis of its suggested novelty. Moreover, this premise which seems to draw strength from the epithets referring to the animal as *mṛga hastin/hasti*, that is the animal with a hand, falls weak when we remind ourselves of the fact that the Indus script remains undeciphered. After all, till we know how the animal was referred to then, it seems somewhat simplistic to harp on its novelty only on the basis of terminology particularly when the content of the references clearly suggests otherwise.

The contention, then, is that not only was the animal fairly well known, but also considerably used. After all, as shall be seen, if the size and strength of the pachyderm were reasons enough for its usage as a common simile to extol the might and virtues of different deities, it is not difficult to envision ways in which this strength would also have been employed. References scattered across the ten *maṇḍalas* can be put together as clues, but before I set out to do that, let us first consider how the *Ṛgveda* perceived the animal, and how closely it appears to have been observed.

The power of the Maruts is likened to that of wild elephants. 'Like the wild elephants ye eat the forests up when ye assume your strength among the bright red flames' (*Ṛgveda* I. 64, 7; Griffith 1963a, I: 88). So is the might of Indra. 'Thou a wild elephant with might invested, like a dread lion as thou wieldest weapons' (*Ṛgveda* IV. 16, 14; Griffith 1963a, I: 412). Similarly, his prowess is likened to that of a wild elephant who rushes on this way and that, mad with heat (*Ṛgveda* VIII. 33, 8; Griffith 1963a, II: 172), referring probably to elephants in *musth*.

Musth is a physiological and behavioural state in male elephants when the animals are known to become temperamental and aggressive. It is characterized by the secretion of a fluid from the *musth* gland behind the eye (Sukumar 2011, 50). Clearly then, the animal, despite its supposed novelty, must have been observed closely and frequently enough to facilitate familiarity with its sexual behaviour.

One can also reflect on the purposes for which the animal would have been trained or the ways in which it was exploited. Curiously, Vedic literature though fairly well-acquainted with elephants, does not mention ivory or objects made of it anywhere (Dwivedi 1976, 17). However, in an indirect reference, a gambler laments that though he would not like to go back to gambling, yet the brown dice enticed him into giving in (*Ṛgveda* X. 34, 5: Griffith 1963a, II: 428). Dwivedi contends that the brown dice in this context could only suggest dice made of ivory which is known to take on a brownish colour with constant handling over time, and can, therefore, be considered as the earliest reference to an ivory object in Vedic literature. The suggestion seems quite plausible if we remind ourselves of the prolific use of ivory in the Indus civilization.

Another clue that can and has been teased out (Singh 1965, 75–6) is in view of a reference where Agni is told to 'go like a mighty king with his attendants. Thou, following thy swift net, shootest arrows: transfix the fiends with darts that burn most fiercely' (*Ṛgveda* IV. 4, 1: Griffith 1963a, I: 398). Though Griffith's translation of the verse does not mention the elephant, the word which has solicited attention is *ibhena*. In view of *ibha* signifying an elephant in Sanskrit literature, the verse, according to Singh (1965, 76), is prophetic of things to come, and marks the debut of the elephant as a royal mount in Vedic literature, and the context seems to be the battlefield. Elsewhere, the Aśvins are likened to 'two mad elephants bending their forequarters and smiting the foe' (*Ṛgveda* X. 106, 6; Griffith 1963a, II: 549). This interestingly has also been treated as an early reference to elephants being used in war (Dikshitar 1944, 167). In yet another passage, Indra is told that the one who earned his enmity 'will battle like a stately elephant on a hill' (*Ṛgveda* VIII. 45, 5; Griffith 1963a, II: 190).

Notwithstanding such allusions, the evidence for elephants being used in war at this time does not seem compelling enough. What can, nevertheless, be surmised from these early references is that people were acquainted with at least wild elephants during the latter half of the second millennium BCE though their familiarity with captive ones also cannot be ruled out if we consider that the Harappans had already tamed them (Sukumar 2011, 37).

I now turn to the three later Vedas generally ascribed to the first half of the first millennium BCE to garner more telling references to the animal. It is interesting to note that gradually, in later language, *mṛga* came to be divorced from both *hastin* and *vāraṇa*, and only the adjectives survived to represent the animal (Macdonell and Keith 1958, II: 172). Whether what was becoming the written word was a manifestation of a growing familiarity with the pachyderm is worth considering.

Elephant-keeping seems to have become a regular profession as the *Vājasaneyī Saṃhitā* (XXX, 11; Griffith 1899, 256) refers to the elephant-keeper, *hastipa*. There is even a mention of the sacrifice of elephants (*Vājasaneyī Saṃhitā* XXIV, 29; Griffith 1899, 221).

The *Taittirīya Saṃhitā* (VI. 4. 5.6; Keith 1914) of the *Kṛṣṇa Yajurveda* states that 'there are three animals which take by the hand, man, the elephant, and the ape.' This seems to be a clear recognition of the importance of the trunk for the animal, echoes of which shall also be seen in Buddhist canonical literature.

Anatomically, as a union of the nose and the upper lip, the trunk's dexterity, and its ability to perform various functions possibly explains the human fascination with elephants. Functions attributed to the organ include feeding, watering, dusting, smelling, touching, communication, and lifting, together with its use as a weapon of defence and offence. These make it an indispensable tool for the animal in its everyday living (Eltringham 1991, 44, 46).

The *Atharvaveda*, considered the latest of the four Vedas, is not far behind in giving us significant glimpses of the animal. A reference to how flies anger an elephant (*Atharvaveda* IV. 36. 9: Whitney 1962, I: 210) seems a comfortable reminder of a sight one can see even today, and assures us of being closer to reality. Another verse describes how 'the elephant strains foot with foot of the she-elephant' (*Atharvaveda* VI. 70. 2; Whitney 1962, I: 333). The reference, though obscure, probably refers to the capture of males with the help of female elephants (Singh 1965, 77). A house presumably raised on four posts is compared to a standing female elephant (*Atharvaveda* IX. 3.17; Whitney 1962, II: 527). The simile is said to have been inspired by the common sight of an elephant standing tirelessly near the poet's residence (Singh 1965, 77). And the observation that all these references adequately indicate a close association with the animal seems to persuasively deliver the verdict that 'it was captured, tamed, used and studied' (Singh 1965, 77).

While there are repeated allusions to the urge to acquire the splendour of the animal (*Atharvaveda* III.22.1; Whitney 1962, I: 126; *Atharvaveda* III.22.5: Whitney 1962, I: 127), far more revealing is another hymn for the taming and training of an elephant for a king to ride on, where the mega herbivore seems to have journeyed its way to becoming the superior (*atiṣṭhāvant*) of the wild beasts (*Atharvaveda* III. 22.6; Whitney 1962, I: 127). The only one of its kind in the Vedic corpus, the hymn is a celebration of the strength and majesty of the animal.

Sukumar (2011, 38) argues that if the elephant had begun to serve as the royal mount then it is also likely to have been trained for use in battle, though a quest for explicit references to such deployment still seems wanting in results. Singh quite reasonably (1965, 78) attributes this gap in later Vedic literature to the character of this corpus, arguing that given its purpose there was not much scope for discussing either the wings of the army or dispositions of battle in the Brāhmaṇas or Upaniṣads.

We, thus, have to wait for a few more centuries, at least till the sixth century BCE for indisputable evidence regarding the arrival of the animal on the battlefield. Till then, glimpses from the remaining later Vedic corpus, Brāhmaṇas, and Upaniṣads await retrieval.

The *Aitareya Brāhmaṇa* (IV.1; Haug 1963, II: 256), for instance, alludes to the docility and obedience of the animal in complying with orders when it mentions an elephant (once tamed) returning to its owner when called out to by him. The same text is also said to furnish us with the earliest allusion to the art of ivory-carving when it says, 'They recite the *Śilpas*. These are the works of art of the gods (*devaśilpānyateṣām*); in imitation of these works of art, here is a work of art accomplished, an elephant (*hastin*), a goblet (*kamso*), a garment (*vāsaḥ*), a gold object (*hiraṇyam*), a mule chariot (*aśvatarī rathaḥ*); a work of art is accomplished in him who knows thus' (cited in Chandra 1957–9, 5). Clearly, argues Dwivedi (1976, 18), since the passage is a direct reference to the arts and crafts of the period such as bronze-casting, weaving, goldsmithing, and chariot-making, *hastin* in the given context would mean the art of ivory-carving.

Of far greater significance is the *Sāma Vidhāna Brāhmaṇa's* (III. 6.11 cited in Singh 1965, 78) allusion to elephants among the four divisions of the army—*hastyaśvarathapadātayaḥ.*

An interesting injunction also occurs in the *Śatapatha Brāhmaṇa* (III. 1, 3, 2–4; Eggeling 1885, II: 12–13), which speaks of the eight sons of Aditi who were born from her body, with seven of whom she went to the gods, and Mārtaṇḍa, who she had brought forth unformed as 'a mere lump of bodily matter, as broad as it was high', was cast off. The seven sons of Aditi, anxious about the fate of their less fortunate sibling, devised a way to ensure his survival. 'That which was born after us must not be lost: come, let us fashion it. They fashioned it as man is fashioned. The flesh which was cut off him and thrown down in a lump became the elephant: hence they say that one must not accept an elephant as a gift, since the elephant has sprung from man.'

Nevertheless, that we are in a milieu where the animal had emerged as a symbol of status, to be owned as well as gifted, is evident when the *Chāndogya Upaniṣad* (VII. 24; Müller 1879, I: 123) spells out the parameters of eminence being the possession of many cows and horses, elephants and gold, slaves and women, fields and houses. Similarly, the *Aitareya Brāhmaṇa* (VIII. 23; Haug 1963, II: 526) refers to Bharata, the son of Duṣyanta, performing the *aśvamedha*, and gifting 10,000 tusked elephants (all decked with gold), and 10,000 slave girls to brāhmaṇs on the occasion, conveying that the practice of gifting elephants was well established in later Vedic times.

GLIMPSES FROM THE PALI *TIPIṬAKA* AND BEYOND

By the sixth century BCE, the elephant had decisively arrived on the landscape of historic India. Inextricably linked to the literature of this period, the animal treads many realms. What follows is not a comprehensive review of all references to the elephant in early Buddhist literature, but an endeavour to weave together strands that underline the multiple images of the animal. Though these images cannot be synthesized into a singular world view, they certainly serve the purpose of conveying the numerous layers of the human–elephant interface.

A fleeting glance at the traditions embedded in the Pali *Tipiṭaka* sufficiently conveys the centrality of the elephant to Buddhist tradition and precept. A deeper perusal, however, reveals that the allusions are often part of larger philosophical discourses. Nevertheless, these can be considered as being broadly reflective of a political, economic, and social milieu where the animal was integrated into nearly every sphere of life. Scattered across this corpus are imageries of the animal acknowledging its might and grandeur as well as recognizing its destructive potential. Contexts vary, but the elephant as a point of reference remains constant.

An obvious entry point would be the oft-recounted narrative of the conception of Buddha. Reclining on her royal couch, Māyā falls asleep, and dreams of the four guardians of the world lifting her in her couch, carrying her to the Himalaya mountains, and placing her under the great sal tree where she is approached by their queens. They take her to the lake of Anotatta, bathe, dress, anoint, and deck her before leading her to a golden mansion where they spread a heavenly couch, and lay her down. The future Buddha, who had taken the form of a white elephant, then approached her with a white lotus in his silvery trunk. Uttering a loud cry, he entered the golden mansion, and after paying obeisance to his mother's couch thrice, struck her on her right side and seemed to enter her womb (*Nidāna Kathā*; Davids 1973, 149–50).

At the same time, the Buddha's association with the pachyderm transcends the conception scene. One is reminded, for instance, of Devadatta's ploy aimed at annihilating him. The fierce elephant Nālāgiri is deployed for the task. As the Lord is seen entering Rājagaha for alms, the mahouts instructed by Devadatta, set the rogue elephant free, which then charges towards its target. People watch with bated breath, some anticipating the worst, while others believing that 'the bull elephant would come into conflict with the elephant (among men)' (*Vinaya Piṭaka*, *Cullavagga*; Horner 1963, V: 273). Eventually, the animal is pacified into submission with loving kindness by the Lord who strokes his forehead saying, 'Do not elephant, strike the elephant (among men), for painful, elephant, is the striking of the elephant (among men), For there is no good bourn, elephant, for a slayer of the elephant (among men) when he is hence beyond' (*Vinaya Piṭaka*, *Cullavagga*; Horner 1963, V: 274).

Implicit here is a clash of might. If Nālāgiri is a mighty elephant, the Lord is its counterpart among men, capable of conquering cruelty with kindness as well as taming the untameable and leading people to marvel at what had transpired: 'Some are tamed by stick, by goads and whips. The elephant was tamed by the great seer without a stick, without a weapon' (*Vinaya Piṭaka, Cullavagga*; Horner 1963, V: 274).

It may also be meaningful to sift through some of the discourses and narratives that provide us with glimpses of the distinct elephant culture that had developed by this time. The *Sāmañña Phala Sutta* of the *Dīgha Nikāya* (Davids 1899, I: 68) furnishes us with crucial evidence of social conditions in the Ganga Valley by listing ordinary occupations which include elephant-riders (*hatthārohā*), cavalry (*assārohā*), charioteers (*rathikā*), archers (*dhanuggahā*), slaves (*dāsakaputtā*), cooks (*āḷārikā*), barbers (*kappakā*), bath attendants (*nahāpakā*), confectioners (*sudā*), garland makers (*mālākārā*), washermen (*rajakā*), weavers (*pesakārā*), basketmakers (*naḷakārā*), potters (*kumbhakārā*), and many others.

Trainers and keepers of elephants appear frequently. The *Majjhima Nikāya* (Horner 1954, I: 223) mentions an elephant-tracker entering an elephant forest, and painstakingly tracking an elephant following its large and broad footprints. He follows the footprints closely, but does not conclude that they belong to a bull-elephant till he actually spots it himself since he is aware that the elephant forest was home also to stunted she-elephants, she-elephants with stumpy tusks, and the footprints could well be theirs. Following the footprints, he finally comes upon the bull-elephant at the root of a tree or in the open, walking or standing or sitting or lying down. In the same *Nikāya*, Pessa, the son of an elephant trainer, asserts his ability to deal with an elephant under training (Horner 1957c, II: 5).

Evidently, the knowledge of elephants must have been sufficiently mastered for elephant training and tracking to be recognized as professions demanding specific skills. Some, we are told, were trained even in elephant lore (*Vinaya Piṭaka, Suttavibhaṅga*; Horner 1949, I: 317; *Aṅguttara Nikāya*; Hare 1961, III: 231). We, thus, find ourselves in a milieu that assigned the care, keep, and treatment of its elephants to specialists.

The elephant and its association with royalty is a theme persistently hinted at. From crowding a king's women's quarters or the royal harem (*Vinaya Piṭaka, Suttavibhaṅga*; Horner 1957b, III: 74), to being the royal mount par excellence, the pachyderm forays into many realms of imperial life. We find King Pasenadi of Kosala riding his bull-elephant Ekapuṇḍarīka (One Lotus) (*Majjhima Nikāya*; Horner 1957c, II: 297), while Seta, another magnificent elephant belonging to the king, is regarded a beauty, a treat for the eyes (*Aṅguttara Nikāya*; Hare 1961, III: 243). Buddha interrogates Prince Bodhi regarding his skills in elephant-riding and handling a goad, and is responded to in the affirmative (*Majjhima Nikāya*; Horner 1957c, II: 281). 'Elephant, horse, chariot, bow and sword skill' qualified a prince to aspire to be the anointed king (*Aṅguttara Nikāya*; Hare 1961, III: 117). Expertise in handling elephants along with horses and chariots, thus, seem virtues commonly associated with and expected of the princes and royalty.

The *Aṅguttara Nikāya* (Woodward 1962, II: 120–1) tellingly expounds on four qualities by virtue of which a rājah's elephant becomes 'worthy of the rājah, a possession of the rājah, is reckoned an attribute of a rājah'. The animal is a listener, a destroyer, a bearer, and a goer. As a listener, he lends his ears to and targets whatever task is assigned to him by the elephant trainer, even if it is a task he has not performed before. In the capacity of a destroyer, he enters the battlefield, and destroys elephant and mahout, horse and rider, chariot and driver, and footman. As a bearer, he enters battle, and bears the blows of spears, swords, arrows, and axes as also the thundering of the drum and kettledrum. As a goer, he takes the direction the trainer turns him to, even if he has not been there before.

Away from specifically royal contexts, the *Piṭaka* discourses often interweave allusions that seem to suggest that the keeping and gifting of animals like the elephant was common enough. That the ownership of elephants was not an exclusively royal prerogative can be presumed from a reference to Soṇa Koḷivisa, a merchant's son, who is said to have given up eighty cartloads of gold, and a herd of seven elephants when he 'went forth from home into homelessness' (*Vinaya Piṭaka, Mahāvagga*; Horner 1962, IV: 245).

Elephants as part of battle scenes are also no longer aberrations. The *Saṃyutta Nikāya* mentions tamed elephants serving both as royal animals as well as beasts of war. The army we are told, means elephants, horses, chariots, and infantry, where an elephant has twelve men, a horse has three men, a chariot has four men, and the infantry has four men (*Vinaya Piṭaka, Suttavibhanga*; Horner 1957a, II: 375). Elsewhere, it is emphasized that 'troops in array means: so many elephants, so many horses, so many chariots, so many infantry . . . a review means a review of elephants, a review of horses, a review of chariots, a review of infantry' (*Vinaya Piṭaka, Suttavibhanga*; Horner 1957a, II: 380). For whatever the statistics mean, what is clearly discernible in these references is an implicit acceptance of elephants as an indispensable wing of the army. The *Majjhima Nikāya* alludes to

> a king's bull-elephant whose tusks are as long as a plough-pole, massive, finely bred, whose home is the battle-field, and who, when going forth to battle, uses his forelegs, uses his hindlegs, uses the forepart of his body, uses the hindpart of his body, uses his head, uses his ears, uses his tusks and uses his tail, protecting only his trunk. (Horner 1957c, II: 88)

Apart from evoking the image of a formidable war animal, it is worth underlining ancient India's familiarity with the morphological features of the animals it interacted with, and put to use. It is known that the trunk is the most sensitive part of an elephant's body, and that it is almost impossible for an animal with a damaged trunk to survive.

Far removed from the royal palace and the din of the battlefield, the *Aṅguttara Nikāya* (Woodward 1962, II: 36–7) stands out in drawing our attention to a dimension scarcely addressed in recent writings—the elephant and lion animosity. It transports us closer to wilderness by alluding to a lion having just emerged from his lair. He stretches himself, and surveys the four quarters in all directions. He then lets out his mighty roar, evoking mortal fear in all creatures and surges ahead in search of prey. Of interest here is the fact that of all the animals intimidated by the lion, the elephant is specifically mentioned: 'whatsoever rājah's elephants in village, town, or rājah's residence are tethered with stout leathern bonds, such burst and rend those bonds asunder, void their excrements and in panic run to and fro. Thus . . . is the lion, king of beasts, over brute creatures: of such mighty power and majesty he is.' Another reference in the same *Nikāya* (Hare 1961, III: 95) refers to a lion striking a blow at an elephant.

The corpus of early Buddhist literature also offers us significant pointers to the economic worth of the animal. Apart from doubtful ones that suggest the use of elephant rugs (*hatthathara* interpreted as an elephant rug since it occurs in a sequence with *assatthara, rathatthara*—horse rug, chariot rug) (*Vinaya Piṭaka, Mahāvagga*; Horner 1962, IV: 257; *Majjhima Nikāya*; Horner 1957c, II: 259, 297) as well as the use of its tusk as a peg in the wall (*Vinaya Piṭaka, Suttavibhaṅga*; Horner 1949, I: 80; *Vinaya Piṭaka, Cullavagga*; Horner 1963, V: 155,161, 214), there are more credible indicators.

Ivory, for instance, is specifically qualified as elephant ivory (*hatthidanta)* in the *Suttavibhaṅga*, which recounts the case of an ivory-worker who offers to provide monks with needle cases. Violating the moderation expected of monastic life, the monks ask for many needle cases so that the ivory-worker ends up being unable to make other goods for sale, and his wife and children suffer. They are eventually rebuked by the Buddha (*Vinaya Piṭaka, Suttavibhaṅga*; Horner 1957b, III: 87–8). Elsewhere, monks are permitted to use ointment boxes made of bone, ivory, horn, reed, bamboo, a piece of stick, lac, crystal, copper, and of the centre of a conch-shell (*Vinaya Piṭaka, Mahāvagga*; Horner 1962, IV: 276).

Similarly, the *Mahāvagga* refers to the death of a king's elephant. Since food was scarce at that time, the people used elephant meat, and gave the same to *bhikkhus* when they came asking for alms. This displeased some, and the matter was reported to the Buddha who forbade the *bhikkhus* from eating elephant meat (*Vinaya Piṭaka, Mahāvagga*; Horner 1962, IV: 298–9). Horner, in this context, reasons that it appears from this passage that people did not as a practice consume elephant meat, and doubts if it was consumed even in times of scarcity, or merely used for offering to monks. Pertinently, she argues that those condemning the monks for it could not have consumed it themselves. Nevertheless, what needs to be kept in mind is that restrictions like these could perhaps only have been imposed in the wake of existing social realities.

The same applied for the flesh of the horse, dog, snake, lion, tiger, panther, bear, and hyena, all of which came to be strictly forbidden (*Vinaya-Piṭaka, Mahāvagga*; Horner 1962, IV: 299–300). It is interesting to note the reasons behind the monks being denied the flesh of these animals. Elephant and horse meat, for instance, were to be abstained from because they were symbols of royalty (*Vinaya-Piṭaka, Mahāvagga*; Horner 1962, IV: 299).

With these aspects of the human–elephant relationship came the growing familiarity with the animal and its behaviour. The well-known parable that revolves around blind men and an elephant deserves mention because of its engagement with the anatomy of the animal though it engages with a larger philosophical position. Buddha recounts how once upon a time the king brought together all the blind men in the city, and also had an elephant brought in. Each of them was asked to touch the animal and describe it. What follows is an interesting jumble, where the one who had felt the head said it was like a water-pot, the one who had felt the ear said it was like a winnowing basket, the tusk was said to be like a ploughshare, the trunk like a plough-handle, the body like a granary, the legs like a pillar, its back like a mortar, its tail like a pestle, and its bristles like a broom. Each of them is so certain regarding his inference that they end up exchanging blows (*Dīgha Nikāya*; Davids 1899, I: 187–8). Notable here is the correlation of each body part with a component of the food-producing and processing sequence suggesting perhaps a symbolic connection between the animal and material prosperity.

The similes and metaphors woven around the pachyderm also clearly give away a keen sense of the animal and its ways. For instance, the similes of the elephant's footprint (*hatthipadopama*) and the elephant's look are the most frequently employed ones. The *Mahāhatthipadopama Sutta* attributed to Sāriputta contends that just as the footprints of all creatures fit into the elephant's footprint, which is pre-eminent for its size, so do all the right states of mind fall within the Four Noble Truths (*Majjhima Nikāya*; Horner 1954, I: 230). Similarly, the *Mahā Parinibbāna Suttanta* of the *Dīgha Nikāya* parallels Buddha's gaze with an 'elephant look' (Davids 1966, II: 131). It is interesting to note the contention that according to Buddhaghoṣa, the Buddhas on looking backwards were accustomed to turning the whole body around like elephants because the bones in their neck were more firmly fixed than those of ordinary men (Davids 1965, III: 131).

Often through the similes it is the anguish of a captive elephant and its longing for the wilderness that is effectively conveyed like 'an elephant that burst each strap and rope' (*Dīgha Nikāya*;

Davids 1966, II: 308), 'an elephant fretted by hook' (*Dīgha Nikāya*; Davids 1966, II: 301), 'as tusker rends his rotten bands' (*Sutta Nipāta*, i, 2; Hare 1947, 5).

We also have the moving allusion to an elephant called Dhanapālaka, who, with his temple running with sap, is difficult to hold and does not eat a morsel when bound (*Dhammapada*; Müller and Fausböll 1881, 77–8). Similarly, emphasizing the virtues of solitude, one is implored to be like the bull-elephant (*Vinaya Piṭaka*, *Mahāvagga*; Horner 1962, IV, 500; *Majjhima Nikāya*; Horner 1959, III: 199):

> Finding none apt with whom to fare,
> None in the well-abiding rapt,
> As rājah quits the conquered realm,
> fare lonely as bull-elephant in elephant jungle.
>
> Better the faring of one alone,
> there is no companionship with the foolish,
> fare lonely, unconcerned, working no evil,
> as bull-elephant in elephant-jungle.

Distant from the cares of service, the discourses also often subtly weave in locales where the animal is found at its natural best, for instance, plunging, crossing over, and trumpeting on the banks of the river Sappinikā near Rājagaha (*Vinaya Piṭaka, Suttavibhaṅga*; Horner 1949, I: 189). The Buddha alludes to a great pond in a stretch of forest with bull-elephants living near it. They plunged into the pond, tugged out lotus fibres and stalks with their trunks, washed and rendered them free of mud, chewed, and swallowed them. Their actions were replicated by young elephant calves (*Vinaya-Piṭaka*, *Cullavagga*; Horner 1963, V: 189).

Again, there is the allegory of a large bull-elephant surrounded by elephants, cow-elephants, calves, and sucklings. The animal ate grass cropped by his herd, while they ate bundles of branches which he broke off. He drank muddied water, and when he crossed over at a ford, cow-elephants went pushing against his body. His desire for solitude makes him leave the herd, and the comfort of being alone dawns upon him. The musings attributed to the bull-elephant in this context could only have emanated from a close observation of the social organization of the animals. The Buddha sums up the animal's reasoning saying, 'Herein agreeth mind with mind, of sage and bull-elephant of plough-pole tusks, since each delights in forest (solitude)'(*Vinaya Piṭaka, Mahāvagga*; Horner 1962, IV: 504).

The Jātaka Tales

The narratives here are again diverse, and abound in references to the pachyderm. The mega herbivore (figuring in twenty-four tales) is almost as popular as the monkey, which appears in twenty-seven of the 550 Jātakas that narrate the stories of the Buddha's former births. We set out in search of these varied glimpses of the human–elephant interface, which clearly seems to have transcended class barriers.

A point of entry can be found in the *Alīnacitta Jātaka* (156), which recounts the bond between an elephant and a group of carpenters who had treated and cured the creature when a splinter of wood had pierced its foot, causing it to swell and fester. Humbled by their care and compassion, the elephant decides to render itself useful to them by offering its services like pulling up trees for them or rolling up the logs they chopped. Having approached old age, he entrusts his responsibilities to his son, a magnificent white elephant who would obediently work for the carpenters, and also play with their children.

Thus it was, till the king of the land was told about the elephant, and he set out to acquire the creature for himself. Having reached his destination, the king expressed his desire and the carpenters readily obliged. The elephant, however, refused to move and on being questioned, demanded that the king pay the carpenters what they had spent on him. This having been done, he still stood firm demanding clothes for the carpenters, their wives, and their children. Then finally, with a last look at the people he had loved so dearly, he bid adieu, and departed with the king.

Chakravarti (1993, 55–6) observes that this narrative brings together animals, humans, and the king in a three-cornered relationship with each other. She reasons that the elephant receives the affection of the poor carpenters 'establishing an enduring reciprocal bond between the high in the animal world and the humble in the human world'. The animal's contestation is also perceived as a critique of kingly power that challenged the injustice of the king's action in appropriating for himself everything that was magnificent in the land. Chakravarti also sees it as a plea for justice and reciprocity between the king and his people 'even as the loving world of reciprocity between the carpenters and the elephant is being destroyed by the king'.

The narrative, however, does not end with the departure of the elephant with the king, and goes on to recount the life of the animal thereafter. Having reached the capital city, the elephant is appointed for the king's own riding, and taken good care of. Sometime later, the queen consort conceived the Bodhisatta, but before she could deliver him, the king died. The news was not broken to the king's elephant, fearing it would break his heart. In the meantime, the ruler of Kosala, having heard of the king's death, planned a conquest, which on request was put on hold till the time the queen delivered. The battle commenced with the birth of the prince, and seeing the tide turning against them, the queen accompanied by the courtiers, visited the state elephant's stable, and placed the baby at its feet. Acquainting him with what had transpired, the queen urged the pachyderm to either kill the child or save the kingdom. With filial affection, the elephant stroked the baby with his trunk, and lifted him on his own head, and then lamenting, brought him down and put him in his mother's arms. With his armour and caparison, the mighty creature emerged from the city gate trumpeting and frightening opponents away, and seized the king of Kosala by his topknot only to drop him at the feet of the young prince (Rouse 1969a, II: 13–17).

More than once, the Bodhisatta is born as the son of an elephant trainer. In the *Saṁgāmāvacara Jātaka* (182), for instance, he is taught everything related to the training of elephants. Having mastered the art, he trained the state elephant to perfection. The king he served was an enemy of the king of Benares, and decided to wage a war against him. Having clad himself in armour, the king mounted his armed state elephant, and challenged his enemy, who rose to the occasion and put up a brave defence that intimidated the state elephant so much that it refused to approach the scene of action. Thereupon, the trainer reminded him, 'Son, a hero like you is quite at home in the battle-field! . . . There stands the gate before thee now: why dost thou turn and yield? Make haste! Break through the iron bar, and beat the pillars down! Crash through the gates, made fast for war, and enter in the town!' The trainer's words appealed, and the elephant complied. 'Winding his trunk about the shafts of the pillars, he tore them up like so many toadstools: he beat against the gateway, broke down the bars, and forcing his way through entered the city and won it for his king' (Rouse 1969a, II: 63–5).

The growing intimacy with the animal, however, brought with it undercurrents of friction. Once, when the Bodhisatta was born as a tree-spirit near the Himalayas, the king put his state elephant in the elephant trainers' hands to be taught to stand firm. They tied the elephant to a post, and with goads set about training him. Unable to bear the agony, the animal broke down the posts, put the trainers to flight, and fled to the Himalayas where he lived in eternal fear of humans, and of being subjected to the same torture if they caught him. It took the Bodhisatta's calming words to take his fears away. 'Fear'st thou the wind that ceaselessly The rotten boughs doth rend alway? Such fear will waste thee quite away!' (*Dubbalakaṭṭha Jātaka* 105; Chalmers 1969, I: 246–7).

This unease and mounting suspicion between humans and the pachyderm seems to run across other tales as well that portray the elephant as a symbol of endurance, gratitude, and self-restraint, often juxtaposed with human treachery and ingratitude. The *Sīlavanāga Jātaka* (72) centres around the Bodhisatta's encounter with a forester from Benares who came to the Himalayas in search of the implements of his craft but ended up losing his way. Weeping in despair, the man is pacified, fed, and shown the way out by the Bodhisatta born as a mighty white elephant. The good deed is, however, paid for through treachery. The forester, having noted the landmarks on his way out, returns to the city, and walking through an ivory-workers' market, and seeing ivory being given diverse shapes and forms, is tempted into finding out the worth of the tusk of a living elephant. Having learnt that it would be worth a lot more than a dead elephant's tusk, he returns repeatedly to the elephant asking him for his tusks till the time he had sawed out even the stumps of his tusks. The tree-fairy's words are a poignant reminder of the ruthlessness of human greed: 'Ingratitude lacks more, the more it gets; Not all the world can glut its appetite' (Chalmers 1969, I: 174–7).

Then we have the kind elephant of the celebrated *Chaddanta Jātaka* (514), which is a poignant tale of revenge woven around an elephant hunt (Francis 1969, V: 20–31). Similarly, the narrative in the *Daḷhadhamma Jātaka* (409) revolves around a she-elephant owned by the king. The animal went a hundred leagues in a day, did the duties of a messenger for the king, and in battle fought and crushed the enemy. As long as she was useful, great honours were bestowed upon her, but as she grew old and weak, the king withdrew all favours, and the animal was left unprotected, thriving on grass and leaves in the forest. She was even given away to a potter for carrying cow dung. It took the Bodhisatta's intervention in the capacity of the king's minister to make the king realize and rectify the folly caused by his ingratitude (Francis and Neil 1969, III: 233–5).

The tales effectively seam together images of the pachyderm in its benevolent as well as malevolent forms. The animal could also be an adversary, particularly the rogue elephant, which knew neither loyalty nor compassion.

In the *Indasamānagotta Jātaka* (161), despite the Bodhisatta's warning against the wisdom of doing so, the headstrong Indasamānagotta goes on to nurture a pet elephant that gradually attains maturity. One day when its master and his accomplices were away in the forest to gather roots and fruits, the animal goes into a frenzy. 'Destruction to this hut! . . . I'll smash the water-jar! I'll overturn the stone bench! I'll tear up the pallet! I'll kill the hermit, and then off I'll go!' And so he does, taking off to the jungle, and waiting there for the group to return. The master carrying food for his pet led the group, but was soon accosted by the elephant, which rushed out from the thicket, seized him in his trunk, dashed him to the ground, and with a final blow on the head crushed him to death. Then madly trumpeting, he scampered into the forest (Rouse 1969a, II: 28–9). Similar is the image of the rampaging elephant in the *Culladhanuggaha Jātaka* (374). Here, however, the animal is dealt with by an archer who strikes him in the head with an arrow that pierces the animal through and through and comes out at the back of his head, killing it on the spot (Francis and Neil 1969, III: 144–8).

More interesting facets emerge in other narratives. The *Kāka Jātaka* (140) recounts as part of a larger narrative how once upon a time when the Bodhisatta was born as a crow, a great fire broke out in the elephant stables. Many elephants were so badly burnt that it was beyond the skills of elephant doctors to cure them (Chalmers 1969, I: 300). The *Susīma Jātaka* (163) emphasizes the knowledge of elephant lore along with that of the three Vedas as a prerequisite for conducting the elephant festival (*hastimaṅgala*). On the day of the festival, 'a hundred elephants were set in array, with golden trappings, golden flags, all covered with a network of fine gold; and all the palace courtyard was decked out' (Rouse 1969a, II: 31–4). Sukumar (2011, 96) pertinently observes that the tradition is reminiscent of the present-day Pooram festival in Thrissur, Kerala.

Apart from references to such elephant festivals, the worship of an elephant figure (*hatthimaha)* is referred to in the *Māti Posaka Jātaka* (455). It recounts the birth of the Bodhisatta as a magnificent white elephant surrounded by a herd of 80,000 elephants. He would send the sweetest fruits for his blind mother through his fellow elephants, but one day learnt that they never reached her, and were all eaten up on the way. Disillusioned, he decided to leave the herd, and take his mother to a place where he could look after her.

One day, a forester lost his way, and as he wailed audibly in desperation, the Bodhisatta rushed to his rescue, and carried him on his back out of the forest. The man, however, intended otherwise, and while on his way out on the kind elephant's back, put markers along the way, only to get to the city and inform the king, who had recently lost his state elephant and had deputed his men to look for a befitting replacement, about this splendid creature. The king sent elephant trainers along with the man to capture the mighty being.

Seeing the troop approaching, the elephant anticipated what lay ahead, but decided to keep his calm. His soliloquy in his hour of distress is telling: 'I can scatter even a thousand elephants; in anger I am able to destroy all the beasts that carry the army of a whole kingdom. But if I give way to anger, my virtue will be marred. So to-day I will not be angry, not even though pierced with knives.' The creature is eventually captured and brought to the king, but freed later when he sees the animal's anxiety over being separated from his blind old mother. The king has a stone image made in the form of the Bodhisatta, and people gathered at the site every year to celebrate the elephant festival (Rouse 1969b, IV: 58–61).

The *Kurudhamma Jātaka* (276) explicates, the association of the animal with rains by recounting how the state elephant of the Kuru kingdom was sought to be acquired in order to give the kingdom of Kaliṅga respite from the long drought it had been experiencing (Rouse 1969a, II: 251–60). The elephant also occurs in the interpretation of omens. A queen dreaming of sitting on the back of a white elephant was taken as a premonition of the king's death, while if she dreamt of touching the moon while riding such an elephant it portended hostile kings attacking her husband (Gokhale 1974, 112–13). Significantly, early Pali literature also refers to a group of people known as *hatthivatikas*, who adhered to a belief system where the elephant was considered sacred (*Mahāniddesa* 89, 92–3 cited in Narain 1991, 34). Thus, transcending purely economic and political roles, the animal had evidently become a part of contemporary folklore as well as assumed quasi-religious implications (Gokhale 1974, 112).

Milindapañha or *The Questions of King Milinda*

The canvas laid out by the *Milindapañha* captures the pachyderm in multiple contexts. King Milinda sets out to review his mighty army with its fourfold array comprising elephants, cavalry, bowmen, and soldiers on foot (Davids 1890, I: 7). References to war elephants recur, and it is emphasized that their management is to be learnt. Princes are urged to know about elephants, horses, and chariots (Davids 1890, I: 247).

However, what is most remarkable are the allusions to the ways of the animal. We have, for instance, elephants, when they are in rut, sucking up water in their trunks, and pouring it out over their towering bodies (Davids 1894, II: 90). Elsewhere, the insinuation is to the games the animal delights in, playing in the water, and plunging into glorious lotus ponds full of clear pure cool water, and covered with lotuses of different hues (Davids 1894, II: 336). Similarly, though devised as metaphors, the similes revolving around the pachyderm serve to provide poignant glimpses regarding the natural behaviour of the animal. 'Like a strong bull who's burst the bonds that bound him, Or elephant who's forced his way through jungle' (Davids 1894, II: 285).

We also encounter diametrically opposite images of the pachyderm in the form of the man-slaying Dhanapālaka (Davids 1890, I: 297), as also the fanciful image of the gentle and all-white elephant king Uposatha showing signs of rutting in three places of his body (Davids 1894, II: 128).

THE JAINA WORLD

Having explored the Buddhist world view regarding the pachyderm, I now move on to examine the position of the animal within the Jaina tradition by cobbling together allusions to it in the four texts chosen. Mulling over the references collated, it does not take much to arrive at the conclusion that despite a strong presence within the tradition, the pachyderm certainly does not seem to have enjoyed the pre-eminence accorded to it by Buddhism.

Nevertheless, as in Buddhism, the animal is appropriated as the harbinger of news regarding the arrival of a great being. The *Kalpa Sūtra*, which contains a biography of Vardhamāna Mahāvīra among other Tirthaṅkaras, recounts the conception of the twenty-fourth *jina* by Devānandā, a brāhmaṇ woman. The event is marked by fourteen auspicious dreams, the first of which is a magnificent elephant followed by a bull, a lion, the anointing of the goddess Śrī, a garland, the moon, the sun, a flag, a vase, a lotus lake, the ocean, a celestial abode, a heap of jewels, and a flame. However, since a *jina* could only be born in a family of illustrious descent, Indra, the bestrider of the elephant Airāvata, deputes his commander Hariṇegamesi to transport the embryo from the womb of Devānandā to the womb of Triśalā of kṣatriya lineage (*Kaḷpa Sūtra*; Jacobi 1884, I: 219–29). The order once executed, Triśalā also sees the fourteen dreams, the first of which is a white four-tusked elephant—an equivalent of Airāvata, the best elephant of the king of gods. The animal once again appears in her dream of Śrī (whose fleshy thighs are likened to the trunk of the pachyderm) being anointed with water from the trunks of elephants. Also her dream of a celestial abode is decorated with pictures of animals including elephants (*Kalpa Sūtra;* Jacobi 1884, I: 231–8).

Apart from being integral to the nativity scene, other allusions occur mostly in the form of similes employed in the context of a mendicant's life. 'Valorous like an elephant, strong like a bull, difficult to attack like a lion', Mahāvīra himself is said to have battled the rigours of mendicancy (*Kalpa Sūtra*; Jacobi 1884, I: 261).

Particularly significant are the similes drawing upon the image of the war elephant within this religious tradition that was perhaps the strongest advocate of *ahiṃsā*. Patiently enduring the assaults that came his way, Mahāvīra proceeded on the path of righteousness like an elephant at the head of the battle (*Ācārāṅga Sūtra*; Jacobi 1884, I: 85). Like a stately elephant at the head of the battle, a monk is not to be deterred by calamitous adversities (*Uttarādhyayana*; Jacobi 1895, II: 110). Similarly, it is said that just as an elephant at the head of the battle kills the enemy, so does a hero in self-control conquer the internal foe (*Uttarādhyayana*; Jacobi 1895, II: 11).

Embedded within such figures of speech are often glimpses of the ways of the animal. For instance, a learned monk is said to have no parallel, just as a strong and irresistible elephant of sixty years surrounded by his females has no equal (*Uttarādhyayana*; Jacobi 1895, II: 48). A person longing for sensual pleasures despite knowing the Law is likened to an elephant sinking in quagmire, which can see the raised ground yet cannot get to the shore (*Uttarādhyayana*; Jacobi 1895, II: 60). Further, elaborating on tender feelings that cannot be easily overcome even by monks, the *Sūtrakṛtāṅga* (Jacobi 1895, II: 264) carries an allusion to keepers always following a newly caught elephant, and a cow which has just calved never going far from the calf.

In more worldly contexts, the paraphernalia accompanying a king out on a hunt included horses, elephants, chariots, and footmen (*Uttarādhyayana*; Jacobi 1895, II: 81). While horses, elephants, subjects, a town, and a seraglio, power, and command are the means enjoyed by a king (*Uttarādhyayana*; Jacobi 1895, II: 101), elephants, horses, chariots, and cars with pleasure trips are also used to lure a monk who leads a pious life.

While there are images of human–animal coexistence in the form of horses, cattle, buffaloes, and elephants housed in stables within villages and towns (*Ācārāṅga Sūtra*; Jacobi 1884, I: 184), encounters with vicious ones are cautioned against. Monks and nuns on their begging tours are told to steer clear if they see fierce animals including the elephant approaching them (*Ācārāṅga Sūtra*; Jacobi 1884, I: 100).

The Jaina cosmos seems to have had its own rationale for classifying animals. The *Uttarādhyayana* (Jacobi 1895, II: 223), for instance, qualifies that quadrupeds and reptiles are the two kinds of terrestrial animals, and that the former are of four kinds: solidungular animals as horses and others, biungular animals as cows and others, multiungular animals as elephants and others, and animals having toes with nails as lions and others. The sense that emerges is that the pedal structures of these animals formed the basis of this classification. What is even more significant is the fact that within a tradition that dogmatically distanced itself from life forms from fear of causing them injury, there was a deeply entrenched awareness of their anatomies.

It is perhaps reasonable to close this section with the mention of a philosophical debate encountered in the *Sūtrakṛtāṅga* (Jacobi 1895, II: 418). It refers to a socio-religious group called the *hastitāpasa* that claimed to kill one big elephant every year, and to live off it in order to spare the lives of other animals. Evidently opposed to the Jaina emphasis on safeguarding all life forms, the reference itself, Sukumar (2011, 113) pertinently contends, is a reminder of contemporary debates over which sections of society are greater champions of conservation.

THE DHARMAŚĀSTRAS

Marshalling references to the pachyderm in this corpus of the 'legal' literature of ancient India leads us to the Dharmasūtras, where Baudhāyana (1:1.10) and Vasiṣṭha (3.11), for instance, compare an uneducated brāhmaṇ with an elephant made of wood, or a deer made of leather, all being so only in name (Olivelle 2003, 197, 363). Prescribing the code of conduct for a student, Baudhāyana restricts a student having attained puberty from greeting his brother's wives and the young wives of his teacher, but does not consider it an offence to sit with them in a boat, on a rock, plank, elephant, terrace, mat, or in a carriage (1:3.34; Olivelle 2003, 203). In the same text, the dust coming from carriages, horses, elephants, grain, and cows is considered auspicious, while the dust of brooms, dogs, goats, sheep, donkeys, and garments is deemed dirty (2:6.34; Olivelle 2003, 265).

The gait of an elephant is used as a simile in the *Mānava Dharmaśāstra* (3.10), which recommends the selection of a bride who walks like a goose or an elephant (Olivelle 2006, 108). A trainer of elephants is an unfit invitee during ancestral offerings (3.162; Olivelle 2006, 116). The text prohibits Vedic recitation on a horse, tree, elephant, donkey, or a camel (4.120;

Olivelle 2006, 130). It gives the king a pre-emptive share in whatever a man wins as war booty—chariot, horse, elephant, parasol, money, grain, livestock, women, all goods, and base metal (7.96; Olivelle 2006, 159)—and enjoins him to fight with chariots and horses on level ground, and with boats and elephants in marshy lands (7.192; Olivelle 2006, 164). It also declares that lost property that is recovered shall remain in the care of competent officials, and the king should have thieves caught in connection with its disappearance executed by an elephant (8.34; Olivelle 2006, 168). The king is also told to kill without hesitation those who break into the treasury, the armoury, and those who steal elephants, horses, or chariots (9.280; Olivelle 2006, 204). Stealing a deer or an elephant entailed rebirth as a wolf (12.67; Olivelle 2006, 233). Similarly, killing donkeys, horses, camels, deer, elephants, goats, sheep, fish, snakes, or buffaloes were sins that caused a man to be of a mixed caste (11.69; Olivelle 2006, 218). The penance for killing an elephant required the giving of five black bulls (11.137; Olivelle 2006, 222).

Keeping in mind the normative character of these texts, the allusions do not convey much. Nevertheless, it may be worthwhile to highlight certain observations which emerge. The animal is a closely guarded form of wealth, stealing or killing of which is an offence entailing severe punishment. At the same time, it is an instrument of war, useful particularly on marshy terrains where horses and chariots are rendered ineffective. It is also employed for executing recalcitrant elements. Ironically, however, though the creature is deemed auspicious, its trainer is not.

ECOLOGY, THERAPEUTICS, AND ELEPHANT MEAT: THE LEGACIES OF CARAKA AND SUŚRUTA

The elephant (*gaja*) (along with the rhinoceros) is one of the animals belonging to the *ānūpa* category in the *Caraka Saṃhitā* (I. 27.39; Sharma 1994, I: 197). Interestingly, the text goes much beyond recommending just the use of elephant meat to the use of its nail, dung, urine, and more.

To begin with, the flesh of the pachyderm (*gajamāṃsa*) is said to cure debility, and as in the case of the rhinoceros, the physician is urged to give the well-spiced meat of the animal in order to cure emaciation (*Caraka Saṃhitā* VI. 8, 154; Sharma 1994, II: 156). The excreta of the animal (*kuñjara-purīṣa*) is used to treat piles through inhalation of the fumes when burnt (*Caraka Saṃhitā* VI.14.51; Sharma 1994, II: 231). Even in the case of *kapha*, the juice of the excrements of the elephant (*gajaśakṛt*) or of the ass, horse, camel, boar, and sheep mixed with honey was to be taken (*Caraka Saṃhitā* VI.17.116; Sharma 1994, II: 297). The text also alludes to a *gajamuktika* or an elephant pearl to be worn as a talisman as an antidote against poison (*Caraka Saṃhitā* VI. 23.252; Sharma 1994, II: 389). Inhalation of the fumes of an elephant's nail when burnt is said to be effective in the treatment of epilepsy (*Caraka Saṃhitā* VI.10.40; Sharma 1994, II: 175). Similarly, the urine of the animal is used in a concoction to treat leucoderma (*Caraka Saṃhitā* VI. 7.169; Sharma 1994, II: 142). Additionally, elephant urine is said to be salty, and beneficial for the retention of urine and faeces, and disorders of *kapha* and piles (*Caraka Saṃhitā* I.1.102; Sharma 1994, I: 12).

Similarly, in the *Suśruta Saṃhitā* (I. 46.49–50; Bhishagratna 1963, I: 487–8), the pachyderm is one of the *kūlacara* quadrupeds belonging to the *ānūpa* category. Its flesh is 'very dry and dessicant, of hot energy and corrupting to bile, sweet, acid, salty, the *gaja* calms phlegm and wind' (cited in Zimmermann 1987, 107).

It is interesting to juxtapose this prescription with proscriptions (discussed earlier) discouraging the consumption of elephant meat. In a sense, then, we have injunctions which are diametrically opposite in their approach towards the animal. On the one hand, we have world views where the animal is not to be consumed or violated since it is a symbol of royalty and a strategic resource; on the other, we have the medical world, which employs its own rationale while sanctioning the use of elephant meat to cure health maladies.

THE ELEPHANT IN THE EPICS

One sets out with a nearly certain sense that compared to our first two heroes—the one-horned rhino and the striped tiger—a search for the pachyderm in the two epics will be a far more daunting enterprise. References abound, and this comes as no surprise given the close association of the animal with human cultures.

The Rāmāyaṇa

We are led into Ayodhyā, the unassailable city. Resplendent in its glory, it is an impregnable fortress with a deep moat impossible to cross, unassailable by its enemies, and filled with horses, elephants, cows, camels, and donkeys (*Bālakāṇḍa* 1.5.13; Goldman 1984, 135). Along with the finest horses, the city has powerful rutting elephants like mountains, born in the Vindhya Hills, and the Himalayas (*Bālakāṇḍa* 1.6.21; Goldman 1984, 137). Full of bull elephants always in rut, elephants were of the *bhadramandra*, *bhadramṛga,* and *mṛgamandra* breeds descended from the cosmic elephants Añjana and Vāmana (*Bālakāṇḍa* 1.6.22–3; Goldman 1984, 137). Robert Goldman (1984, 291) points out that commentators associate these breeds with the major mountain ranges like the Himalaya, Vindhya, and Sahya respectively. They also describe the elephants according to their physical features, Bhadras having contracted limbs, Mandras being stocky, and Mṛgas, lean and long-limbed. The animals mentioned are, therefore, of mixed breeds.

As the narrative progresses, the princes complete their education, and just when old Daśaratha is contemplating their marriage, sage Viśvāmitra arrives urging the king to give him Rāma to kill the two menacing *rākṣasas*, Mārīca and Subāhu, obstructing his sacrificial ritual. Daśaratha is anguished by the thought of letting his most beloved son, still a minor, walk into a situation fraught with peril. He pleads with the seer to spare Rāma, and instead offers himself along with all the four branches of his army—an obvious reference to the presence of elephants in the line of battle (*Bālakāṇḍa* 1.19.9; Goldman 1984, 164). Eventually, the king relents, and grants leave to Rāma and Lakṣmaṇa, who set out with the sage traversing difficult terrains including a forbidding forest infested with elephants along with lions, tigers, and boars (*Bālakāṇḍa* 1.23.13; Goldman 1984, 170). The task first set for Rāma is to eliminate Tāṭakā, the mother of Mārīca, who lives in the forest and is a terrifying *yakṣa* (a class of benevolent/malevolent nature spirits regarded as the attendants of Kubera, the god of wealth) woman with the strength of a thousand elephants (*Bālakāṇḍa* 1.23.24; Goldman 1984, 171).

In the company of the seer, the princes hear many tales including one about the sacrifice performed by their ancestor Sagara, where the sacrificial horse gets stolen. While tearing up the earth in search of the horse thief, the sons of king Sagara encounter Virūpākṣa, one of the mountainous elephants supporting the earth, and the guardian of the East (*Bālakāṇḍa* 1.39.12, 15; Goldman 1984, 200). As they proceed to tear up the South, West, and North, they encounter the great elephants Mahāpadma, Saumanasa, and the snow-white Bhadra respectively supporting the earth (*Bālakāṇḍa* 1.39.16–17, 19, 21; Goldman 1984, 200–1).

Viśvāmitra's negotiation with Vasiṣṭha for the acquisition of the latter's cow Śabalā involves an offer of 14,000 elephants with gold chains for girth and neck along with goads of gold (*Bālakāṇḍa* 1.52.17; Goldman 1984, 225). Lavish gifts including beautifully adorned troops of elephants, horses, chariots, and foot soldiers are bestowed upon the daughters of Janaka as they embark on their marital journeys with the sons of Daśaratha (*Bālakāṇḍa* 1.73.3–4; Goldman 1984, 263).

Displaying his valour, the able Rāma, proficient in training and riding horses and elephants (*Ayodhyākāṇḍa* 2.1.23; Pollock 1986, 80), accomplishes the mission assigned to him, and returns to Ayodhyā. Exuding all the makings of a righteous ruler, the scion of the Ikṣvāku dynasty while returning from battle on chariot or elephant, would inquire about the welfare of the people of the city (*Ayodhyākāṇḍa* 2.2.25–6; Pollock 1986, 83). With long arms, and immense strength, he carried himself like a bull-elephant in rut (*mattamātaṅga*) (*Ayodhyākāṇḍa* 2.3.11; Pollock 1986, 85).

Just as preparations begin for the consecration of Rāma, fortune takes a turn in the form a heartless intervention by Kaikeyī. The hunchback Mantharā instigated the queen and urged her to save prince Bharata, who according to her was threatened by Rāma like the leader of an elephant herd attacked by a lion in the forest (*Ayodhyākāṇḍa* 2.8.25; Pollock 1986, 98). Once aware of Kaikeyī's demands, Daśaratha tries to persuade her to relent, caressing her like a great bull-elephant in the wilderness would caress his cow wounded by the poisoned arrow of a hunter lurking in the forest (*Ayodhyākāṇḍa* 2.10.4; Pollock 1986, 102).

Meanwhile, unaware of the turn of events, arrangements for the consecration continue, and we find a white bull, a white horse, and a majestic rutting elephant fit to be a king's mount, ready for the occasion (*Ayodhyākāṇḍa* 2.13.10; Pollock 1986, 109). For Rāma, however, the decision is made—to comply with the order that he be exiled, and Bharata be made king to rule the land with its treasures of horses, chariots, and elephants (*Ayodhyākāṇḍa* 2.16.26; Pollock 1986, 115).

He proceeds to his mother Kausalyā's chamber heaving sighs like an elephant (*Ayodhyākāṇḍa* 2.17.1; Pollock 1986, 118). As Rāma speaks, Lakṣmaṇa keeps his head lowered, and shaking his hand as an elephant shakes its trunk from side to side, lets his head fall on his chest (*Ayodhyākāṇḍa* 2.20.4; Pollock 1986, 126). Raging at the misery that had befallen his faultless sibling, he resolves to turn back fate which was 'running wild, like a careering elephant beyond control of the goad, in a frenzy of rut and might' (*Ayodhyākāṇḍa* 2.20.15; Pollock 1986, 127). Resolute about undoing the revocation of his rightfully deserving brother's consecration, he decides to take on his enemies in a way that would make the earth impassable, knee-deep in the trunks, flanks, and heads of elephants along with horses and men hacked off by his sword (*Ayodhyākāṇḍa* 2.20.28; Pollock 1986, 128). Elephants would drop to the ground with the blows of his sword, and he would ply arrows in their vitals as also those of men and horses (*Ayodhyākāṇḍa* 2.20.29, 31; Pollock 1986, 128).

Undeterred by all the lamenting, Rāma proceeds to execute his father's orders, and as he does so, anguished Kausalyā bestows her blessings on his journey. Among prayers for warding off other perils, she wishes that huge elephants may not harm her tenderly raised son, nor should lions, tigers, bears, boars, or the ferocious horned buffalo (*Ayodhyākāṇḍa* 2.22.7; Pollock 1986, 131).

Rāma takes leave of his mother, and returns to his residence. Sītā, oblivious of what had transpired, notices the pallor on his face, and is puzzled by the absence of the royal elephant leading his procession (*Ayodhyākāṇḍa* 2.23.15; Pollock 1986, 133). Before embarking on his journey into the forest, Rāma gives away his riches along with his elephant named Śatruṃjaya, 'worth a thousand others'—*gajasahasreṇa* (*Ayodhyākāṇḍa* 2.29.9; Pollock 1986, 144). He refuses the amenities his grieving father wishes to send with him reasoning thus, 'Would a man who gives away a prize elephant cling to the cinch-belt? Why cherish the rope once the animal is gone?' (*Ayodhyākāṇḍa* 2.33.3; Pollock 1986, 153).

Gloom takes over as Rāma leaves Ayodhyā. The heroic prince himself is tormented by the sight of his wailing father and mother. It is like a goad tormenting an elephant (*Ayodhyākāṇḍa* 2.35.31; Pollock 1986, 159). In the city, afflicted by turmoil and confusion, elephants become wild and unruly, and horses raise a loud clangour (*Ayodhyākāṇḍa* 2.35.16; Pollock 1986, 158). Women cry and despair like the wailing of cow-elephants when their great bull is captured (*Ayodhyākāṇḍa* 2.35.25; Pollock 1986, 158). Elephants let their fodder drop, while cows would not suckle their calves (*Ayodhyākāṇḍa* 2.36.9; Pollock 1986, 160).

As they travel on with Sītā, they come across enchanting mountain landscapes, forests, rivers, lotus-covered ponds, thronged by water-birds, herds of antelopes, rutting horned buffaloes, boars, and elephants butting at trees (*Araṇyakāṇḍa* 3.10.2–4; Pollock 1991, 104). They notice woodland trees by the hundreds, all in flower, their barks rubbed raw by elephants' trunks (*Araṇyakāṇḍa* 3.10.72–4; Pollock 1991, 109).

While the wonders of nature in the forest enthral, its perils in the form of dreaded beasts and demons spell doom. Spurning the advances of Śūrpaṇakhā invites the wrath of the dreadful *rākṣasas*. Dūṣaṇa is slain, and is like a spirited elephant with shattered tusks (*Araṇyakāṇḍa* 3.25.9; Pollock 1991, 140), while Rāma and Triśiras clash like lion and elephant (*Araṇyakāṇḍa* 3.26.10; Pollock 1991, 142).

Eventually Rāvaṇa himself sets out to avenge his sister's humiliation with the evil intention of abducting Sītā. Approaching her in disguise, he praises her beauty comparing her smooth thighs to an elephant's trunk (*Araṇyakāṇḍa* 3.44.18–19; Pollock 1991, 180). He asks her how she is all alone in the deep forest, and how is it that she does not fear the mighty elephants which run wild, maddened by rut (*Araṇyakāṇḍa* 3.44.29; Pollock 1991, 181). Revealing his identity, he tries to lure Sītā by boasting about the opulence of Laṅkā, full of beautiful gardens, and crowded with elephants, horses, and chariots (*Araṇyakāṇḍa* 3.46.12; Pollock 1991, 186). Having carried her away, he forces her to see his lavish dwelling, which had magnificent windows made of ivory and silver (*Araṇyakāṇḍa* 3.53.10; Pollock 1991, 203). Firm-willed Sītā is then handed over to horrifying *rākṣasa* women who are instructed to break her will as is done with a cow-elephant (*Araṇyakāṇḍa* 3.54.28; Pollock 1991, 207).

In the meantime, Rāma returns to the hermitage along with Lakṣmaṇa to find his beloved gone. He searches frantically asking all and sundry, and failing to find her, sinks in despair like an elephant in deep mud (*Araṇyakāṇḍa* 3.59.12; Pollock 1991, 216). Finally, it is through their encounter with the giant Kabandha, who with his long arms could pull in and devour creatures of the forest like lions, tigers, elephants, and deer, that Rāma and Lakṣmaṇa are shown the way to securing an alliance with Sugrīva (*Araṇyakāṇḍa* 3.67.14; Pollock 1991, 234).

Having slain Vālin, Rāma secures an ally in Sugrīva. However, the act (of striking when Vālin was not looking) itself earns the unblemished epic hero the epithet of a 'mad elephant' that had broken the fetters of good conduct (*Kiṣkindhākāṇḍa* 4.17.38; Lefeber 1994, 90). Wounded Vālin, like an elephant mired in mud, cries out in distress (*Kiṣkindhākāṇḍa* 4.18.45; Lefeber 1994, 94). Killed by a shaft strung from Rāma's bow, he is like an elephant struck by an arrow (*Kiṣkindhākāṇḍa* 4.20.1–3; Lefeber 1994, 97).

The pachyderm, at times, also serves to make descriptions more picturesque. Rāma's description of the rainy season frequently employs the animal to create a dramatic effect. The clouds resembling majestic mountain peaks are said to emit deep rumblings like maddened elephants trumpeting excitedly in battle (*Kiṣkindhākāṇḍa* 4.27.20; Lefeber 1994, 113). The clouds adorned with lightning banners are likened to elephants ready for battle (*Kiṣkindhākāṇḍa* 4.27.28; Lefeber 1994, 113). 'Wandering in mountain forests, the majestic elephant in rut who has set out on his way eager for battle, runs back upon hearing the roar of the clouds, thinking he hears a rival elephant' (*Kiṣkindhākāṇḍa* 4.27.29; Lefeber 1994, 113). Similarly, as the rainy months pass by, the clouds are like elephants no longer in rut, their violence having been calmed (*Kiṣkindhākāṇḍa* 4.29.24; Lefeber 1994, 119). Very often elephants are associated with mountains, trumpeting, and grazing around in a broad and well-watered place (*Kiṣkindhākāṇḍa* 4.41.14; Lefeber 1994, 148).

Kiṣkindhā is crowded with monkeys as big as elephants, some with the strength of ten elephants, some ten times that, while some had the valour of a thousand elephants (*Kiṣkindhākāṇḍa* 4.30.25; Lefeber 1994, 123). Kubera's royal mount is the bull-elephant Sarvabhauma (*Kiṣkindhākāṇḍa* 4.42.34; Lefeber 1994, 153), while the king of elephants (*gajendra*) is Airāvata, Indra's elephant (*Kiṣkindhākāṇḍa* 4.11.15; Lefeber 1994, 75).

Elephants could be playful as well as fierce. The wilderness around sage Mataṅga's hermitage is a site where one hears the loud trumpeting of elephant calves coming to play in lake Pampā. At the same time, there were also fierce bull-elephants with hides dark as storm clouds which wandered in herds, or all alone, running with streams of ichor. They came to drink the cool water there, and then plunged back into the forests which were their homes (*Araṇyakāṇḍa* 3.69.27–9; Pollock 1991, 239).

Hanumān, the elephant among monkeys, sets out in search of Vaidehī (*Sundarakāṇḍa* 5.1.61; Goldman and Goldman 1996, 105). As he flew across, he was like the sun coursing through the sky, or like a mighty elephant with a gird bound around it (*Sundarakāṇḍa* 5.1.62; Goldman and Goldman 1996, 105). He is struck by the splendour of Rāvaṇa's palace, which dazzled with exquisite horses and elephants, and with great four-tusked bull-elephants resembling masses of white clouds (*Sundarakāṇḍa* 5.3.35–7; Goldman and Goldman 1996, 121). Filled with mahouts mounted on elephants, it was constantly traversed by chariots covered with the skins of lions and tigers, and adorned with ivory, gold, and silver (*Sundarakāṇḍa* 5.5.5–6; Goldman and Goldman 1996, 124). In the compound, he saw well-bred war elephants, trained like Airāvata himself in battle, and destroyers of enemy forces (*Sundarakāṇḍa* 5.5.29–30; Goldman and Goldman 1996, 126). With streams of rut fluid flowing, they were like rain clouds or mountains with their cascades. Trumpeting with a sound like thunder, they were invincible in battle (*Sundarakāṇḍa* 5.5.31; Goldman and Goldman 1996, 126).

Observing the magnificence of the palace surrounded by two, three, and four-tusked elephants (*Sundarakāṇḍa* 5.7.4; Goldman and Goldman 1996, 129), Hanumān proceeds to a beautiful hall embellished with panels inlaid with ivory (*Sundarakāṇḍa* 5.7.19; Goldman and Goldman 1996, 131). It was full of beautiful women in slumber with their garlands crushed and torn like flowering creepers crushed by mighty elephants in the great forest (*Sundarakāṇḍa* 5.7.44; Goldman and Goldman 1996, 132). Resting on the most magnificent bed in the hall, mighty Rāvaṇa seemed like an indomitable bull-elephant asleep on it (*Sundarakāṇḍa* 5.8.12; Goldman and Goldman 1996, 135). His arms, like iron beams, resembled elephant trunks (*Sundarakāṇḍa* 5.8.16; Goldman and Goldman 1996, 136), while he himself was like an elephant sleeping in the waters of the Ganges (*Sundarakāṇḍa* 5.8.26; Goldman and Goldman 1996, 137). Surrounded by his women, the lord of the *rākṣasas* is a reminder of a great bull-elephant surrounded by his cows in the forest (*Sundarakāṇḍa* 5.9.9; Goldman and Goldman 1996, 139).

When his search does not lead to Vaidehī, Hanumān turns to the *aśoka* grove as his last hope. Here he finally spots her surrounded by hideous looking *rākṣasa* women with faces, ears, noses, and feet like those of elephants among other animals (*Sundarakāṇḍa* 5.15.9–17; Goldman and Goldman 1996, 160). Bereft of her companion, Sītā herself was like an elephant cow captured by a lion, and separated from her herd (*Sundarakāṇḍa* 5.15.22; Goldman and Goldman 1996, 161). She was like a lotus pond fouled by the trunks of elephants (*Sundarakāṇḍa* 5.17.14; Goldman and Goldman 1996, 165), and heaving deep sighs like an elephant lord's captured mate, bound fast to a post, cut off from the leader of the herd (*Sundarakāṇḍa* 5.17.17; Goldman and Goldman 1996, 165).

Rebuking Rāvaṇa for his immoral advances, she urges to be reunited with Rāma like a young elephant cow with the lord of the elephants in the forest (*Sundarakāṇḍa* 5.19.17; Goldman and Goldman 1996, 170). Warning him of a miserable fate, she likens Rāma to a proud bull-elephant and Rāvaṇa to a rabbit (*Sundarakāṇḍa* 5.20.16; Goldman and Goldman 1996, 172).

Tormented by the fearsome *rākṣasa* women, hapless Sītā gets some reprieve from old Trijaṭā who narrates her dream as a forewarning of the impending doom. She recounts having seen Rāma moving through the sky in a celestial palanquin made of ivory (*Sundarakāṇḍa* 5.25.10; Goldman and Goldman 1996, 185). Then she saw the heroic princes mounted on a four-tusked elephant, which was later also mounted by Sītā (*Sundarakāṇḍa* 5.25.12, 14; Goldman and Goldman 1996, 185). Eventually, that grand elephant with Rāma, Lakṣmaṇa, and Sītā aboard is seen standing above Laṅkā (*Sundarakāṇḍa* 5.25.16; Goldman and Goldman 1996, 186).

Terrified by Rāvaṇa's sharp retort to her resilience to yield to his passion, Sītā is like the daughter of an elephant king attacked by a lion in the depths of the forest (*Sundarakāṇḍa* 5.26.1; Goldman and Goldman 1996, 187). Hanumān's encounter with her is the harbinger of joy as well as doubt. The mighty monkey, however, tries to dispel her suspicions by talking at length about Rāma. He tells Sītā that her husband can find no more peace than can an elephant harried by a lion (*Sundarakāṇḍa* 5.34.35; Goldman and Goldman 1996, 209).

The elephant and the lion are persistently pitted against each other in the *Yuddhakāṇḍa*, which is the longest episode in the epic capturing the tumultuous clash between the forces of Rāma and Rāvaṇa. The latter's son Atikāya threatens to assault Lakṣmaṇa with an arrow that would drink his blood the way a raging lion, the king of beasts, would drink the blood of a king among elephants (*Yuddhakāṇḍa* 6.59.55; Goldman et al. 2009, 317). Similarly, Sugrīva falls upon Kumbhakarṇa's son Kumbha like a swift lion upon an elephant roaming the mountain slopes (*Yuddhakāṇḍa* 6.63.29–30; Goldman et al. 2009, 339).

Also pervasive are allusions to war elephants that are distinct from the ones which along with other animals flee out of terror created by the clamour of the battlefield (*Yuddhakāṇḍa* 6.30.17; Goldman et al. 2009, 195). The presence of the pachyderm in the combat zone is conveyed in different ways. While Rāvaṇa's line-up comprises rutting war elephants, chariot horses, and armed foot soldiers, once fighting commences, the enraged monkeys drag about elephants with riders and chariots, slashing them with their fangs (*Yuddhakāṇḍa* 6.34.8; Goldman et al. 2009, 210). Striking the mount ensured access to the rider. Aṅgada leaps up to strike the mighty elephant of Mahodara (Rāvaṇa's brother) violently with his open hand, knocking out its eyes, and ripping out one of its tusks, making the bull-elephant trumpet loudly (*Yuddhakāṇḍa* 6. 58. 13–14; Goldman et al. 2009, 309). The field is soon transformed into a gruesome sight with rivers of blood flowing and carrying corpses. Elephants and horses formed their banks, warhorses their fish, and flagstaffs their trees (*Yuddhakāṇḍa* 6.81.9; Goldman et al. 2009, 391). Slain Rāvaṇa is mourned by his *rākṣasa* women like elephant cows when the leader of their herd is slain (*Yuddhakāṇḍa* 6.98.5; Goldman et al. 2009, 440).

Interwoven in the narrative are more dramatic and diverse glimpses. While it takes a force of a thousand elephants to finally rouse Kumbhakarṇa from his legendary slumber (*Yuddhakāṇḍa* 6.48.47;

Goldman et al. 2009, 265), struck by Rāvaṇa's arrows, the monkeys flee like elephants being burnt by the flames of a forest fire surrounding them (*Yuddhakāṇḍa* 6.84.3; Goldman et al. 2009, 401). Elsewhere, there is an allusion to an elephant being snared by men carrying ropes (*Yuddhakāṇḍa* 6. 10.6; Goldman et al. 2009, 142).

The Mahābhārata

Once again the pachyderm is inextricably woven into the narrative. Vengeance against the Pāṇḍavas is harboured as Duryodhana is mocked for being tricked by the marvels in Yudhiṣṭhira's palace, struck by wonders he had never seen in the City of the Elephant, Hastināpura (*Sabhāparvan* 2.43.2; Buitenen 1975, 2: 109). Envious of the opulence he had witnessed, he sets out to describe the tributes brought to the Pāṇḍavas by kings. Together with other riches such as cloth, silk, slaves, horses, the Vaṅgas and the Kaliṅga chieftains, the Tāmraliptas, Puṇḍrakas brought 'gold-caparisoned, cloth-decked elephants with pole-long tusks, lotus-dotted, towering like mountains, always rutting . . . each gave a thousand tuskers covered with armor, patient and well-bred'. Similarly, the king of the Śukaras brought hundreds of elephants, and so did Virāṭa, the Matsya king, who offered 2,000 gold-caparisoned rutting elephants. From the domain of Pāṃśu came twenty-six elephants, and so did twenty-six elephant-pulled chariots come from Yajñasena. Even the Siṃhalas offered hundreds of elephant trappings (*Sabhāparvan* 2.48.16–30; Buitenen 1975, 2: 118–19).

Along with the elephant, its ivory was also considered a gift befitting kings. Bhagadatta of Prāgjyotiṣa brought a jade vase, and swords with hilts of pure ivory (*Sabhāparvan* 2.47.14; Buitenen 1975, 2: 117). Costly seats, palanquins, and beds, gem-studded chariots inlaid with and mostly made of ivory were some of the other tributes brought (*Sabhāparvan* 2.47.29; Buitenen 1975, 2: 118).

In the decisive game of dicing, Yudhiṣṭhira stakes among other assets, a thousand *must* (implying *musth*) elephants with golden caparisons. Expounding their qualities, he declares, 'They are crowned with chaplets, hung with garlands, and spotted with lotus dots. They are well-trained mounts, fit for a king, and deaf to any noise on the battlefield. Their tusks are as long as poles, and each big-bodied bull has a herd of eight elephant cows. They are all bastion-battering tuskers, huge like mountains and monsoon clouds' (*Sabhāparvan* 2.54.8–10; Buitenen 1975, 2: 129).

Eventually all is lost, including Draupadī, the last stake for Yudhiṣṭhira. To amplify their indignation, the humiliated and despairing woman is invited by imprudent Duryodhana, who shows her his left thigh (an act impelling doom, as Bhīma vows to break the same thigh with his club) auspiciously marked by an elephant trunk and a thunderbolt (*Sabhāparvan* 2.63.11; Buitenen 1975, 2: 151).

Defeated and exiled, the Pāṇḍavas looking for abode decide to go to lake Dvaitavana. The landscape canopied with trees, and resonating with the voices of peacocks, *cakoras,* and cuckoos is enchanting. In the forest are to be found great herds of mighty elephants, leaders of herds flowing with rut together with herds of elephant cows (*Āraṇyakaparvan* 3.25.19; Buitenen 1975, 2: 269–70).

Though enchanting, the forest is also replete with perils in the form of the deadly beasts it harboured. When a tired group of merchants decides to spend the night there, little do they anticipate their fate. A herd of elephants on its way for a drink at a mountain stream found the caravan obstructing their way and ended up trampling the sleeping people. Some were killed by tusks and trunks, while others were trampled underfoot (*Āraṇyakaparvan* 3.62.8; Buitenen 1975, 2: 342).

Allusions capturing the physical features and natural behaviour of the pachyderm are not infrequent. A lotus pond with withered flowers, its birds chased away, and its mud stirred up and perturbed by elephant trunks conjures up a familiar image of the way elephants dabble with water (*Āraṇyakaparvan* 3.65.14–15; Buitenen 1975, 2: 346). Also, the near obsession with 'rutting' elephants can scarcely be missed, and is used multiple times to denote contexts involving might. The reference here is to the phenomenon of *musth* in elephants. As he sets out to fetch the lotus Draupadī coveted, Bhīma is said to proceed with 'the prowess of an elephant in rut, the vehemence of an elephant in rut, the copper-red eyes of an elephant in rut, able to ward off an elephant in rut' (*Āraṇyakaparvan* 3.146.31; Buitenen 1975, 2: 500). Again, in his onslaught on the *rākṣasa* Kirmīra, he is compared to an elephant whose temple glands have burst (*Āraṇyakaparvan* 3.12.56; Buitenen 1975, 2: 243). Elsewhere, we encounter elephants drinking at a mountain stream that was muddied by their flowing ichor (*Āraṇyakaparvan* 3.62.6; Buitenen 1975, 2: 342).

The pachyderm also serves as a frequently employed *upamāna* in similes. Extolling Kuntī's beauty, the Sun compares her to a woman who strides like an elephant—*gajagāminī* (*Āraṇyakaparvan* 3.290.14; Buitenen 1975, 2: 788). Elsewhere, Draupadī implores Bhīma to annihilate the smitten Kīcaka by pulling him out as an elephant pulls out a reed (*Virāṭaparvan* 4.21.28; Buitenen 1978, 3: 58). In the clash that ensues, they wrestle like two mighty bull-elephants over a cow (*Virāṭaparvan* 4.21.49; Buitenen 1978, 3: 59). The picture evoked can also be poignant. As Arjuna in the guise of Bṛhannaḍā makes his way like a rutting elephant to assist princess Uttarā's brother, the large-eyed girl followed him as a baby elephant follows its mother (*Virāṭaparvan* 4.35.9; Buitenen 1978, 3: 81). The battlefield is often the stage where the animal is used to allude to might or clashes between warriors. Launching arrows at each other, Droṇa and Arjuna are likened to two elephants goring each other with the points of their tusks (*Virāṭaparvan* 4.53.41; Buitenen 1978, 3: 106).

Outstanding, however, in the epic are the bold and vivid descriptions of elephants as war animals and the fantastic figures enumerating their strength in armies. Their presence on the battlefield is a given. The idea of the four wings of the army clearly finds expression, for instance, when we find the Pāṇḍavas sitting in the magnificent assembly hall. Nārada arrives, and is welcomed by Yudhiṣṭhira, who is then questioned on matters of policy, one of which is if their army was equipped with four kinds of troops (*Sabhāparvan* 2.5.53; Buitenen 1975, 2: 42). The reference here is obviously to elephants, chariots, cavalry, and infantry. We find the army of the Matsyas yoked with a large number of chariots, elephants, and horses, the assemblage also having foot soldiers (*Virāṭaparvan* 4.30.8; Buitenen 1978, 3: 74). 'Terrifying rutting elephants with riven temples, well-tusked sixty-year olds like gliding clouds, well-mounted by expertly trained riders', followed the Matsya king like moving mountains. Behind him followed his men with 8,000 chariots, 1,000 elephants, and 60,000 horses (*Virāṭaparvan* 4.30.26–8; Buitenen 1978, 3: 75).

Dhṛtarāṣṭra foretells the doom awaiting his sons pitted against the Pāṇḍavas: 'When you see the mountainous war elephants brought down by Bhīma, their tusks in pieces, their temples cut and dripping with blood, when you see them on the battlefield like pulverized mountains . . . you shall remember my words' (*Udgyogaparvan* 5.57.24–5; Buitenen 1978, 3: 325). The line-up of Kaurava warriors has among others Suyodhana with the emblem of an elephant on a field of gold (*Virāṭaparvan* 4.50.12; Buitenen 1978, 3: 102) and Karṇa with a red elephant's girth (*Virāṭaparvan* 4.50.15; Buitenen 1978, 3: 102).

Lances and hooks were used to goad the elephants (*Virāṭaparvan* 4.31.3; Buitenen 1978, 3: 75). Striking the pachyderm was an effective way of dislodging and targeting the mount (*Virāṭaparvan* 4.60.7–11; Buitenen 1978, 3: 115). Their trumpeting compounded the din of the battlefield (*Virāṭaparvan* 4.57.3; Buitenen 1978, 3: 111), while the ground itself was strewn with dead bodies, and elephants, horses, and other animals struck by shafts (*Virāṭaparvan* 4.57.8; Buitenen 1978, 3:111).

Much more can be garnered from the expanse of the epics, but the need to sum up makes us reflect on the images which emerge. What crucially sets the two epics apart is the 'more sober account of the use of elephants and more graphic descriptions of the forests and the denizens of the region' in the *Rāmāyaṇa* (Sukumar 2011, 40). Though Indra rides the elephant Airāvata (as against the horse in the Vedas), it is pointed out that the role of the animals in the final battle is not as pronounced as that in the *Mahābhārata*. Moreover, war elephants are mentioned frequently as a part of Rāvaṇa's forces, possibly suggesting a more regular use south of the Vindhyas (Sukumar 2011, 45).

On the other hand, the *Mahābhārata* has more elaborate descriptions of the use of elephants in war. Yet, it is interesting to see that the great heroes still ride chariots, and not elephants, and that the warriors who ride on elephants, like Bhagadatta on the side of the Kauravas and Ghaṭotkach on the side of the Pāṇḍavas, are not the main characters of the narrative (Sukumar 2011, 47). Though the origins of the war elephant are traced back to the early part of the first millennium BCE, the epic is perceived as the transitional period in ancient Indian warfare during the first millennium BCE from the predominance of horses and chariots to the decline in chariotry and the strengthening of elephant corps (Sukumar 2011, 47–8).

I would argue that while the treatment of the war elephant as pointed out by Sukumar serves as a point of departure between the two epics, the animal itself binds them in ways more than one. We find the elephant as *gaja*, *kuñjara*, *nāga*, *mātaṅga*, *vāraṇa*, *hastin*, *dvipa* straddling across the epic canvas as formidable war animals, harbingers of good as well as bad times, items of gift, symbols of status, wealth, and most frequently in the form of figures of speech (similes woven around the elephant are common to both the epics). There is a consciousness of elephant ecologies as well as behaviour. What also stands out is the massive preoccupation with 'rutting' elephants within the overall epic tradition. This perhaps emanates from the wonder evoked by the phenomenon of *musth*, which was generally not understood by biologists even until very recently.

ELEPHANTS AND EMPIRES: THE TESTIMONY OF THE *ARTHAŚĀSTRA*

'With redness formed, covered, with smoothened sides, with an even girth, with flesh spread evenly, level with the back-bone, and with a trough formed, these are appearances (of an elephant)' in the *Arthaśāstra* (2.31.17; Kangle 1963, II: 203). The animal occupies pride of place in this text as a critical resource in the economic and political edifice of the kingdom.

The text itself goes on to elucidate the reasons behind according such importance to the animal when it asserts, 'Victory (in battle) for a king depends principally on elephants. For, elephants, being possessed of very big-sized bodies and being capable of life-destroying activities, pound the troops, fortresses and camps of enemies' (*Arthaśāstra* 2. 2. 13–14; Kangle 1963, II: 68).

Further, the pact for unsettled land also reinforces the same while spelling out the economic utility of elephant forests, 'As between the usefulness of a material forest and an elephant forest, the use of a material forest is the source of all undertakings and able to secure plenty of stores, the reverse is the use of an elephant forest', say the teachers. 'No', says Kauṭilīya. 'It is possible to plant many material forests in many tracts of land, not so an elephant forest. For, the destruction of an enemy's forces is principally dependent on elephants' (*Arthaśāstra* 7.11.13–16; Kangle 1963, II: 412).

In these contexts, the motive behind the importance assigned to elephants is obvious. Clem Tisdell (2005, 9) argues that the political treatise mentions no religious or empathetic reasons for conserving and protecting elephants though some may possibly have existed. Sukumar (2011, 66) similarly underlines the lack of sentiment in the Mauryan desire for elephants, arguing that these creatures were invaluable in the army, and, hence, had to be protected. More recently, we have Trautmann (2015, 305) asserting that it 'was not out of a sentiment favouring wildlife, but purely for reasons of state, or king-centered self-interest of the most direct kind'. Undoubtedly, the impression looming large is that the centrality accorded to the pachyderm was primarily in view of its utility to kings particularly on the battlefield.

This, however, does not mean that one escapes contexts that go beyond purely political and strategic ones. For instance, despite being a treatise which was principally a manual of statecraft, compassionate perspectives emerge, for instance, when we take into account a reference where the text contends that an elephant calf may be caught for play (*Arthaśāstra* 2.31.16; Kangle 1963, II: 203). Also interesting is the directive regarding the creation of an animal park for the king's recreation. Stocked with tamed deer and other animals, the park was also to have wild animals with their claws and teeth removed, and male and female elephants and calves useful for hunting. On its border, the treatise also envisioned another animal park where all animals were welcomed as guests and given full protection (*Arthaśāstra* 2.2.3–4; Kangle 1963, II: 67).

The care and concern for the species begins from the level of the king, finding a place in the day's schedule prescribed for him. Instructed to divide his day into eight parts, the king is told to devote the seventh part of it to the review of elephants, horses, chariots, and troops (*Arthaśāstra* 1.19.6, 15; Kangle 1963, II: 51–2).

It is, however, the superintendent of elephants (*hastyadhyakṣa*) who looked into the nitty-gritty of elephant affairs from guarding elephant forests (*hastivana*) to overseeing food, fodder, and shelter for the animals (*Arthaśāstra* 2.31.1–18; Kangle 1963, II: 201–3). The physician, the trainer, the rider, the driver, the guard, the decorator, the cook, the fodder-giver, the foot-chainer, the stall-guard, the night-attendant, and so on formed the entourage looking after the animals (*Arthaśāstra* 2.32.16; Kangle 1963, II: 206).

Sukumar (2011, 66) contends that unlike in the later *Gajaśāstra* attributed to Palakāpya, the biology of elephants is not dealt with in the *Arthaśāstra.* While the exclusion is understandable in the case of a treatise which was principally a manual of statecraft, I would underline that the prescriptions, nevertheless, convey a thorough familiarity with the animal and its ways. This could only have been the result of very close association with it. One can take, for instance, the distinction between elephants inhabiting riverbanks and those from mountainous regions, when the text enjoins that a part of the tusks of elephants living on riverbanks should be cut every two and a half years, while in the case of those from mountainous regions, every five years (*Arthaśāstra* 2.32.22; Kangle 1963, II: 207). The awareness of such a distinction is significant, and emanates from an obvious difference underlined by Sukumar (2011, 66) in the growth rate of tusks in elephants inhabiting the plains, and those from the hills where the former were the larger elephants.

Similarly, the time for catching elephants according to the text is in summer (*Arthaśāstra* 2.31.8; Kangle 1963, II: 202). Water is a critical resource for elephants, so the assertion probably hints at the more likely presence of the animals around waterbodies particularly in summer months.

While a twenty-year-old elephant is advised to be caught (*Arthaśāstra* 2.31.9; Kangle 1963, II: 202), a forty-year-old is considered best, a thirty-year-old is middling, while a twenty-five-year old is the lowest (2.31.11; Kangle 1963, II: 202). On the other hand, the catching of a calf (*vikka*), an elephant with small tusks (*moḍha*), one without tusks (*makkaṇa*), a diseased elephant (*vyādhita*), a pregnant or a suckling female (*garbhiṇī, dhenukā hastinī*) was to be abstained from (*Arthaśāstra* 2.31.10; Kangle 1963, II: 202).

Here, we should take cognizance of Sukumar's (2011, 66) intervention that considering that one could not catch a juvenile elephant or a tuskless bull, and that a mature cow was likely to have a calf at heel or a suckling offspring practically throughout her reproductive lifespan suggests that the *Arthaśāstra* was recommending the selective capture of male tuskers that had just attained adulthood. The capture of young adult tuskers, he contends, best served the needs of an elephant force in the king's army, for the latter were the most sought after for their psychological effect on the battlefield.

Moreover, according to the text, the animals were to be assigned tasks in conformity with their appearance, and also keeping in mind the season (*Arthaśāstra* 2.31.18; Kangle 1963, II: 203). That elephants were used for various purposes comes forth in the way they were classified on the basis of the ones used in training, ones used in war, ones for riding, and the rogue elephant. Each of these was further categorized (*Arthaśāstra* 2.32.1; Kangle 1963, II: 204).

Curiously enough, amidst all these directives, most of which suggest a sound knowledge regarding the animal, one stands out as an anomaly. This is where the ration prescribed for the elephant is said to consist of, among other things, fifty *palas* of meat (*Arthaśāstra* 2.31.13; Kangle 1963, II: 202), when the animal, we know, is a herbivore feeding on grasses and other plant material. I would argue that rather than attributing this incongruity to ignorance regarding the animal, the inclusion of meat in the diet of this celebrated mega mammal was possibly more metaphoric than literal. Meat being a strength-giving food item, its inclusion in the diet of an elephant was perhaps considered apt keeping in mind its size and might.

Overall, there is little to defy the sense of a close acquaintance with the animal, with the treatise also suggesting a pressing concern for the protection and conservation of the species by envisioning elephant sanctuaries. This, for instance, clearly comes forth in the section dealing with the disposal of non-agricultural land where it is urged,

> On the border (of the kingdom), he should establish a forest for elephants guarded by foresters. The superintendent of the elephant-forest should, with the help of guards of the elephant forest protect the elephant-forest (whether) on the mountain, along a river, along lakes or in marshy tracts, with its boundaries, entrances and exits (fully) known. They should kill anyone slaying an elephant. To a person bringing in the pair of tusks of an (elephant) dying naturally, a reward of four paṇas and a quarter (shall be given). Guards of elephant forests, aided by elephant keepers, foot-chainers, border guards, foresters and attendants . . . moving with five or six female elephant decoys should ascertain the size of the herds of elephants, by means of indications provided by sleeping places, foot-prints, dung and damage caused to river-banks. They should maintain a record in writing of (every) elephant, (whether) moving in a herd, moving alone, lost from a herd, lord of a herd, and (whether) wild, intoxicated, cub or released from captivity. They should catch elephants

> whose outward marks and behaviour are excellent in the judgment of elephant-trainers. (*Arthaśāstra* 2.2.6–12; Kangle 1963, II: 68)

Having mentioned the battlefield, I now move on to take up the war elephant specifically, and follow the treatise in analysing the strategic role that it accorded to the animal in the combat line where it was supposed to stand in attendance, go round, march together, kill and trample, fight with elephants, assault towns, and fight in battle (*Arthaśāstra* 2.32.4; Kangle 1963, II: 204).

That the animal's efficacy on the battlefield had been amply realized by now clearly comes forth in the modes of fighting listed for the *caturaṅg* (four arms) of the army. Spelling out the functions of the four wings, the *Arthaśāstra* expounds as follows:

> Marching in the van, making new roads, halting places and fords, repelling as with arms, crossing and descending in water, remaining steadfast, marching forward and descending, entering difficult and crowded places, setting fire and extinguishing it, securing victory single-handed, reuniting broken ranks, breaking up unbroken ranks, protecting in a calamity, assault, frightening, causing terror, showing magnificence, capturing, setting free, breaking ramparts, gates and towers, bringing in and carrying away treasury, these are the functions of elephants. (X. 4.14; Kangle 1963, II: 514)

Apart from the animal's role in determining the strength of an army, we can also remind ourselves of the inclusion of the elephant in the list of wild animals whose skin, bones, bile, tendons, eyes, teeth, horns, hooves, and tails are considered valuable forest produce (*Arthaśāstra* 2.17.13; Kangle 1963, II: 149). Talking of war, it is equally important to refer to the mention of *nistrimśa*, *maṇḍalāgra*, and *asiyaṣṭi* as swords whose hilts were fashioned out of the tusk of the elephant along with the horn of the rhinoceros and buffalo, wood and bamboo-root (*Arthaśāstra* 2.18.12; Kangle 1963, II: 152). Additionally, combinations of skin, hooves, and horns of dolphin, rhinoceros, *dhenuka*, elephant, and bull served as armours (*Arthaśāstra* 2.18.16; Kangle 1963, II: 152).

What is also fascinating is the sense this manual of statecraft conveys of the range of habitats inhabited by elephants. Perhaps the first account of its kind, it names eight *gaja vanas* as sources of wild elephants, and grades them according to their quality.

Though the treatise does not mention the geographical boundaries of these forests (which are mentioned only in later texts), it qualifies that elephants from the Kaliṅgas and the Aṅgaras were the best, those from Cedi and Karūṣa and those from the Daśārṇas and the Aparāntas were of medium quality, and those from the Surāṣṭras and the Pañcanadas were of the lowest quality (*Arthaśāstra* 2. 2.15–16; Kangle 1963, II: 68–9).

Trautmann (1982, 263) contends that the list seems to have been fashioned from a north Indian perspective, for it does not include the elephant forests of Kerala or of the Karnataka–Tamil Nadu border where wild elephants are found to this day. The doubtful case of south India apart, he argues, what the list of the eight *gaja vanas* tells us is that central and eastern India were well stocked with wild elephants, and that the Indus Valley (probably the Punjab only, though this is uncertain) and the western coast, where they are no longer found, also had some. He also asserts that the *Arthaśāstra* and other ancient texts are unanimous in declaring the elephants of the Indus basin (the *Pañcanada Vana*), and the Kathiawar peninsula (the *Surāṣṭra Vana*) as being the most inferior. These, he further observes, are the very regions where they are now extinct, and in which, as one can infer, they were less abundant in ancient times than those in the other forests (Trautmann 1982, 265–6). Sukumar (2011, 70) adds that these two western regions are located in comparatively arid zones where elephant numbers could have been naturally low, and the animals were of a physique not suited for use in war.

THE WEST BEHOLDS THE WONDER

In view of what is known about elephants today, let us now turn to the classical accounts to see how accurately the animal was portrayed in the ancient Western world. Since early classical writers have much to tell us about the animal, a convenient way to organize the discussion would be to arrange our sources chronologically as far as possible, keeping in mind the order in which the works were produced. However, within that we have to remember that we are dealing mostly with works which are lost, fragments of their contents being preserved in the form of citations found in later works. Moreover, since these were Western authors writing about the 'Indian' elephant, some engagement with the probable sources behind these writings could also help us determine the credibility that might be accorded to these accounts.

I begin with Scullard's (1974) epic tome, *The Elephant in the Greek and Roman World*, which not only serves as an indispensable source for any discussion on Westerly encounters with the pachyderm, but also provides an entry point when it tells us that the elephant made a somewhat late arrival on the stage of classical history. We are similarly informed that the word *elephas* meant to Homer (700 BCE), and later to Hesiod and Pindar, not the animal but ivory, and that the animal itself is first mentioned in Greek literature by Herodotus (Scullard 1974, 32).

Significant as these and other early references put together by Scullard (1974, 32–3) are in suggesting early Western knowledge regarding the animal, we have to exercise caution in distinguishing between contexts. The allusions to the ones discussed so far, are clearly to the African elephant, and not to the Asian species which forms the subject of my enquiry.

The search for Asian elephants leads us to Ktesias, the first Greek author to tell us about elephants in India as well as to describe them. His original work being lost, we know that much of what Ktesias wrote comes down to us through later writings that drew from his work. We would do well by reminding ourselves of the long career he had as the royal physician at the court of Persia, and how that enabled him to gather information about India. It has been argued that the courtly setting possibly provided him with the opportunity of seeing elephants at first hand (Scullard 1974, 34). This assertion was made on the basis of his remark regarding the semen of the elephant turning so hard on drying that it became like amber (cited in Scullard 1974, 43–4). I shall return subsequently to Aristotle's scepticism regarding this remark. For the moment, it would suffice to say that even if untrue, what the reflection suggests is that Ktesias had possibly studied the behaviour of the animals at first hand, or may have interacted with men who knew the animals well (Scullard 1974, 34).

Weaving together fragments of the *Indika* surviving in the works of other writers, one learns, for instance, about Ktesias's perception of war elephants from Aelian, who wrote that based on hearsay Ktesias reported that when the king of the Indians went on a campaign, 100,000 war elephants marched on before him, while 3,000 more of superior size and strength marched behind him, the animals being trained to demolish the walls of the enemy. This they achieved by rushing against them at the king's instruction, and throwing them down by the overwhelming force with which they pressed their chests against them (McCrindle 1882, 35). Though the strength of the mentioned elephant force seems vastly exaggerated, the observations certainly serve the purpose of confirming the large-scale deployment of Indian elephants on the battlefield in the fifth century BCE. Aelian also tells us that Ktesias claimed to have seen elephants tear up palm trees and roots with furious violence, and they did this whenever they were instigated to the act by their drivers (McCrindle 1882, 35).

Further, the *Indika* of Ktesias abridged by Patriarch Photios contains an incidental reference to elephants in a passage describing a mysterious man-eating beast called the *martikhora* (discussed in Chapter 3). Having mentioned that the *martikhora* could kill all animals it attacked except the elephant, he goes on to record that to kill these beasts which were numerous in India, the natives rode on elephants, and shot darts at them (McCrindle 1882, 11–12). Again from Photios, one fathoms that Ktesias mentioned elephants which demolished walls (McCrindle 1882, 8). The reference was possibly a pointer to the use of the animals in war.

Scullard (1974, 34–5), in fact, also draws on the *Persika* by Ktesias for more accounts of elephants in battle. In one, Ktesias relates that when Amoraius, king of the Derbikes (Scythians to the east of the Caspian) was attacked by the elder Cyrus of Persia, he placed some elephants in an ambush where they routed the cavalry of Cyrus. The passage becomes relevant to us since the Derbikes are said to have received the elephants from Indians who were fighting on their side.

Another anecdote, though far less credible, also demonstrates the wide employment of elephants in war. Cited at length by Scullard (1974, 35), it revolves around the campaign against India, of Semiramis, the daughter of a Syrian goddess, and the queen of Assyria. Recorded by Diodorus, who closely followed the account of Ktesias, the narrative recounts how the Indian king had many elephants, 'equipped in an extremely splendid fashion with things which would strike terror in war'. Diodorus also remarked that India had an unbelievable number of elephants, which far surpassed those of Libya in strength and courage, a clear recognition of the existence of elephants in both India and North Africa.

We are further told that to make up for her own lack of elephants, Semiramis decided to make dummy elephants of ox-hide stuffed with straw round a light frame, moved by a camel and driver inside, and two men on each animal's back. She also trained her horses to get used to these dummies. In the meantime, the Indian king organized a hunt of wild elephants with the aim of increasing his force further, resulting in a multitude that seemed irresistible. Notwithstanding this, things went awry for him on the battlefield when his cavalry approached the dummy elephants of Semiramis, and finding their smell and other aspects unfamiliar, was left baffled. The situation was, however, saved, when the king's own elephants advanced in front of his infantry, and smashed through the ranks of Semiramis. King Strabrobates, we are told, was eventually victorious, and Semiramis had to withdraw to Bactria (Scullard 1974, 35).

The legend has been attributed to a historical figure Sammuramat, regent of the Assyrian throne between 810–805 BCE after the death of her husband. The veracity of the anecdote, however, remains uncertain since we also have Megasthenes later telling us that Semiramis, the Assyrian queen, undertook an expedition against India, but died before she could execute her design (McCrindle 1877, 109). Nevertheless, whether fact or fiction, the account reinforces a recognition of the decisive role of elephants in war.

Ktesias is, thus, indispensable to our narrative for two reasons. First, in view of being the first Greek writer to impart information about Indian elephants, including 'their ability to uproot trees, their use in tiger hunting, and their employment in war, both to overthrow walls and in open battle, together with the fact that cavalry had to be trained to face them' (Scullard 1974, 36). Second, notwithstanding later apprehensions regarding the veracity of what Ktesias wrote, his *Indika* served as a source for many later writers. However, despite Ktesias, contends Scullard (1974, 36), the animal remained 'beyond the horizon of the average Greek', awaiting the breakthrough made by Aristotle (fourth century BCE), a contemporary of Alexander of Macedon, who has much to tell us about elephants in his works *De Partibus Animalium* and *Historia Animalium*.

Aristotle's zoological writings furnish us with numerous details about the elephant, most, though not all of them, being remarkably accurate (Bigwood 1993, 537). Scullard (1974, 37–52) offers a succinct summary of his observations regarding the animal. It has been argued that unlike Ktesias, who presumably did not know African elephants, Aristotle was aware of both the species, but from his emphasis on the similarities between the two, it seems that he was perhaps ignorant of the characteristics which set them apart (Scullard 1974, 35, 49).

In a compelling essay, J.M. Bigwood (1993, 537–55) points out that apart from concerns regarding the origins of Aristotle's intense knowledge regarding the animal, an assumption commonly made is that he was discussing the Indian elephant. Following are a few highlights from the essay which according to him enable us to identify the subject of Aristotle's observations.

Bigwood (1993, 550) contends that much of what Aristotle writes regarding the diet, diseases of elephants in captivity, their temperament, functions they were trained to perform, including their use in war, and in the hunting of wild elephants, all seem to point to the Indian elephant. For instance, Aristotle (cited in Scullard 1974, 48) tells us that the Indians employed these animals for purposes of war, irrespective of sex, and that the females were inferior both in size and spirit. Additionally, a great deal of Aristotle's description is of the 'domesticated' animal, and in the absence of firm evidence for the domestication of the African elephant before about 280 BCE, or a little earlier, his description on the whole, seems to apply to the Indian, and not the African animal (Bigwood 1993, 550–1).

Coming to the sources that Aristotle seems to have based his observations on, Bigwood emphasizes that his knowledge of zoology came not only from personal observation and the opinions of experts, but also from what he read. It is not surprising then, that though in none of his works Aristotle has much to say about the Indian subcontinent or its inhabitants, he is remarkably well informed about the Indian elephant, and talks at length about its structure and habits. For instance, his trepidation about the observation of Ktesias regarding the hardening of the elephant's semen, mentioned earlier, certainly suggests the latter as a source, and Ktesias we know, was writing about the Indian elephant. Despite his misgivings about Ktesias, Aristotle's description of the elephant betrays his influence. Like Ktesias, Aristotle's elephants also participated in battle, tore down walls, and uprooted palm trees (Bigwood 1993, 544).

Other probable sources include the Athenian physician Mnesitheus, Eudoxus of Cnidus (c. 391–342 BCE), and Aristotle's nephew Callisthenes. Though Mnesitheus remains a problematic figure because of the uncertainty regarding his dates, Eudoxus, a doctor by profession, and known to Aristotle, stands out as a more reliable informer. Remnants of his work show that like other descriptive geographies, it discussed the fauna of the areas described. Significantly, along with other eastern parts of the Persian Empire, India was treated in his work. Callisthenes, on the other hand, not only accompanied Alexander's expedition to India, but was also present at Gaugamela in 331 BCE to witness for the first time, along with other members of the expedition, the fifteen Indian elephants in the forces of Darius. Given his close association with Aristotle, it only seems natural that he would have enlightened him about elephants and other affairs (Bigwood 1993, 544–8).

Based on a survey of such probable sources, Bigwood (1993, 550) argues that while some of Aristotle's descriptions may pertain to the African elephant, much of what he says clearly relates to the Indian species. After all, like Ktesias, Callisthenes was also writing about Indian elephants.

It is, therefore, crucial to underline that much before Alexander's invasion of India, and the oft-repeated account of his epic encounter with the mighty elephants of Porus on the banks of the Hydaspes, writers like Ktesias and Aristotle had already introduced the Indian elephant to Western consciousness.

Taking the story further, I now proceed to retrace Alexander's encounters with the mighty animals as he embarked on his expedition to the East. However, before transporting ourselves to the banks of the Hydaspes (Jhelum), where his troops for the first time witnessed the terror unleashed by elephants in action on the battlefield, let us go further back in time to remind ourselves that the episode was not their first encounter with the Indian elephant. Not only this, but by the time Alexander and Porus confronted each other, the former had assembled a significant number of elephants in his camp. It is, however, a different matter that he never put them to use in war for reasons I shall subsequently weave into my narrative.

The Greeks first encountered elephants on the battlefield in the Battle of Gaugamela, also called the Battle of Arbela, in 331 BCE, when the forces of Alexander met those of Darius III, thereby deciding the fate of the Persian Empire. It was an unusual line-up that awaited Alexander, for the forces of Darius in this battle included some fifteen Indian elephants. The animals, however, seem to have played no role in the actual fighting. Possibly, they had joined the forces of Darius recently and, thus, had not been properly trained to act with his cavalry (Scullard 1974, 64). Nevertheless, despite their marginal role in the actual battle, they were probably the first elephants that Alexander had seen in war, and their appearance, if not their performance, must have set him thinking at a time when he was contemplating an advance as far as India, the land of elephants (Scullard 1974, 65).

After the battle, the animals were captured, and presumably taken to Babylon, though their fate, thereafter, remains unknown (Bigwood 1993, 548). But these fifteen elephants were not all that Alexander acquired, for we are told that when he was approaching Susa, the local satrap sent gifts which included swift dromedaries and twelve elephants Darius had imported from India (Curtius cited in Scullard 1974, 65). Hence, even before he turned his eyes towards India, Alexander already had twenty-seven elephants. Many more were to become a part of his entourage in the course of his Indian campaigns.

I begin by following the trail set by the 'Alexander historians' (his contemporaries) whose works come to us in fragments through later authors such as Strabo, Pliny, and Arrian who used their accounts. In this context, we consider the works of Onesicritus, the chief navigator of Alexander's fleet, and Nearchus (mentioned in Chapter 3). Though an exact dating of Onesicritus is not possible, his work is likely to have preceded that of Nearchus, and both perhaps appeared after the deaths of Alexander and Aristotle (Scullard 1974, 52).

For instance, though Strabo is writing much later, he cites the remarks of Onesicritus about the longevity of the animals (300 years, and though rarely, even 500) as well as their gestation period, ten years (McCrindle 1901, 50). Though these are clearly magnified, what deserves attention is Strabo's assertion that Onesicritus and others stated that Indian elephants were larger and stronger than their African counterparts (a view which became common from the time of Onesicritus), and that they could 'pull down battlements with their trunks, and tear up trees by the roots, standing erect on their hind legs' (McCrindle 1901, 50). Subsequently, Pliny cites the observation of Onesicritus that the elephants of Taprobane (Ceylon) were larger and more aggressive than those of India (McCrindle 1901, 102).

Fragments of the *Indike* of Nearchus, on the other hand, are available to us through Strabo and Arrian, and contain among other things, a description of India's animals, particularly elephants. From Strabo, we gather Nearchus's version of what was to become a popular theme in later writings, namely, the elephant hunt. Nearchus mentioned the use of foot-traps for the hunt as well as the use of tame elephants to overcome wild ones, also adding that the animals once tamed were docile enough to be taught to throw a stone at an assigned mark, to use weapons of war, and to swim admirably (McCrindle 1901, 50). While some of this finds corroboration in the account of the hunt given later by Megasthenes, what seems slightly out of place is the simplistic understanding Nearchus seems to have had of the ease with which elephants could be trained.

Far more interesting is an excerpt from Strabo, where he cites the observation by Nearchus which says that 'a chariot drawn by elephants is considered a very great possession, and they are driven without bridles. A woman is signally honoured who receives from her lover the present of an elephant' (McCrindle 1901, 50). But this statement, adds Strabo, does not agree with what has been said by another writer that a horse and an elephant are the property of kings only (McCrindle 1901, 50). The writer Strabo mentions is undoubtedly Megasthenes, whose *Indika* I will turn to after reconstructing Alexander's encounters with the animals in the course of his Indian campaigns.

Equally important are passages Arrian (who admits having based his *Indika* on the annotations of Nearchus and Megasthenes) derives from Nearchus, which are significant in terms of their implications for the value assigned to the animal:

> But Indian women, if possessed of uncommon discretion, would not stray from virtue for any reward short of an elephant, but on receiving this a lady lets the giver enjoy her person. Nor do the Indians consider it any disgrace to a woman to grant her favours for an elephant, but it is rather regarded as a high compliment to the sex that their charms should be deemed worth an elephant. (McCrindle 1877, 222)

Not only this, we are also told that the Indians wore earrings of ivory, but only the wealthy did so since all Indians did not wear them (Arrian cited in McCrindle 1877, 220). Elsewhere, it is qualified that if a man is rich, he uses pricks made of ivory (Arrian cited in McCrindle 1877, 220–1).

From piecing together the fragments available from the aforementioned historians of Alexander, I now move on to narratives of his Indian campaigns compiled several centuries after his death from the works of writers who either witnessed the events they described or were living when they occurred. We know that these works though themselves lost, served as sources for the accounts left by Diodoros Siculus, Quintus Curtius Rufus, Plutarch, Arrian, and Justinus Frontinus.

From the time Alexander entered India, narratives attest to his acquisition of elephants either by way of capture or as gifts received. Let us approach our sources, which, though scattered and seldom rid of ambiguities, give us an inkling of the number of elephants Alexander acquired en route.

As Alexander entered India, chiefs of various tribes and kingdoms came to meet him. Considered the best of Alexander's historians, Arrian mentions Taxiles and other chiefs who brought him gifts esteemed most by the Indians, also offering the twenty-five elephants they had with them (McCrindle 1896, 59). Curtius in his narrative, however, mentions a gift of fifty-six elephants made by him to Alexander (McCrindle 1896, 202). The siege of Ora, as part of the campaign against the Assakenians (who are said to have possessed thirty elephants) gave him possession of more of the animals, though Arrian does not specify the number (McCrindle 1896, 70).

While these are just some of the statistics scattered across the corpus of classical Western writings, Scullard (1974, 66) estimates that by the time Alexander faced Porus, he had mustered an elephant force of more than 126, yet 'made no attempt to experiment in this new arm'. His contention is that the acquired herd was possibly integrated with his army not for actual fighting, but for purposes of transport, or to impress tribes through whose territory he advanced. Quite convincingly, he further argues that Alexander evidently had not had time to coordinate his elephants into a cohesive force nor had he been able to train his horses to cooperate with them. However, as the ensuing trial of might with Porus testified, what he had managed to do was to devise ways of dealing with the enemy's elephants (Scullard 1974, 66).

We can now set out to reconstruct what transpired on the banks of the Hydaspes from the time Porus was seen on the opposite side surrounded by his army and array of elephants, which as all accounts seem to suggest, were a crucial concern for Alexander as he formulated his stratagem for crossing the river. The enormity of the task in hand was unmistakable and the perils obvious. Alexander clearly foresaw the impracticality of attempting to cross where Porus had encamped himself largely because of his elephants. His horses, he knew, would refuse to mount the opposite bank where the elephants would at once charge at them, terrifying them by their very appearance and thunderous roaring. He anticipated that even the sight of the large animals would make his horses frantic enough to leap into the water. As he manoeuvred his moves to cross the river unobserved, Arrian tells us, what loomed large on Alexander's mind was the difficulty posed by the elephants on the other bank. The rest of the army, he was certain, could cross over without difficulty (McCrindle 1896, 95–8). The bank, according to Curtius, presented a far more formidable sight, for as far as vision reached, it was covered with cavalry and infantry amidst which huge elephants stood like massive structures emitting frightful roars when provoked by their drivers (McCrindle 1896, 204).

Arrian recounts that as tidings reached Porus regarding Alexander having crossed the river with the strongest division of his army, he marshalled forces for battle, taking with him a cavalry strength of 4,000, 300 chariots, 200 elephants, and 30,000 infantry (McCrindle 1896, 102). Other accounts, however, differ regarding the strength of the forces of Porus. Curtius gives the same number of infantry and chariots, but puts the number of elephants at eighty-five (McCrindle 1896, 204). For Diodorus, it is 130 elephants along with 50,000 infantry, 3,000 cavalry, and more than 1,000 chariots (McCrindle 1896, 274). The confusion, observes Scullard (1974, 266), may have arisen from the fact that Porus did not take all his infantry and elephants, but left some as a covering force in his rear against Craterus. Notwithstanding the divergence in statistics, it is easy to envision the psychological impact these massive animals must have had on the Macedonian army as it prepared to take on the forces awaiting them.

The battle array that then awaited Alexander was a formidable one. According to Arrian, Porus had posted his elephants in the front line at intervals of at least 100 feet so that the animals ranged in front of the infantry could spread terror among Alexander's cavalry. He was certain that neither the enemy's cavalry nor its infantry would dare to push in between the gaps separating the elephants. Behind the giant beasts, Porus positioned his infantry so that the units not only covered intervals in the line of elephants in front, but also formed a second line in their rear. Troops of infantry were posted on wings stretching beyond the elephants, and on both sides of the infantry, the cavalry was stationed with chariots in front (McCrindle 1896, 103).

Curtius and Diodorus are unanimous in confirming the impact that the sight of the army of Porus had on the Macedonians. The beasts, according to Curtius, from a distance appeared like towers, while the majestic Porus, dressed in shining armour, stood taller than other men mounted on his elephant, which towered over all other elephants (McCrindle 1896, 209). Diodorus, on the other hand, tells us that the overall disposition of the army gave it the appearance of a city, with the elephants resembling its towers, and the infantry positioned between them resembling the lines intervening between towers (McCrindle 1896, 275).

It can scarcely be denied that the animals decided the course of the battle. Because of them, Alexander had an entirely novel problem to deal with. His cavalry could not take on the main line of the enemy since his horses would not charge against the formidable row of elephants. On the other hand, he was also apprehensive about sending his phalanx against this 'untested opponent' anticipating the havoc the animals might wreak against men inexperienced in dealing with them (Scullard 1974, 68–9). Consequently, Arrian tells us, he resolved not to let Porus reap the benefit of the layout, and decided not to advance against the centre in front of which the elephants had been posted, but instead march along towards the left wing of the enemy, and attack from that quarter (McCrindle 1896, 104).

Even amidst such machinations, the impact of what he beheld, writes Curtius, led Alexander to remark,

> I see at last a danger that matches my courage. It is at once with wild beasts and men of uncommon mettle that the contest now lies . . . The formidable length and strength of our pikes will never be so useful as when they are directed against these huge beasts and their drivers. Hurl, then, their riders to the ground, and stab the beasts themselves. Their assistance is not of a kind to be depended on, and they may do their own side more damage than ours, for they are driven against the enemy by constraint, while terror turns them against their own ranks. (McCrindle 1896, 209)

In the action that ensued, the brute force of the animals unleashed terror, with the Macedonians sometimes chasing, and sometimes escaping from the huge beasts as they went about trampling men under their feet, and killing friend and foe indiscriminately. The deaths they inflicted were, according to Diodorus, terrible, crushing the armour and bones of men, twining their trunks around some, and dashing them to the ground, while tearing apart many others with their tusks (McCrindle 1896, 275). The most dismal of all sights, according to Curtius, was of elephants grasping men with their trunks, and handing them over to their riders to deal with. In fact, the wrath of the animals evoked such fear in the Macedonians that they not only left no means untried for killing them, but also resorted to unprecedented forms of cruelty, hacking the feet of the animals with axes prepared for the purpose, and slashing their trunks with swords called choppers (McCrindle 1896, 211). For the Macedonians undoubtedly, Arrian emphasizes, this was a kind of warfare different from what they had experienced in former contests (McCrindle 1896, 106).

Gradually, the elephants were pushed into a narrow space where exhausted and spent with the wounds they had been afflicted with, they began to spread havoc among their own ranks. What unfolded, thereafter, was a sordid tale of ally turning adversary. We learn from Curtius that maddened with terror, they threw their drivers to the ground, and trampled them to death (McCrindle 1896, 211). The Indians being in the midst of these animals, according to Arrian, consequently suffered far more the effects of their rage (McCrindle 1896, 106).

It would not be out of context to mention a little of what has been written about the elephant that carried Porus. In describing it, Curtius leaves us with the image of a compassionate animal and a loyal aide fighting valiantly for its collapsing master. The setting undoubtedly is that of the violent battlefield, but the picture is poignant. A badly wounded Porus begins to slip from his elephant. The driver thinking that the king wanted to alight, orders the elephant to kneel down. Other elephants follow suit, having been trained to lower themselves when the royal elephant did so. Porus and his men are, thus, brought within the reach of Alexander, whose men advance towards the wounded king. Itself severely wounded, the royal elephant, nevertheless, makes a brave bid to defend its master, and turns upon the men while lifting Porus and placing him once more on its back. The endeavour costs it dearly for it soon becomes the recipient of darts from all sides, and is eventually stabbed to death (McCrindle 1896, 213).

Writing later, Aelian gives us a similar account, but takes it a little further by recounting how the wounded animal gently and cautiously pulled out with its trunk, the darts that had pierced its master's body, and continued doing so till it saw him becoming weak and ready to pass out due to the excessive loss of blood. He then lowered him slowly and gently with its knees bent in a way to ensure that Porus did not hit the ground with force (McCrindle 1901, 139).

However, despite Porus's inexorable spirit of fighting till the finish, the day eventually belonged to Alexander and his men. The Greeks had outdone their formidable enemies—the elephants, eighty of which Diodorus tells us, survived the battle to be eventually captured (McCrindle 1896, 276). What facilitated this triumph can be speculated. Richard Glover (1944, 264–5), for instance, is uncertain whether this victory can be attributed to Alexander's genius or Porus's blundering. The latter, he asserts, did blunder. By allowing Alexander to seize the initiative and rout his cavalry, Porus lost the opportunity to use his elephants in a mass assault upon the Macedonian infantry, which instead grabbed the chance, attacked the animals, and proved that an impactful offensive could make them run amok, destroying their own ranks rather than those of the enemy, and tragically turn victories into calamitous defeats.

Elucidating the military characteristics of elephants, Glover (1944, 258) attributes to the animals a highly-strung nervous system that could be easily upset. Nevertheless, he argues that their shock value was far greater than the horse's not only because of their size, but also because they could trample men and gore them with their tusks. In addition to bearing armed riders, the pachyderm's kick was enough to send a man flying!

The encounter at Hydaspes can, thus, be taken as an early, and a classic instance of psychological warfare. Huge trumpeting beasts lined up in the front row of battle were undoubtedly an intimidating sight as well as sound, particularly for men and cavalry horses new to the experience (Shelton 2006, 7). Though Alexander never used them in war, the grim battle with the elephants of Porus had a lasting imprint on the mind of his general, Seleucus, who made Indian elephants his special arm in war, and the emblem of his house.

Armed with victory, Alexander's insatiable thirst for conquest spurred him east towards the Ganges. From Curtius we know that Alexander learnt about the kingdoms of Gangaridai and Prasii in the east, whose king Agrammes was said to have the most formidable troop of 3,000 elephants, apart from 20,000 cavalry, 200,000 infantry, and 2,000 four-horsed chariots (McCrindle 1896, 222). In the same context, Diodorus cites an elephant force of 4,000 (McCrindle 1896, 282). Plutarch, claiming to derive his particulars from Alexander's own letters, puts the figure at 6,000 (McCrindle 1896, 310). Notwithstanding the variations in these accounts, the picture that emerges is of a vast elephant corps inhabiting the land east of the Ganges.

A sense of the lessons learnt from the Battle of Hydaspes can be had from a speech Curtius attributes to Alexander as he cajoled his fatigued soldiers into embarking on a 'noble spoil' of 'the much-rumoured riches of the East' with its profusion of pearls, precious stones, gold, and ivory (McCrindle 1896, 215). Elephants no longer seemed to agonize the conqueror as he addressed his men imploring,

> Who but till the other day believed that it was possible for us to bear the shock of those monstrous beasts that looked like so many ramparts . . .

> which after all were more formidable to hear of than they proved to be in actual experience . . . Can you suppose that the herds of elephants are greater than of other cattle when the animal is known to be rare, hard to be caught, and harder still to tame? . . . Is it then, I ask, the monstrous size of the elephants or the number of the enemy that you dread? As for the elephants, we had an example of them before our eyes in the late battle when they charged more furiously upon their own ranks than upon ours, and when their vast bodies were cut and mangled by our bills and axes. What matters it then whether they be the same number as Porus had, or be 3000, when we see that if one or two of them be wounded, the rest swerve aside and take to flight. Then again, if it be no easy task to manage but a few of them, surely when so many thousands of them are crowded together, they cannot but hamper each other when their huge unwieldy bodies want room either to stand or run. For myself, I have such a poor opinion of the animals that, though I had them, I did not bring them into the field, being fully convinced they occasion more danger to their own side than to the enemy. (McCrindle 1896, 223–4)

Persuasive as Alexander's discourse was, it failed to move his men, who clearly found themselves too spent to pursue the imperial ambitions of their king. Curtius recounts that entreating him to see their bloodless bodies pierced with wounds and gashed with scars, they put forth the perils of exposing such an army to the mercy of savage beasts, whose numbers even if exaggerated, were likely to be considerable (McCrindle 1896, 229).

Apart from the sources that tell us about Alexander's anxiety regarding the difficult nature of the terrain ahead as well as the turbulent nature of the rivers to be crossed en route, it seems perfectly reasonable to concur with Scullard's (1974, 73) contention that elephants were equally instrumental in forcing Alexander to give up his march of conquest. From the men who actually fought them at Hydaspes to the inestimable force that awaited them in the east, the animals to a large extent determined Alexander's capitulation in the face of opposition from his dispirited men.

I now turn to the *Indika* of Megasthenes, which survives only in fragments scattered in classical works spanning centuries. Though most of what Megasthenes has to offer in terms of insights regarding the Indian elephant has already been put together by Scullard (1974, 55–9) in the form of entire quotations, I will run through the *Indika* once again, not with the aim of following it verbatim, but to weave in the implications of its observations with the sociopolitical milieu it represented. It is also crucial to establish as far as possible, the veracity of what Megasthenes writes since his is the first eyewitness account of the Gangetic Plain by any Westerner.

I begin with a fragment retrieved from Diodorus, where Megasthenes describes India as a country prolific in well-fed and monstrous elephants, far exceeding in size the ones bred in Libya. He attributes the might of the Indian elephant to the abundant food the soil of the land provided. Caught in great numbers by the Indians and trained for war, the animals were considered momentous for victory (McCrindle 1877, 31).

Further, Diodorus cites Megasthenes mentioning Gangaridai as a nation with a vast force of the largest-sized elephants due to which the country had never been conquered by any foreign king since all other nations dreaded the overwhelming number and strength of these animals. Megasthenes seems to be reminiscing the past, when he reasons that after conquering all of Asia, and having arrived at the Ganges with all his troops, Alexander did not make war upon the Gangaridai having learnt that they possessed 4,000 elephants, well-trained and equipped for war (McCrindle 1877, 33–4).

From Strabo, we pick up a fragment where spelling out the administration of public affairs, Megasthenes mentions after the city magistrates, a governing body consisting of six divisions directing military affairs. The charge of war elephants was attributed to the sixth division. Royal stables housed the esteemed animals, and a soldier had to return his arms to the magazine and his horse and elephant to the stables. Elephants were used without bridles (McCrindle 1877, 88–9). As an instrument of war, the animal carried four men either in a tower or on its bare back, two of them shooting from the side, and one from behind, while the fourth in the capacity of a driver, steered the animal with a goad in his hand (Aelian cited in McCrindle 1877, 90).

What ensues seems to further reinforce the above passage suggesting a royal monopoly of elephants. We are told that a private person was not allowed to keep a horse or an elephant, these animals being the special property of the king, with persons being appointed to take care of them (Strabo cited in McCrindle 1877, 90).

Let us digress a little to juxtapose the unambiguous assertion by Megasthenes of a royal monopoly of elephants with the observation earlier made by Nearchus regarding the issue. Trautmann (1982, 255–6) offers a persuasive resolution to this apparent contradiction pointed out by Strabo. Nearchus, Trautmann contends, was referring to what he encountered in that part of India which Alexander penetrated, namely the Punjab to the Beas River and Sindh in 327–326 BCE. The private ownership of elephants, horses, camels, and asses prevailed then in the north-western sector of the Indian subcontinent which Candragupta Maurya established his sway over after overthrowing the Nandas in c. 324 BCE. Megasthenes, on the other hand, was describing what he found two or more decades later in the Mauryan Empire, probably at Pāṭaliputra where he resided. He was, therefore, referring to this empire, and its monopoly of the sinews of war, which must have contributed to its success against the loosely ordered states of the north-west.

Away from the affairs of state, also striking is Megasthenes' description of the royal hunt in which the king leaves his palace in Bacchanalian fashion with crowds of women surrounding him. Other details of the hunt notwithstanding, significant is the picture of the king hunting in open grounds, shooting from the back of an elephant. The scene comprising women mounted on chariots, horses, and even on elephants is paralleled with that of an approaching campaign (Strabo cited in McCrindle 1877, 72).

The hunt of the animal itself was an arduous enterprise, a fact Megasthenes acknowledges as well as painstakingly demonstrates in a passage dealing exclusively with the subject. Mission accomplished, the violence and brutality of the hunt is supplanted by soothing, coaxing, and pacifying.

> Few of them are found difficult to tame, for they are naturally so mild and gentle in their disposition that they approximate to rational creatures. Some of them take up their drivers when fallen in battle, and carry them off in safety from the field. Others, when their masters have sought refuge between their forelegs, have fought in their defence and saved their lives. If in a fit of anger they kill either the man who feeds or the man who trains them, they pine so much for their loss that they refuse to take food, and sometimes die of hunger. (Strabo cited in McCrindle 1877, 92)

In Pliny's list of Indian tribes, which he borrows mostly from Megasthenes, he mentions a half-wild class engaged in the hunting and taming of elephants, which are said to have been employed in ploughing and for riding on, and regarded as forming the main part of their stock in cattle. They were also employed in war and in fighting for their country, and their choice as war animals was determined by their age, strength, and size (McCrindle 1877, 136–7).

The same list enumerates a number of kingdoms with sizeable elephant corps. For instance, the king of the royal city of Calingae is said to have been watched over by 60,000 foot soldiers, 1,000 horsemen, and 700 elephants. Further, there is a mention of tribes said to have been chiefly located in the regions between the left bank of the Ganges and in the Himalayas. Their king, we are told, was armed with 50,000 foot soldiers, 4,000 cavalry, and 400 elephants. A still more powerful race was that of the Andarae, which supplied its king with an army of 100,000 infantry, 2,000 cavalry, and 1,000 elephants. However, the Prasii, with their capital at Palibothra, surpassed all in power and glory, with their king having a standing army of 600,000 foot soldiers, 30,000 cavalry, and 9,000 elephants (McCrindle 1877, 135–9).

After a general account of the basins of the Indus and the Ganges, Pliny goes on to enumerate the tribes of north India. We learn from McCrindle's (1877, 143) postscript that the tribes first mentioned in the list occupied the country extending from the Jamuna to the western coast about the mouth of the Narmada. The mention of Megallae, whose king mastered 500 elephants together with an infantry and cavalry of unknown strength, is followed by a reference to the Chrysei, the Parasangae, and the Asangae, who commanded a force of 30,000 infantry, 300 elephants, and 800 horses (McCrindle 1877, 142).

From Aelian also, we have details purportedly borrowed from Megasthenes. Having spelt out the arduous task of capturing a full-grown elephant, he describes the difficulties in taming one, if caught. The animal, he says, longs for its freedom and, if confined in chains, is even more exasperated and resists submission. The Indians pacify it with food, music, and other things of its liking, but the animal does not yield. Enthralled by the music, the animal, gradually settles down though not without occasional outbursts of resistance. Though freed from its bonds now, it shows no desire to escape and eagerly consumes the food offered to it (McCrindle 1877, 93).

Another fragment recounts the affection between an elephant-trainer and a young white elephant calf which he brought home, reared, tamed gradually, and even used as a mount. The king of the Indians having heard of the elephant wished to acquire it, but the elephant-trainer, distraught at the idea of having to let go of his beloved pet, made off to the desert mounted on the pachyderm. Enraged, the king sent his men in pursuit with orders to capture the elephant, and bring back its master for punishment. Overtaken by the king's men, the elephant-trainer put up stiff resistance, perched on the back of his pet which faithfully fought on his side. However, wounded badly, the trainer soon slipped to the ground, but his trusted ally took charge, standing astride, and shielding his master. Using his trunk, he put him on his back, and carried him home only to stay with him forever as a faithful friend (Aelian cited in McCrindle 1877, 118–19).

Apart from the aforementioned gleanings, Aelian, drawing from Megasthenes, also furnishes us with details regarding how the diseases of elephants were cured by the Indians (McCrindle 1877, 93–4). Additionally, we are told about the animal's love for flowers and sweet perfumes, as also that Indian elephants were nine cubits in height and five in breadth, and that the largest elephants were those called the Praisian followed by the Taxilan (McCrindle 1877, 117–18).

After piecing together these extracts from Megasthenes, I move on to underline some glimpses of the animal that we derive from the writers we have been citing so far. Amidst particulars which he attributes to 'historians', Strabo writing his *Geography* in the first century CE, mentions processions at the time of Indian festivals where numerous elephants adorned with gold and silver were lined up along with four-horsed chariots and yokes of oxen (McCrindle 1901, 75). There is also mention of a marriage alliance between Seleukos Nikator and Sandrokottos, and a gift of 500 elephants by the latter to the former (McCrindle 1901, 89).

India, according to Pliny the Elder, produced the largest elephants and dragons which could entwine themselves around the gigantic bodies of the former, with the fight ending in the death of both the combatants, as the vanquished elephant in falling crushed with its weight the reptile coiled around it. He also relates that in India, wild elephants found alone or separated from the herd were caught with the help of tame ones. Once caught, the animals were beaten to exhaustion and then mounted. He also recounts that elephants mad with rage were tamed by hunger and blows, and that the animals when in heat were in the worst of temper and known to demolish the huts of the Indians with their tusks in that condition (McCrindle 1901, 114–15).

Aelian, who flourished about the middle of the second century CE, has much to say about animals in general, and the elephant in particular. Drawing upon Greek authors and some personal observation, his work *On the Peculiarities of Animals* is a collection of facts and beliefs concerning the habits of animals.

Aelian tells us that the Indians cannot easily capture a full-grown elephant; hence, hunters prefer the swamps adjoining a river where they can catch the young ones. A sound sense of the ecology of the animal reverberates in his description of the elephant as an animal loving moist and soft grounds, enjoying being in water, and preferring to 'spend time in haunts of this nature, so that one may say he is a creature of the marsh'. What follows is an empathetic dimension of the human relationship with the pachyderm. He contends that once caught at a young age, they are brought up with care by the Indians who supply them with food they relish the most, groom them carefully, and talk to them in soothing tones, for elephants understand the native tongue. The animals, he emphasizes, are reared like children with great care and attention, and then subjected to a long course of training (McCrindle 1901, 137).

Aelian also mentions elephants in the act of pulling down trees, and records that when set to the task by Indians, the animals began by giving the tree a violent shake so as to ascertain the viability of the job assigned (McCrindle 1901, 139). He also reports having been told about the antipathy between the snake and the elephant in India, and elucidates how the pachyderm is attacked by the reptile waiting ambushed in trees (McCrindle 1901, 139).

'Horses and elephants being animals of great use in arms and warfare' were held in the highest esteem by the Indians according to Aelian. If unhappy with the services rendered, the king severely punished the men in charge of elephants and horses. Next to men, elephants were considered the most faithful of sentinels, guarding the king's person, and serving in other ways they were trained for. Strokes of the goad by the keeper served as reminders of lessons taught to the animals (McCrindle 1901, 142–3).

Interestingly, in another passage he leaves us with a glimpse of the animal as part of a public spectacle organized by the king, where before the end of the show, elephants were brought in to fight and would often end up killing each other by inflicting fatal wounds with their tusks (McCrindle 1901, 145).

As mentioned before, one treads with caution while dealing with this corpus of literature, which is evidently a mix of fact and fiction. Nevertheless, amidst the maze of images oscillating between fanciful descriptions to credible ones, what is reinforced are the multiple ways of perceiving the animal. The Western world, in this sense, was no exception as it grappled with putting in words what it had heard and seen of this magnificent animal.

6

CONCLUSION

STRADDLING ACROSS MILLENNIA WITH OUR three protagonists, we reach the end of the road we had set out to traverse. As their stories approach closure (within the chronological limits of this book), it is time for them to return to their natural haunts—the marshes, forests, and woods of ancient India—and to the lives and cultures of the people who voluntarily or involuntarily forged associations with them. After all, it is from there that we picked up traces of their footprints in time, and it is from there that they beckon us now to connect those vestiges to give them the histories they deserve.

Having strung together their individual trajectories in the preceding chapters, I move on to consider issues which emerge when our three heroes are juxtaposed in the light of an archive comprising archaeology, a gamut of literary texts as well as visual depictions. The evidence collated, what I now look for are not neat conclusions, but meaningful indicators that equip us to reflect on the ecological and cultural images these mega mammals evoke.

I began this journey with the testimony of archaeology, hence it would be meaningful to put together the faunal records for these animals. A quick overview of the evidence available not only gives us a sense of the limitations inherent in the faunal record, but also enables us to correlate our data with Figures 6.1, 6.2, and 6.3 mapping the same.

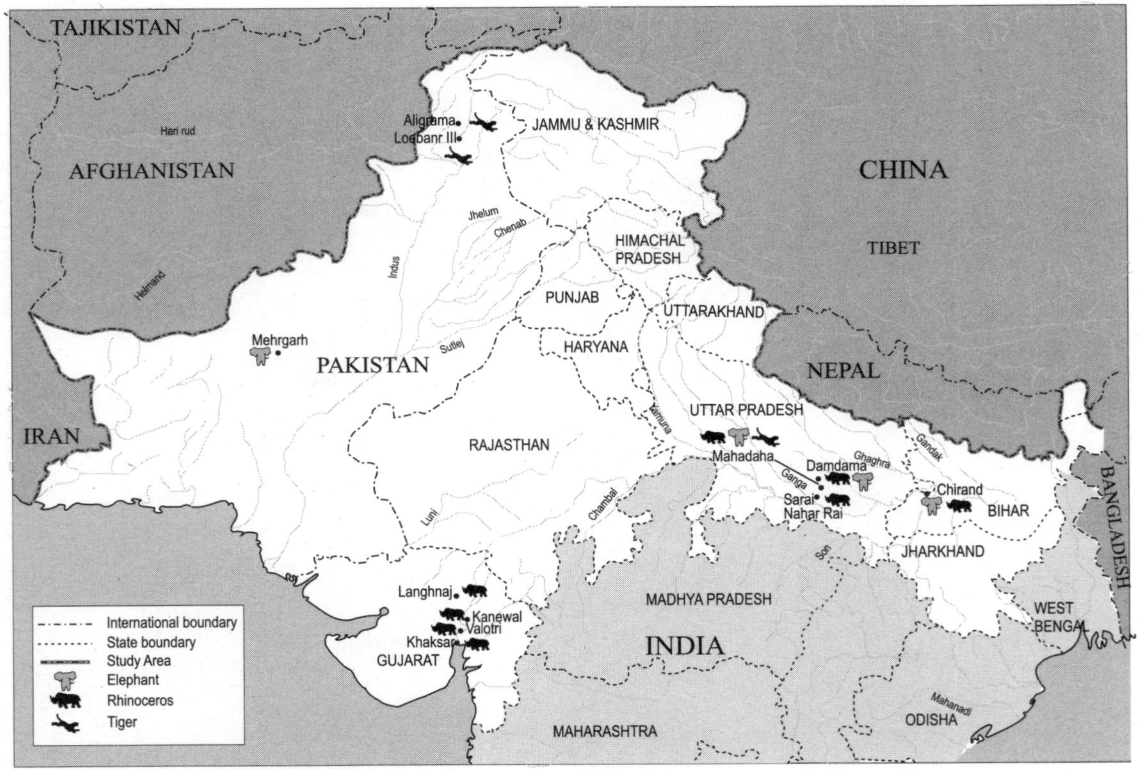

Figure 6.1 Megafauna from mesolithic/microlithic and neolithic sites (for representative purposes only)
Source: Courtesy of the author.

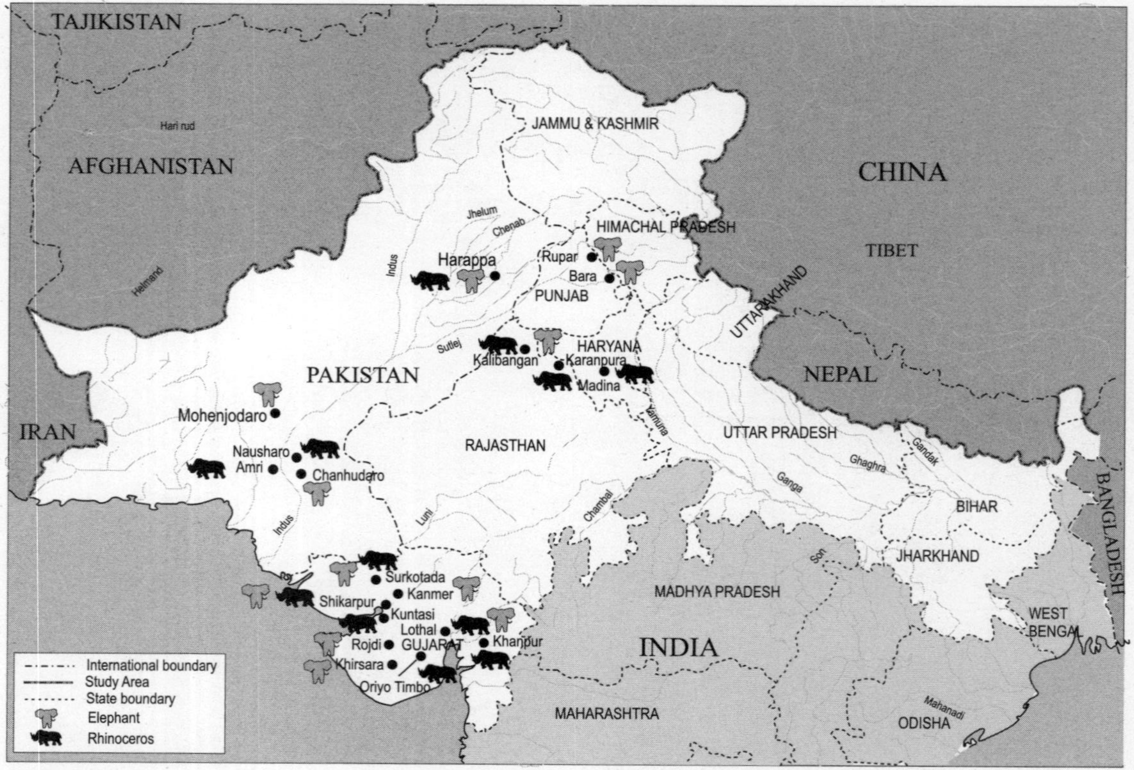

Figure 6.2 Megafauna from Harappan sites (for representative purposes only)
Source: Courtesy of the author.

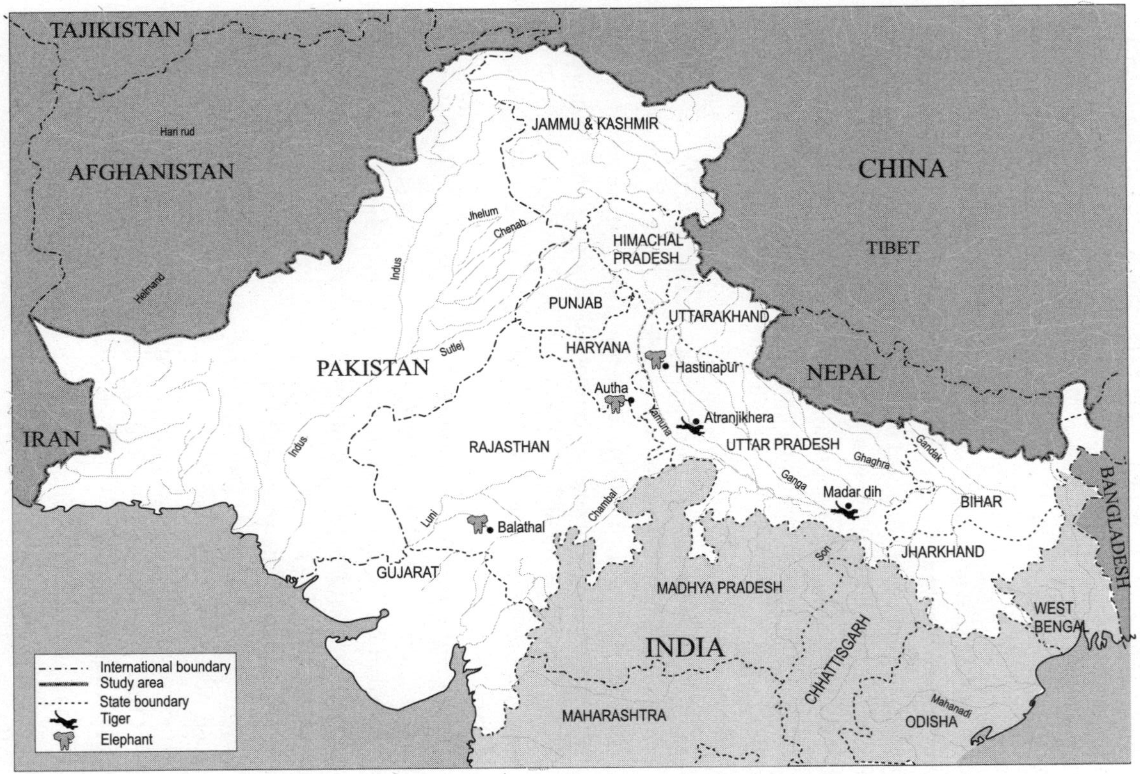

Figure 6.3 Megafauna from chalcolithic, iron age, and early historic sites (for representative purposes only)

Source: Courtesy of the author.

Beginning with the rhinoceros, from its first appearance in the fossil record in a terminal Pleistocene context in the Middle Son Valley (maxillary fragment bearing three cheek teeth), the animal steadily makes its way to the early mesolithic sites of Damdama (no details of remains), Sarai Nahar Rai (no details of remains), and Mahadaha (teeth, mandible, and vertebra) in the Ganga Valley as well as later mesolithic/microlithic contexts at Khaksar (two pieces of cervical vertebrae), Valotri (a cervical vertebra), Kanewal (no details), and most tellingly at Langhnaj (left scapula, right humerus, a talus, and a fragment of molar tooth) in the western state of Gujarat. While neolithic Chirand in Bihar also yielded remains (proximal fragment of ulna, left tibia without proximal and distal epiphysis, fragment of upper molar tooth, and a complete left humerus) of the animal, rhino bones at Nausharo (no details of remains), Amri (no details), Harappa (scapula), and Madina (astragalus) attested the presence of the animal in Harappan and Late Harappan contexts.

However, what is actually remarkable is that at Kalibangan in Rajasthan, there is a profusion of bones (distal fragment of radius, one broken medial and one broken lateral condyle of left humerus, fragment of the shaft of left ulan-3, the broken right ramus of a mandible, second phalanx of right hind limb, fragment of right pelvic girdle with acetabulum, broken rib-7, left lunate, left trapezoid magnum, left external cuneiform, right metacarpal, right unciform, distal portion of left femur, lumbar vertebra, right astragalus, proximal portion of left ulna, proximal end of right radius, fragment of a rib and broken left pelvic girdle with pubis). The evidence from Karanpura (shoulder bone, a part of pelvis, one radius, and ulna) serves to reinforce the presence of the animal in Rajasthan. Faunal remains from Surkotada (no details of remains), Shikarpur (no details of remains), Kuntasi (the third phalanx), Lothal (fragment of rhinoceros mandible), Khanpur (a complete humerus), and possibly also Oriyo Timbo firmly placed the animal on the proto-historic map of western India.

The tiger has a scant presence in the fossil record, but figures in the archaeozoological record at mesolithic Mahadaha (phalanx), neolithic Loebanr III and Aligrama in Swat (no details), in the OCP level at Atranjikhera (one fragment of ulna), and in a NBPW context at Madar Dih (complete maxillary canine).

Notwithstanding the paucity of fossil remains, definite evidence for the presence of the Asian elephant comes forth from mesolithic contexts at Damdama (no details of remains) and Mahadaha (two ivory pieces, three rib fragments—charred and one with a cut mark, charred fragment of right scapula of adult elephant and an almost complete vertebra of a young elephant). While Mehrgarh (fragment of long bone and part of tusk) and Chirand (fragment of upper molar tooth and unciform bone of manus) yielded remains in neolithic contexts, Harappan sites such as Mohenjodaro (caput of femur), Chanhudaro (tusk), Harappa (no details), Rupar (proximal fragment of the left scapula), Bara (first phalanx of the left second digit of the forefoot), Kalibangan (broken phalanx of the left forelimb and a fragment of rib), Lothal (three fragments of right femurii, the first phalanx of the fourth digit of its right hind limb and one fragment of radius), Rojdi (rib), Surkotada (tusk), Shikarpur (two fragments of ivory), Kanmer (two fragments of ivory), and Khirsara (ivory fragment) testified to the presence of the pachyderm in north-western and western India. Remains from early historic sites are, however, scarce and come from Hastinapur (tip of tusk, pisiform bone of the left forefoot), Autha (fragmentary skeleton), and Balathal (a rib) within the study area.

Admittedly, there are gaps in evidence due to the uneven and often patchy character of the archaeozoological record itself. As can be seen, in many cases, what is affirmed is the mere presence of a species at an archaeological site without any mention of the details of the remains found or other telltale signs on them. As more sites are excavated and their faunal assemblages examined, we hope to know much more about human interactions with these animals. Till then, even on the basis of the evidence available, it is possible to ask certain questions, the first relating to the exploitation of these animals.

While not much can be said about early human interactions with the tiger because of the paucity of its remains in archaeological deposits, the picture is somewhat different when one considers the rhinoceros and the elephant, where it is clear that their economic utilization goes back to ancient times. At the mesolithic site of Damdama, for instance, we have seen that the carcasses or isolated bones of these big mammals were suggested to have been collected and utilized for making bone tools. At Langhnaj, deliberate pits on the shoulder blade of a rhinoceros indicated its use as an anvil for making microlithic tools. Similarly, in the context of the elephant, evidence from Mahadaha and Mehrgarh suggests that the antiquity of ivory working in India can be traced back to prehistoric times. Additionally, and notwithstanding doubts expressed by most faunal experts, this book considers the possibility of the inclusion of such big mammals in the diet, and underlines the need to look at the evidence in totality including rock paintings illustrating the hunting and even butchering (rhinoceros, Kerwaghat, Mirzapur) of such mega species.

It is equally or perhaps more crucial to think about what archaeology tells us regarding the ecological landscape of early India. A case in point would be the presence of the rhinoceros in areas where it is no longer found. Now popularly associated with eastern India, I have already underlined the presence of the animal at prehistoric as well as proto-historic sites in the semi-arid and arid states of Gujarat as well as Rajasthan. This invites some thought about past landscapes which evidently must have been able to carry marshy-swampy patches for moisture-loving species such as the rhinoceros. We should also perhaps transport ourselves back to the iconographic representation on seal no. 420 at Mohenjodaro. The animals associated with the central figure include the elephant, the tiger, the rhinoceros, and the buffalo—animals suggesting a tree-dotted tall grassland ecology. In fact, we only need to remind ourselves of the presence of the mega herbivore in the study area (as shown in Chapter 2) until fairly recent times. Evidently then, as landscapes changed, the animal retreated. From the point of view of environmental history, charting the journey of the rhinoceros also calls for considering the role of human activities like agriculture and hunting, which eventually brought pressure on rhino habitats leading to local extinctions.

Taken together, the faunal evidence not only helps us map the past distribution of these mega mammals, but also makes us wonder if an absence in record (mostly in the case of the tiger) necessarily implies an absence in real life.

No perhaps, because even when bones are not available, the animals assert their presence on the landscape of early India through visual representations both in the form of rock paintings and terracottas. And in most cases, it is the fidelity of the representations that tells us that they must have been adequately seen around to have entrenched themselves in human consciousness, which then sought to express itself through tangible portrayals.

Whether it is the art of early hunter-gatherers in the form of rock engravings or paintings, or the glimpses derived from Harappan iconography on seals, tablets, amulets, and terracottas, in most cases, the qualitative details and graphic fidelity of the representations is remarkable. Particularly in the Harappan context, what deserves to be underlined, and is clearly borne out by the illustrations accompanying the chapters is a spectacular sense of familiarity with these animals. It is remarkable that a civilization known for its urban character shows such close acquaintance with the natural world.

As far as the rhinoceros is concerned, the prominent single horn unambiguously certifies it as the one-horned species as does the well-represented thick tuberculed hide with the nearly real wrinkles and folds of the skin. Depictions of the tiger and the elephant reiterate the same impression. While the former is clearly delineated by its stripes, brawny, and often ready-to-pounce sinuous body, representations of the pachyderm show even greater details such as the wrinkles on its back, the fold of the trunk terminating at times in two distinct processes, bristles along the outline of the body to show a younger animal, or the positioning of the legs in order to capture an animal in motion. The outstandingly close observation of the elephant could possibly be explained by the fact that the Harappans were actively engaging with the animal as can be inferred from painted elephant heads in terracotta as well as depictions of the animal apparently draped.

This then links up to a larger point that seems to run across the narrative. It relates to the observation that early India reflects a deep connect with the natural world, evident in allusions capturing the anatomies, habitats, and temperaments of the animals they knew and interacted with. This clearly comes forth not only through visual representations as already elucidated, but also in the ways in which these animals were being thought of and written about. Not only are they interwoven in lyrical descriptions of natural beauty (most strikingly in the epics), their ecology and behaviour are also observed and referred to.

What should perhaps also be emphasized is that this consciousness permeates nearly all genres of the literary texts dealt with. Whether it is the ritualistic corpus of Vedic literature, the philosophical discourses of the Buddhist and Jaina canons, texts of jurisprudence (the Dharmaśāstras), the medical treatises of Caraka and Suśruta, the intrigues and battle scenes intrinsic to the two epics, a political treatise like the *Arthaśāstra*, or the often warped classical Western accounts, placing animals in their natural settings with subtle allusions to their ways, is endemic within textual traditions. What comes to mind as a striking example is Suśruta's placing of the rhinoceros and the elephant amidst the *kūlacara* (living on the banks) quadrupeds of the *ānūpa* (marshy land) category, while the lion and the tiger are *guhāśaya* (living in a lair) animals falling within the *jāṅgala* (dry land) category. Implicit in this classification (which is actually more elaborate) is a deep sense of the ecologies of different animals.

On the basis of more specific examples already cited in the respective chapters, it can then safely be said that ancient Indian literature knows the rhinoceros for its love of marshy, swampy habitats, its armour-like hide, one-horned identity, and solitary character. It is also aware of the tiger as a denizen of deep forests living in the vicinity of water, its predatory skills, and as a key player in maintaining ecological balance. The elephant, on the other hand, is known for the dexterity of its trunk as well as the phenomenon of *musth* (explaining the evident fixation with 'rutting' elephants). Overall, the sense looming large is of ancient India's reasonable accuracy in depicting and writing about its animals.

However, even as I emphasize this realism, what comes to mind is an element which consistently surfaces, and which is its polar opposite—where the ancient mind distanced itself from reality which was maneuvered to conjure up images invested with their own symbolism.

This argument can be substantiated if we mentally assemble the visual evidence, particularly in the context of the Harappan civilization. What immediately comes to mind are images of the horned tiger, horned elephant, intriguing depictions of wild animals such as the rhinoceros and tiger feeding out of troughs the way tamed animals do, the portrayal of the rhinoceros with a collar, or even the large number of composite figures combining parts of various animals often with a human face. In the absence of a deciphered script, it may not be possible to access the definite meanings of such portrayals, yet what can be said with a fair amount of conviction is that these had their own symbolic rationale, which perhaps reflected a larger world view that used (at least the elephant, and perhaps also the rhinoceros), feared, and possibly even deified these animals. Thus, portraying them in ways other than the actual was perhaps part of a vocabulary of shared meanings.

While these illustrations perhaps give us the most spectacular sense of the use of animal symbolism in proto-historic times, it is not to say that the former is the first efflorescence of such imageries in the domain of art. One only needs to go back in time and mull over some of the prehistoric rock paintings. Of the ones I have looked at in my chapters, most seem to be quite explicit in their portrayals of the animals, yet there are some which call for attention. For instance, one wonders at the rather abstract renderings of the rhinoceros at Kerwaghat and Tarsang. In both cases, it is only the horn which gives the animal some semblance of being a rhinoceros. Evidently, the animals were known, yet chosen to be depicted with an element of abstraction. We can be far from certain about the implications of such garbled representations. The reasons could be stylistic as well as symbolic, but the fact that they occur, gives us a sense of the parallel existence of an imagined world. This, then, also underlines the fact that the tendency to use animals as symbols actually predates a trend that conspicuously manifested in later textual traditions.

Coming to the latter, it may be meaningful to remind ourselves that literary representations in ancient India are replete with allusions to mythical animals. Most pertinent from the perspective of my narrative are images of the surreal multiple-tusked white elephant. We have seen that in Buddhist and Jaina traditions the animal heralds the birth of both Buddha as well as Mahāvīra. While Queen Māya dreams of a six-tusked white elephant entering her womb, Triśalā's first dream amongst the fourteen auspicious ones was that of a four-tusked white elephant. Similarly, more than once, the Jātaka Tales (*Chaddanta* and many others) recount the birth of the Bodhisatta as a white elephant. The king of gods, Indra, rides the four-tusked white Airāvata in the *Rāmāyaṇa*, while Rāvaṇa's palace is guarded by two-, three-, and four-tusked elephants resembling masses of white clouds.

Thus, when it came to playing up the significance of an event or the eminence of a being, the elephant was chosen as a symbol, but since an ordinary one would not suffice, a white elephant was roped in. This magnificent and celestial creature then cut across traditions, and recurrently served as a symbol of exclusivity and auspiciousness. Apart from the colour of the animal itself, what makes it more distinctive is the addition of extra pairs of tusks (elephants are normally known to have a single pair; however, it may be worthwhile to point out that though albino elephants as well as four-tusked elephants are known, these are rare). The aim, clearly, was to create an animal extraordinaire for beings and settings extraordinary.

The point one is trying to make here is that where required, and as the following examples will also elucidate, pragmatism was defied, and the terrains created were ones where the imagined met the real.

In the Dharmasūtras, for instance, we stumble upon another anomaly in the form of the three-toed rhinoceros in the list of the edible *panca pañcanakha* animals. An oddity due to its size as well as its pedal structure, the inclusion does not emanate from a lack of awareness regarding the animal. Rather, it is an awareness superseded by the logic of assimilating a tradition which already extolled the use of rhino skin on ritual occasions.

Here, it is crucial to underline that even in death (as in life) an animal could be a powerful metaphor. This is evident, for instance, in prescriptions relating to the use of the skin, horn, and meat of the rhinoceros in ritual contexts, where the idea was to imbibe the might of the animal through the use of its body parts. In fact, the horn of the rhinoceros is celebrated for its exclusivity across the literary genres dealt with in this book. Similarly, the centrality assigned by the *Śatapatha Brāhmaṇa* to the skin of the tiger in the consecration ceremony of the king also serves to illustrate the point. The sovereign stepping on tiger skin during the *rājasūya* was not just symbolic of his imbibing the prowess of the animal, but perhaps also suggested his subjugation of the animal which represented the land. Himself a tiger here, the king treads upon tiger skin which then immediately turns into a symbol of supremacy since it signifies a dead animal which has been overpowered and reduced to a lifeless form.

Where required, the animal could also be divorced from its fundamental character. As seen in Jātaka Tales like the *Kuntani Jātaka* and the *Tittira Jātaka,* the tiger, itself a perpetrator of fatalities, is transformed into an instrument for delivering justice. In the *Mahābhārata,*

Damayantī's anguish for her lost husband, Nala, in the heart of the forest is once again a reminder of the tiger's association with royalty. As she wails, it is in 'a four-tusked, broad-jowled tiger . . . sovereign of animals the master of this jungle' that she finds befitting audience. The king of the jungle, only he could bring her tidings regarding her companion Nala, also a king. The image of the animal is then inverted, and from a dreaded adversary it is transformed into a passive audience to Damayantī's pleadings to eat her, despite which the mighty carnivore moves on without harming her.

Similarly, we are reminded of Rāma's frantic search for Sītā, when he chooses a deer, an elephant, and a tiger to enquire about his beloved. While the choice of the deer and the elephant are symbolic of the characteristics Sītā shares with them (beautiful fawn-like eyes and smooth thighs like an elephant's trunk), the tiger's presence in the scene is somewhat perplexing. There is no attempt to compare Sītā with the predator, nor is there an apprehension regarding the latter having harmed the former. Rather, the animal needs to be reassured of its safety. Reducing the animal to a submissive role is, then, also a means of underlining the prowess of Rāma who could subdue all.

The realm of the imagined could also extend to hierarchies operating in the wild. A tendency running across textual traditions is to juxtapose these big mammals, and the imageries constructed are not always in adherence with what actually transpires in the wild. For instance, while the *Milindapañha* pits the rhinoceros against the elephant, a similar tendency can be seen in the context of the elephant and lion which are repeatedly pitted against each other, with the lion being the inevitable victor (in the wild, though lions are known to attack elephants when food is scarce, when they cross paths, the maned predator will not easily take on an adult elephant getting the better of which is in any case an arduous task). Given that the theme is so recurrent, it would be pertinent to ponder over the significance of such allusions. While the reasons may be more complex, in view of the wide distribution of both lions and elephants until fairly recent times, a possibility which seems plausible is that real life overlap of ranges may have generated such imageries.

A far more pronounced tendency, however, is to repeatedly juxtapose the two big cats, the tiger and the lion. The purpose of putting these animals together also needs some reflection. Evidently, both inspired awe, and were frequently employed as metaphors for power, yet were made to jostle for supremacy. I would argue that such references should be seen as cultural resonances endorsing perceptions that alternately chose the two as the animals par excellence, and making them interact within the same physical space may have been a ploy devised to reiterate the same.

Apart from embedded cultural implications, these allusions also subtly recreate the ecological landscape for us. In a land, where in recent centuries, habitat destruction has threatened both these big cats, and there have been almost no opportunities for them (with the lion now confined to the Gir forest in Gujarat) to cross paths in the wild, it is enthralling to visualize a landscape where the two were close enough to lead to encounters.

Not just this, there are also glimpses in these texts of an environmental consciousness that has a strikingly modern resonance. What immediately strikes us is the ecologically evocative *Vyaggha Jātaka* emphasizing the role of big predators in keeping the forest intact. An analogous reverberation comes through in the *Udyogaparvan* of the *Mahābhārata*, where one is eloquently reminded of the symbiotic relationship between the forest and the tiger, each protecting the other. Entrenched herein are fairly obvious pointers to the ecological scheme which functions on the premise of the interaction and interdependence of the diverse units within it.

In attempting to trace the human interface with the animals under study, it may also be worthwhile to highlight the multiplicity of contexts witnessing the coming together of the human and animal worlds. Within these, what runs across as a connecting thread is the theme of conflict as well as coexistence. Here our entry point would be the early rock paintings showing these animals being hunted. While as pointed out, tiger portrayals are rare in the rock art regions within the study area, depictions of rhinoceros and elephant hunts are elaborate, the animals being generally taken on and wounded by armed human figures. However, it is not always conflict that is depicted. Though there are graphic representations of attempts at asserting the dominance of humans over nature, there are also poignant and compassionate glimpses of pregnant elephants or a mother and baby rhinoceros which captured the attention of these early artists.

Moving on, we have movable art in the form of the Harappan seals and tablets, which conjure up a maze of images when we think of these animals. While in the absence of a deciphered script, Harappan iconography remains a mystery, the special importance attributed to animals within it has already been emphasized.

More specifically, we are reminded of the human–tiger contest motifs that convey a sense of the animal being treated as a force to be reckoned with, and, thus, to be guarded against. On the other hand, images of caparisoned elephants could perhaps be perceived as early beginnings of a close cultural association with the pachyderm, which seems to have been kept, used, and looked after.

In fact, when we consider the elephant, what instantly comes to mind is also the extensive repertoire of elaborately decorated terracotta elephants of the later periods. The popular elephant and rider motif depicted in rock paintings and terracottas of the early historic period are telling testimonies of a close interface between humans and the pachyderm.

Ancient Indian literature amplifies such imageries of conflict and coexistence with the faunal world. Though the elephant is perhaps the best exemplar of a relationship fraught with harmony as well as discord, it is possible to tease out similar glimpses for the rhinoceros as well as the tiger. Broadly speaking, their characteristics are considered worthy of emulation, yet their destructive potential is recognized and underlined.

Scattered across these early texts are glimpses embodying diverse facets of human interactions with these animals. The rhinoceros, for instance, is assigned a relatively passive role in the *Sudhābhojana Jātaka*, where it is found along with other brute creatures in the vicinity of a hermitage sans any conflict. However, in the *Vessantara Jātaka*, it is one of the forces to be reckoned with along with lions and tigers. Similarly, while the *Atharvaveda* alludes to lions and tigers going about man-eating, and the *Śatapatha Brāhmaṇa* refers to a tiger killing milk-giving cattle, we also have the imagery of the infant son of Duhṣanta and Śakuntalā grappling with tigers, lions, and other wild beasts, and playing, riding, and taming them in the *Mahābhārata*. Equally fascinating are references in the classical Western accounts to the presence of tamed rhinos and tigers.

It is, however, the elephant that is exceptional in terms of the intimacy it has shared with humans across millennia. In the docile and obedient animal which comes when called in the *Aitareya Brāhmaṇa*, or the poignant tale of the bond between an elephant and a group of carpenters in the *Alīnacitta Jātaka*, or the compassionate and loyal elephant of wounded Porus in the Battle of Hydaspes, we have images of an animal which was lovingly engaged with on terms of reciprocity. However, things could well get awry if it was a rogue elephant on rampage, like the one which eventually had to be killed as in the *Culladhanuggaha Jātaka* or the elephant turning upon its trainers for the torture they had subjected it to in the *Dubbalakaṭṭha Jātaka*. Even elephants running amok in battle with calamitous results for the camp which had set out to use them to its advantage, exemplifies how an ally could become an adversary.

Having assimilated some points of intersection (harmonious as well as discordant) in human interactions with these mega mammals, it may be worthwhile to also highlight the glimpses texts provide regarding using the body parts of these mega mammals. References occur in contexts ranging from the ritual to the political, and in this sense, these animals permeate the world of gods as well as humans. While the Vedic interest in rhinoceros skin graduates to an interest in its meat for appeasing ancestors in the Dharmasūtras, the centrality attributed to tiger skin by Vedic texts in the *rājasuya* ceremony establishes the striped carnivore as an uncontested exemplar of royal authority. Similarly, ancient Indian texts consistently underline the interest in ivory even as the animal itself is perceived as a critical resource in the *Arthaśāstra*.

Integrating our evidence, it may be worthwhile to underline that while it may not always be possible to establish a connection between the fate of an animal and historical processes transpiring over millennia, each of the three animals under study has its own distinct story to tell. The impression looming large is that when it comes to weaving historically connected narratives, it is the elephant which takes the lead by virtue of its close and consistent interactions with human cultures. The trajectories seem somewhat different in the case of the rhinoceros and the tiger, particularly when we consider the evidence offered by the visual records we sift through. The waning of depictions after the Harappans may possibly be explained by the fact that unlike the elephant, these two animals were not encountered or engaged with on a regular basis. While this may partly explain their not being chosen as popular motifs for portrayal, it also reiterates the point we started out with, that is, an absence or a break in record does not always imply absence in real life. Notwithstanding the dearth of visual evidence (though one keeps in mind that portrayals of both persist even as late as the Gupta period in the form of the tiger-slayer and the rhinoceros-slayer type coins of Samudragupta and Kumāragupta I), textual sources attest to continuing interactions with these animals. This book then also makes a strong plea for the need to corroborate material sources with literary ones for any historical reconstruction.

In view of the rich material and textual evidence as well as the contemporary interest in megafauna and its conservation, it is only natural to ask why faunal histories in general have not received as much attention as other areas of historical enquiry. We have to, perhaps, take into account historiographical shifts, and the gradual coming into being of environmental history as a field engaging with historical records of human interfaces with forests, rivers, fauna, and flora.

The importance of long-ranging faunal histories emanates from a growing recognition of the implications these have for our own world, particularly when it comes to policymaking and generating consciousness regarding the need to protect our faunal heritage in the face of impending threats. For instance, the *Arthaśāstra*'s idea of creating protected spaces for animals in general, and elephants in particular, along with Aśoka prohibiting the killing of selective species are cogent pointers to the ecological sensibilities and conservation concerns of ancient India. Here it may be relevant to examine whether such directives came in the wake of an actual decline in faunal resources, or whether it was simply a royal endeavour to seize control of the forest and its inhabitants. While a definite answer would require more evidence, what can be said for the moment is that any attempt to write an environmental history of India needs to take cognizance of these early voices.

This book presents enough evidence to unsettle the idea of an idyllic ecological past, and in that sense, it takes forward Kathleen

Morrison's (2014, 41) contention that even prior to colonization, several parts of India witnessed environmental transformations that were as significant as later ones. While Morrison has worked on forests, water, and agricultural change, my work envisions early India and its interactions with the natural world by focusing on megafauna as an index for chronicling landscape and ecological changes in ancient India. It argues that animals have histories which can be retrieved using multiple prisms, and that these histories cannot be divorced from the interface humans have had with them. It also reiterates that when it comes to working out a long-term ecological history, owing to their vulnerability (discussed in Chapter 1), megafauna can serve as crucial if not always adequate indices of environmental quality.

The trails of these mega mammals since prehistoric times present a kaleidoscope of shifts in their fortunes that fluctuated with changing forms of human settlement and production as also with the ebb and flow of kingdoms and cultures. Not only do these animals have captivating stories to tell of their journeys across millennia, they also help us map past ecologies. The cultural as well as ecological questions these animals pose may not always be neatly answered. Nevertheless, asking them is integral to understanding aspects of the environmental history of ancient India.

REFERENCES

Agrawala, V.S. 1947–8. 'Terracotta Figurines of Ahichchhatrā, District Bareilly, U.P.'. *Ancient India*, 4: 104–79.

Ali, Salim. 1983. 'The Moghul Emperors of India as Naturalists and Sportsmen'. In J.C. Daniel, ed., *A Century of Natural History*, pp. 1–16. Bombay: Bombay Natural History Society.

Altekar, A.S., and Vijayakanta Mishra. 1959. *Report on Kumrahar Excavations 1951–1955*. Patna: K.P. Jayaswal Research Institute.

Alter, Stephen. 2004. *Elephas Maximus: A Portrait of the Indian Elephant*. Orlando: Harcourt.

Alur, K.R. 1980. 'Faunal Remains from the Vindhyas and the Ganga Valley'. In G.R. Sharma, V.D. Misra, D. Mandal, B.B. Misra, and J.N. Pal, eds, *From Hunting and Food Gathering to Domestication of Plants and Animals: Beginnings of Agriculture*, pp. 201–30. Allahabad: Abinash Prakashan.

Amundson, Ronald, and Elise Pendall. 1991. 'Pedology and Late Quaternary Environments Surrounding Harappa: A Review and Synthesis'. In Richard H. Meadow ed., *Harappa Excavations 1986–1990: A Multidisciplinary Approach to Third Millennium Urbanism*. Monographs in World Archaeology No. 3, pp. 13–27. Madison, Wisconsin: Prehistory Press.

Anand, Kumar. 1991–2. 'Rock Paintings of Kaimur'. *The Journal of the Bihar Purāvid Pariṣad*, 15 and 16: 55–64.

Anderson, Kenneth. 1967. *Tiger Roars*. London: George Allen & Unwin Ltd.

Ashfaque, Syed M. 1974–86. 'Constellations in the Harappan Seals'. *Pakistan Archaeology*, 10–22: 135–67.

Atre, Shubhangana. 1985–6. 'Lady of Beasts'. *Puratattva*, 16: 7–14.

———. 1987. *The Archetypal Mother: A Systemic Approach to Harappan Religion*. Pune: Ravish Publishers.

———. 1990. 'Harappan Seal Motifs and the Animal Retinue'. *Bulletin of the Deccan College Post-Graduate and Research Institute*, 49: 43–9.

Auboyer, Jeannine. 1972. 'Animals in India'. In Alan Houghton Brodrick, ed., *Animals in Archaeology*, pp. 115–45. New York: Praeger.

Badam, G.L. 1979. *Pleistocene Fauna of India with Special Reference to the Siwaliks*. Pune: Deccan College Postgraduate and Research Institute.

———. 1985. 'Pleistocene Fossil Vertebrates in India'. In K.N. Dikshit, ed., *Archaeological Perspective of India Since Independence*, pp. 123–37. New Delhi: Books & Books.

Badam, G.L., and S.C. Jayakaran. 1993. 'Pleistocene Vertebrate Fossils from Tamil Nadu, India'. In Nina G. Jablonski, ed., *Evolving Landscapes and Evolving Biotas of East Asia Since the Mid-Tertiary*, pp. 241–64. Hong Kong: Centre of Asian Studies, University of Hong Kong.

Badam, G.L., and V.G. Sathe. 1991. 'Subsistence Economy of the Indus Civilization'. In C. Margabandhu, K.S. Ramachandran, A.P. Sagar, and D.K. Sinha, eds, *Indian Archaeological Heritage* (*Sh. K.V. Soundara Rajan Felicitation Volume*), vol. I, pp. 135–43. Delhi: Agam Kala Prakashan.

Bala, Madhu. 2003. 'Minor Antiquities'. In B.B. Lal, Jagat Pati Joshi, B.K. Thapar, and Madhu Bala, eds, *Excavations at Kalibangan: The Early Harappans (1961–1969)*, pp. 223–42. Memoirs of the Archaeological Survey of India No. 98. New Delhi: Archaeological Survey of India.

Ball, V. 1879–88. 'On the Identification of the Animals and Plants of India which Were Known to Early Greek Authors'. *Proceedings of the Royal Irish Academy. Polite Literature and Antiquities*, 2: 302–46.

———. 1889–91. 'Further Notes on the Identification of the Animals and Plants of India which Were Known to Early Greek Authors'. *Proceedings of the Royal Irish Academy*, 1: 1–9.

Banerjee, S., and S. Chakraborty. 1973. 'Remains of the Great One-horned Rhinoceros, *Rhinoceros unicornis* Linnaeus from Rajasthan'. *Science and Culture*, 39 (10): 430–1.

Banerjee, S., R.N. Mukherjee, and B. Nath. 2003. 'Identification of Animal Remains'. In B.B. Lal, Jagat Pati Joshi, B.K. Thapar, and Madhu Bala, eds, *Excavations at Kalibangan: The Early Harappans (1961–1969)*, pp. 267–339. Memoirs of the Archaeological Survey of India No. 98. New Delhi: Archaeological Survey of India.

Banerji, Arundhati. 1994. *Early Indian Terracotta Art circa 2000–300 BC (Northern and Western India)*. New Delhi: Harman Publishing House.

Banerji, Sures Chandra. 1962. *Dharma Sūtras: A Study in Their Origin and Development*. Calcutta: Punthi Pustak.

———. 1980. *Flora and Fauna in Sanskrit Literature*. Calcutta: Naya Prokash.

Bapat, G.V. 1985–7. 'The Elephant in the Rāmāyaṇic Simile'. *Bhāratī*, 16: 37–74.

Barua, Maan. 2006. 'The Road Ahead for the Indian One-horned Rhinoceros'. *Sanctuary Asia*, 26 (1): 48–9.

Baskaran, N., Surendra Varma, C.K. Sar, and Raman Sukumar. 2011. 'Current Status of Asian Elephants in India'. *Gajah*, 35: 47–54.

Bautze, Joachim Karl. 1985. 'The Problem of the Khaḍga (*Rhinoceros unicornis*) in the Light of Archaeological Finds and Art'. In J. Schotsman, and M. Taddei, eds, *South Asian Archaeology 1983*, pp. 405–33. Naples: Istituto Universitario Orientale.

———. 1995. *Early Indian Terracottas*. Leiden; New York; Koln: E.J. Brill.

Bhardwaj, Sudarshan. 1997. 'A Terracotta Cylindrical Seal from Rakhigarhi'. In Manmohan Kumar, ed., *Numismatic Studies*, vol. V, pp. 153–5. New Delhi: Harman Publishing House.

Bhishagratna, Kaviraj Kunjalal. 1963. *An English Translation of the Sushruta Samhita Based on Original Sanskrit Text*. Vol. I, *Sutra-Sthāna*. Varanasi: The Chowkhamba Sanskrit Series Office.

Bigwood, J.M. 1993. 'Aristotle and the Elephant Again'. *The American Journal of Philology*, 114 (4): 537–55.

Bisht, R.S. 1982. 'Excavations at Banawali 1974–77'. In Gregory L. Possehl, ed., *Harappan Civilization: A Contemporary Perspective*, pp. 113–24. New Delhi: Oxford & IBH Publishing Co.

Biswas, S.S. 2012. *Rock Art: A Catalogue*. New Delhi: Indira Gandhi National Centre for the Arts and Indian Archaeological Society.

Blakiston, J.F., ed. 1927. *Annual Report of the Archaeological Survey of India 1924–1925*. Calcutta: Government of India Central Publication Branch.

Blanford, W.T. 1888–91. *The Fauna of British India, Including Ceylon and Burma. Mammalia*. London: Taylor and Francis.

Blumenschine, R.J., and U.C. Chattopadhyaya. 1983. 'A Preliminary Report on the Terminal Pleistocene Fauna of the Middle Son Valley'. In G.R. Sharma and J.D. Clark, eds, *Palaeoenvironments and Prehistory in the Middle Son Valley*, pp. 281–4. Allahabad: Abinash Prakashan.

Böhtlingk, Otto, and Rudolph Roth. 1990. *Sanskrit Wörterbuch*, vol. IV. Delhi: Motilal Banarsidass Publishers Private Ltd.

Bose, Shibani. 2014. 'From Eminence to Near Extinction: The Journey of the Greater One-Horned Rhino'. In Mahesh Rangarajan and K. Sivaramakrishnan, eds, *Shifting Ground: People, Animals, and Mobility in India's Environmental History*, pp. 65–77. New Delhi: Oxford University Press.

———. 2018. 'Before the Written Word'. In Divyabhanusinh, Asok Kumar Das, and Shibani Bose, eds, *The Story of India's Unicorns*, pp. 32–43. Mumbai: Marg Foundation.

Bowles, Adam, trans. 2006. *Mahābhārata. Book Eight: Karna*, vol. I. Clay Sanskrit Library. New York: New York University Press: JJC Foundation.

Briggs, G.W. 1931. 'The Indian Rhinoceros as a Sacred Animal'. *Journal of the American Oriental Society*, 51: 276–82.

Brockington, J.L. 1984. *Righteous Rama: The Evolution of an Epic*. Delhi: Oxford University Press.

———. 1998. *The Sanskrit Epics*. Leiden; Boston: Brill.

van Buitenen, J.A.B., trans. and ed. 1973. *The Mahābhārata. Book 1: The Book of the Beginning*, vol. 1. Chicago and London: The University of Chicago Press.

———, trans. and ed. 1975. *The Mahābhārata. Book 2: The Book of the Assembly Hall, Book 3: The Book of the Forest*, vol. 2. Chicago and London: The University of Chicago Press.

———, trans. and ed. 1978. *The Mahābhārata. Book 4: The Book of Virata, Book 5: The Book of the Effort*, vol. 3. Chicago and London: The University of Chicago Press.

Burton, R.G. 1933. *The Book of the Tiger*. London: Hutchinson & Company.

Casal, Jean-Marie. 1964. *Fouilles D'Amri*, vol. I, *Texte*. Paris: Librairie C. Klincksieck.

Casson, Lionel. 1989. *The Periplus Maris Erythraei: Text with Introduction, Translation, and Commentary*. Princeton, N.J.: Princeton University Press.

Chhabra, B.Ch. 1973. 'Elephant in Indian Art'. *Journal of Indian History*, 51(3): 485–9.

Chakrabarti, Dilip K. 1999. *India: An Archaeological History*. New Delhi: Oxford University Press.

———. 2006. *The Oxford Companion to Indian Archaeology: The Archaeological Foundations of Ancient India: Stone Age to AD 13th Century*. New Delhi: Oxford University Press.

Chakravarti, Monmohan. 1906. 'Animals in the Inscriptions of Piyadasi'. *Memoirs of the Asiatic Society of Bengal*, 1: 361–74.

Chakravarti, Uma. 1993. 'Women, Men and Beasts: The *Jataka* as Popular Tradition'. *Studies in History*, 9 (1): 43–70.

Chalmers, Robert , trans. 1969. *The Jātaka or Stories of the Buddha's Former Births*, edited by E.B. Cowell, vol. I. London: Luzac & Company.
Chandra, Moti. 1957–9. 'Ancient Indian Ivories'. *Prince of Wales Museum Bulletin*, 6: 4–63.
Chandra, Pramod. 1970. *Stone Sculpture in the Allahabad Museum: A Descriptive Catalogue*. Poona: American Institute of Indian Studies.
Chandra, Rai Govind. 1973. *Studies of Indus Valley Terracottas*. Varanasi: Bhartiya Publishing House.
Chattopadhyaya, U.C. 1991. 'A Study of Subsistence and Settlement Patterns during the Late Prehistory of North-Central India'. Unpublished PhD dissertation. University of Cambridge.
———. 1996. 'Settlement Pattern and the Spatial Organization of Subsistence and Mortuary Practices in the Mesolithic Ganges Valley, North-central India'. *World Archaeology*, 27 (3): 461–76.
———. 1999. 'The Ganges Neolithic'. In M. Ember, and P.A. Peregrine, eds, *Encyclopaedia of Prehistory*, pp. 127–32. New Haven: Human Relations Area Files Inc.
———. 2001. 'Complementary Partitioned Network System: A Regional Model of Post-Pleistocene Human Adaptations in the Vindhya-Ganga Complex'. *The Oriental Anthroplogist*, I (1): 16–34.
———. 2002. 'Researches in Archaeozoology of the Holocene period (Including the Harappan Tradition in India and Pakistan)'. In S. Settar and R. Korisettar, eds, *Archaeology and Interactive Disciplines: Indian Archaeology in Retrospect*, vol. III, pp. 365–422. New Delhi: Manohar.
Chaudhari, Sibadas. 1952–4. 'Concordance of the Fauna in the Rāmāyaṇa'. *Indian Historical Quarterly*, 28: 135–41, 249–56, 350–9; 29: 56–63, 121–8, 276–85, 378–86; 30: 148–53.
Chitalwala, Y.M. 1990. 'The Disappearance of Rhino from Saurashtra: A Study in Palaeoecology'. *Bulletin of the Deccan College Post-Graduate and Research Institute*, 49: 79–82.
Chitalwala, Y.M., and Thomas P.K. 1977–8. 'Faunal Remains from Khanpur and Their Bearing on the Culture, Economy and Environment'. *Bulletin of the Deccan College Post-Graduate and Research Institute*, 37 (1–4): 11–14.
Chowdhury, K.A., and S.S. Ghosh. 1951. 'Plant-remains from Harappa 1946'. *Ancient India*, 7: 3–19.
Clutton-Brock, Juliet. 1965. *Excavations at Langhnaj*. Part II: *The Fauna*. Poona: Deccan College Postgraduate and Research Institute.
Cockburn, John. 1883. 'On the Recent Existence of Rhinoceros Indicus in the North-Western Provinces and a Description of a Tracing of an Archaic Rock Painting from Mirzapore Representing the Hunting of this Animal'. *Journal of the Asiatic Society of Bengal*, 52 (1): 56–64.

Compagnoni, Bruno. 1979. 'Preliminary Report on the Faunal Remains from Protohistoric Settlements of Swat'. In Maurizio Taddei, ed., *South Asian Archaeology 1977*, vol. 2, pp. 697–700. Naples: Istituto Universitario Orientale.

Constable, Archibald, and Vincent A. Smith, trans. 1914. *Travels in the Mogul Empire A.D. 1656–1668 by Francois Bernier*. New York: Oxford University Press.

Cowell, E.B., and W.H.D. Rouse, trans. 1969. *The Jātaka or Stories of the Buddha's Former Births*, edited by E.B. Cowell, vol. VI. London: Luzac & Company.

Crosby, Kate, trans. 2009. *Mahābhārata Book Ten: Dead of Night; Book Eleven: The Women*. Clay Sanskrit Library. New York: New York University Press: JJC Foundation.

Dales, George F., and J. Mark Kenoyer. 1989. *Preliminary Report on the Fourth Season (January 15–March 31, 1989) of Research at Harappa, Pakistan*. Berkeley/Madison: University of California and University of Wisconsin.

Dani, Ahmad Hasan. 1965–6. 'Shaikhan Dheri Excavation (1963 and 1964 Seasons)'. *Ancient Pakistan*, 2: 17–214.

———. 1967. 'Timargarha and the Gandhara Grave Culture'. *Ancient Pakistan*, 3: 1–407.

———. 1983. *Chilas: The City of Nanga Parvat (Dyamar)*. Islamabad: Ahmad Hasan Dani, Director Centre for the Study of the Civilisations of Central Asia, Quaid-i-Azam University.

Daniel, J.C. 1998. *The Asian Elephant: A Natural History*. Dehradun: Natraj Publishers.

———. 2001. *The Tiger in India: A Natural History*. Dehradun: Natraj Publishers.

Danino, Michel. 2015. 'Climate and Environment in the Indus-Sarasvati Civilization'. In Arundhati Banerji, ed., *Ratnaśrī: Gleanings from Indian Archaeology, Art, History and Indology: Papers Presented in Memory of Dr. N.R. Banerji*, pp. 39–47. New Delhi: Kaveri Books.

Das Gupta, Charu Chandra. 1961. *Origin and Evolution of Indian Clay Sculpture*. Calcutta: University of Calcutta.

Davids, Rhys T.W., trans. 1890. *The Questions of King Milinda*. Part I: *Sacred Books of the East*, edited by F. Max Müller, vol. XXXV. Oxford: Clarendon Press.

———. 1894. The *Questions of King Milinda*. Part II: *Sacred Books of the East*, edited by F. Max Müller, vol. XXXVI. Oxford: Clarendon Press.

———. 1899. *Dialogues of the Buddha*. Part I. London: Pali Text Society.

———. 1950. *The Book of the Kindred Sayings (Sanyutta-Nikāya) or Grouped Suttas*. Part I: *Kindred Sayings with Verses (Sagāthā-Vagga)*. London: Luzac & Company Ltd.

———. 1952. *The Book of the Kindred Sayings (Sanyutta-Nikāya) or Grouped Suttas*. Part II: *The Nidāna Book (Nidāna-Vagga)*. London: Luzac & Company Ltd.

———. 1973. *Buddhist Birth-Stories (Jātaka Tales): The Commentarial Introduction Entitled Nidāna-Kathā: The Story of the Lineage*. Varanasi; Delhi: Indological Book House.

Davids, Rhys T.W., and C.A.F. Rhys Davids. 1965. *Dialogues of the Buddha*. Part III. London: Luzac & Company Ltd.

———, trans. 1966. *Dialogues of the Buddha*. Part II. London: Luzac & Company Ltd.

Davids, T.W. Rhys, and Stede William. 1975. *Pali–English Dictionary*. New Delhi: Oriental Books Reprint Corporation.

De Terra, Helmut, and Thomas Thomson Paterson. 1939. *The Ice Age in the Indian Subcontinent and Associated Human Cultures (With Special Reference to Jammu, Kashmir, Ladakh, Sind, Liddar and Central & Peninsular India)*. Washington D.C.: Carnegie Institution of Washington.

Delort, Robert. 1992. *Life and Lore of the Elephant*. New York: Harry N. Abrams.

Dhavalikar, M.K. 1988. 'Early Bronzes'. In Asha Rani Mathur, ed., *The Great Tradition—Indian Bronze Masterpieces*, pp. 12–21. New Delhi: Brijbasi Printers Pvt. Ltd.

Dikshitar, V.R. Ramachandra. 1944. *War in Ancient India*. Madras: Macmillan.

Dinerstein, Eric. 1992. 'Effects of Rhinoceros Unicornis on Riverine Forest Structure in Lowland Nepal'. *Ecology*, 73(2): 701–4.

———. 2003. *The Return of The Unicorns: The Natural History and Conservation of the Greater Indian One-Horned Rhinoceros*. New York: Columbia University Press.

Divyabhanusinh. 1995. *The End of a Trail: The Cheetah in India*. New Delhi: Banyan Books.

———. 2005. *The Story of Asia's Lions*. Mumbai: Marg Publications.

Dubey, Sitaram, Ashok Kumar Singh, and Gautam Kumar Lama. 2012. *Pakkākot: Some New Archaeological Dimensions of Mid-Ganga Plain*. Delhi: Rishi Publication.

During Caspers, E.C.L. 1979. 'Caricatures, Grotesques and Glamour in Indus Valley Art'. In M. Taddei, ed., *South Asian Archaeology 1977*, pp. 345–74. Naples: Istituto Universitario Orientale.

———. 1987. 'Singular Aspects of Indus Valley Artistry'. In B.M. Pande and B.D. Chattopadhyaya, eds, *Archaeology and History: Essays in Memory of Sh. A. Ghosh*, vol. I, pp. 219–33. Delhi: Agam Kala Prakashan.

———. 1989. 'Magic Hunting Practices in Harappan Times'. In K. Frifelt and P. Sorensen, eds, *South Asian Archaeology 1985*, Occasional

Papers No. 4, pp. 227–36. London: Scandinavian Institute of Asian Studies.

———. 1992. 'Rituals and Belief Sytems in the Indus Valley Civilization'. In A.W. Van den Hoek, D.H.A. Kolff, and M.S. Oort, eds, *Ritual, State and History in South Asia: Essays in Honour of J.C. Heesterman*, pp. 102–27. Leiden, The Netherlands: E.J. Brill.

Dutta, Arup Kumar. 1991. *Unicornis: The Great Indian One-Horned Rhinoceros*. Delhi: Konark Publishers Pvt. Ltd.

Dwivedi, Vinod P. 1976. *Indian Ivories: A Survey of Indian Ivory and Bone Carvings from the Earliest to the Modern Times*. Delhi: Agam Prakashan.

Edgerton, Franklin. 1970. *Buddhist Hybrid Sanskrit Grammar and Dictionary*, vol. II. Delhi. Motilal Banarsidass.

Eggeling, Julius, trans. 1882. *The Śatapatha Brāhmaṇa*. Part I. *Sacred Books of the East*, edited by F. Max Müller, vol. XII. Oxford: Clarendon Press.

———, trans. 1885. *The Śatapatha Brāhmaṇa*. Part II. *Sacred Books of the East*, edited by F. Max Müller, vol. XXVI. Oxford: Clarendon Press.

———, trans. 1894. *The Śatapatha Brāhmaṇa*. Part III. *Sacred Books of the East*, edited by F. Max Müller, vol. XLI. Oxford: Clarendon Press.

———, trans. 1897. *The Śatapatha Brāhmaṇa*. Part IV. *Sacred Books of the East*, edited by F. Max Müller, vol. XLIII. Oxford: Clarendon Press.

———, trans. 1900. *The Śatapatha Brāhmaṇa*. Part V. *Sacred Books of the East*, edited by F. Max Müller, vol. XLIV. Oxford: Clarendon Press.

Eltringham, S.K. 1991. *The Illustrated Encyclopedia of Elephants: From Their Origins and Evolution to Their Ceremonial and Working Relationship with Man*. New York: Crescent Books.

Enzel, Y., L.L. Ely, S. Mishra, R. Ramesh, R. Amit, B. Lazar, S.N. Rajaguru, V.R. Baker, and A. Sandler. 2000. 'High Resolution Holocene Environmental Changes in the Thar Desert, Northwestern India'. In Nayanjot Lahiri, ed., *The Decline and Fall of the Indus Civilization*, pp. 226–38. New Delhi: Permanent Black.

Fairservis, Walter A., Jr. 1976. *Excavations at the Harappan Site of Allahdino: The Seals and Other Inscribed Material*. New York: American Museum of Natural History.

Fentress, M.A. 1976. *Resource Access, Exchange Systems, and Regional Interaction in the Indus Valley: An Investigation of Archaeological Variability at Harappa and Mohenjodaro*. PhD dissertation. Michigan, Ann Arbor: University Microfilms International.

Fitzgerald, James L., trans. and ed. 2004. *The Mahābhārata. Book 11: The Book of the Women, Book 12: The Book of Peace*, Part I, vol. VII. Chicago: The University of Chicago Press.

Francfort, H.P. 1984. 'The Harappan Settlement of Shortughai'. In B.B. Lal and S.P. Gupta, eds, *Frontiers of the Indus Civilization*, pp. 301–10. New Delhi: Books & Books.

Francis, H.T., trans. 1969. *The Jātaka or Stories of the Buddha's Former Births*, edited by E.B. Cowell, vol. V. London: Luzac & Company Ltd.

Francis, H.T., and R.A. Neil., trans. 1969. *The Jātaka or Stories of the Buddha's Former Births*, edited by E.B. Cowell, vol. III. London: Luzac & Company Ltd.

Frankfort, H. 1939. *Cylinder Seals: A Documentary Essay on the Art and Religion of the Ancient Near East*. London: Macmillan and Co.

Frenez, Dennys, and Massimo Vidale. 2012. 'Harappan Chimaeras as "Symbolic Hypertexts". Some Thoughts on Plato, Chimaera and the Indus Civilization'. *South Asian Studies*, 28(2): 107–30.

Gadgil, Madhav, and Ramachandra Guha. 1992. *This Fissured Land: An Ecological History of India*. New Delhi: Oxford University Press.

Gairola, Vachaspati., ed. and trans. 1977. *The Arthaśāstra of Kauṭilya and the Cāṇakya-Sūtra*. Varanasi: Chowkhamba Vidyabhawan.

Ganguli, Kisari Mohan, trans. 1970. *The Mahabharata of Krishna-Dwaipayana Vyasa*, vol. XI: *Anusasana Parva*. Part II. New Delhi: Munshiram Manoharlal Publishers Pvt. Ltd.

Gaur, R.C. 1983. *Excavations at Atrañjīkherā: Early Civilization of the Upper Gaṅgā Basin*. Delhi: Motilal Banarsidass.

van der Geer, Alexandra Anna Enrica 2008. *Animals in Stone: Indian Mammals Sculptured through Time*. Leiden: Brill.

Ghosh, Manoranjan, Rai Sahib. 1932. *Rock-Paintings and Other Antiquities of Prehistoric and Later Times*. Memoirs of the Archaeological Survey of India No. 24. Calcutta: Government of India, Central Publication Branch.

Glover, Richard. 1944. 'The Elephant in Ancient War'. *The Classical Journal*, 39(5): 257–69.

Gokhale, Balkrishna G. 1974. 'Animal Symbolism in Early Buddhist Literature and Art'. *East and West*, nos 1–2: 111–20.

Goldman, Robert P., trans. 1984. *The Rāmāyaṇa of Vālmīki: An Epic of Ancient India*, vol. I: *Bālakāṇḍa*. Princeton, N.J.: Princeton University Press.

Goldman, Robert P., and Sally J. Sutherland Goldman, trans. 1996. *The Rāmāyaṇa of Vālmīki: An Epic of Ancient India*, vol. V: *Sundarakāṇḍa*. Princeton, N.J.: Princeton University Press.

Goldman, Robert P., Sally J. Sutherland Goldman, and Barend A. van Nooten, trans. 2009. *The Rāmāyaṇa of Vālmīki: An Epic of Ancient India*, vol. VI: *Yuddhakāṇḍa*. Princeton, N.J.: Princeton University Press.

Gole, Susan. 1988. *Maps of Mughal India Drawn by Colonel Jean-Baptiste-Joseph-Gentil, Agent for the French Government to the Court of Shuja-ud-daula at Faizabad, in 1770*. New Delhi: Manohar Publications.

Gorakshar, Sadashiv. 1979. *Animal in Indian Art*. Bombay: Prince of Wales Museum of Western India.

Goyal, Pankaj, and P.P. Joglekar. 2012. 'Archaeozoological Remains from the Site of Kanmer'. In J.S. Kharakwal, Y.S. Rawat, and Toshiki Osada, eds, *Excavation at Kanmer 2005–06–2008–09*, pp. 767–94. Japan, Kyoto: Indus Project, Research Institute for Humanity and Nature; India, Gandhinagar: Gujarat State Department of Archaeology; India, Udaipur: Institute of Rajasthan Studies, JRN Rajasthan Vidyapeeth.

Griffith, Ralph T.H., trans. 1899. *The Texts of the White Yajurveda*. Benaras: E.J. Lazarus & Co.

———, trans. 1963a. *The Hymns of the Ṛgveda, Translated with a Popular Commentary*, 2 vols. Varanasi: The Chowkhamba Sanskrit Series Office.

———, trans. 1963b. *The Ramayan of Valmiki*. Varanasi: The Chowkhamba Sanskrit Series Office.

Guggisberg, C.A.W. 1966. *S.O.S. Rhino*. London: Andre Deutsch.

Gupta, H.P. 1976. 'Holocene Palynology from Meander Lake in the Ganga Valley, District Pratapgarh, U.P.'. *The Palaeobotanist*, 25: 109–19.

Gupta, S.K. 1983. *Elephant in Indian Art and Mythology*. New Delhi: Abhinav Publications.

Gupta, S.P. 1980. *The Roots of Indian Art*. Delhi: B.R. Publishing Corporation.

Hare, E.M., trans. 1947. *Woven Cadences of Early Buddhists (Sutta-Nipāta)*. London: Oxford University Press.

———, trans. 1961. *The Book of the Gradual Sayings (Aṅguttara-Nikāya) or More-Numbered Suttas*, vol. III. London: Luzac & Company Ltd.

Härtel, Herbert. 1993. *Excavations at Sonkh: 2500 Years of a Town in Mathura District*. Berlin: Dietrich Reimer Verlag.

Haug, Martin., trans. 1863. *The Aitareya Brahmanam of the Rigveda*, vol. II. Bombay: Bombay Central Government Book Depot.

Heesterman, J.C. 1957. *The Ancient Indian Royal Consecration: The Rājasūya Described According to the Yajus Texts*. The Hague, The Netherlands: Mouton & Co. Publishers.

Hilton, Julie. 1996. 'The Good, the Bad and the Ugly: Birds, Animals and Omens in Ancient Indian Literature'. In Sarva Daman Singh, ed., *Culture through the Ages (Prof. B.N. Puri Felicitation Volume)*, pp. 59–83. Delhi: Agam Kala Prakashan.

Hooja, Reema, and Vijay Kumar. 1997. 'Aspects of the Early Copper Age in Rajasthan'. In R. Allchin and B. Allchin, eds, *South Asian Archaeology 1995*, pp. 323–39. New Delhi: Oxford & IBH.

Horner, I.B. 1945. 'Early Buddhism and the Taking of Life'. In D.R. Bhandarkar, K.A. Nilakanta Sastri, B.M. Barua, B.K. Ghosh, and P.K. Gode, eds, *B.C. Law Volume*, Part I, pp. 436–55. Calcutta: The Indian Research Institute.

———, trans. 1949. *The Book of the Discipline (Vinaya-Piṭaka)*, vol. I (*Suttavibhaṅga*). London: Luzac & Company Ltd.

———, trans. 1954. *The Collection of the Middle Length Sayings (Majjhima-Nikāya)*, vol. I: *The First Fifty Discourses (Mūlapaṇṇāsa)*. London: Luzac & Company Ltd.

———, trans. 1957a. *The Book of the Discipline (Vinaya-Piṭaka)*, vol. II (*Suttavibhanga*). London: Luzac & Company Ltd.

———, trans. 1957b. *The Book of the Discipline (Vinaya-Piṭaka)*, vol. III (*Suttavibhaṅga*). London: Luzac & Company Ltd.

———, trans. 1957c. *The Collection of the Middle Length Sayings (Majjhima-Nikāya)*, vol. II: *The Middle Fifty Discourses (Majjhimapaṇṇāsa)*. London: Luzac & Company Ltd.

———, trans. 1959. *The Collection of the Middle Length Sayings (Majjhima-Nikāya)*, vol. III: *The Final Fifty Discourses (Uparipaṇṇāsa)*. London: Luzac & Company Ltd.

———, trans. 1962. *The Book of the Discipline (Vinaya-Piṭaka)*, vol. IV (*Mahāvagga*). London: Luzac & Company Ltd.

———, trans. 1963. *The Book of the Discipline (Vinaya-Piṭaka)*, vol. V (*Cullavagga*). London: Luzac & Company Ltd.

Hultzsch, E. 1969. *Corpus Inscriptionum Indicarum Inscriptions of Aśoka*, vol. I. Delhi: Indological Book House.

Indian Archaeology 1963–64: A Review. Available at http://nmma.nic.in/nmma/nmma_doc/Indian%20Archaeology%20Review/Indian%20Archaeology%201963-64%20A%20Review.pdf.

Indian Archaeology 1964–65: A Review. Available at http://nmma.nic.in/nmma/nmma_doc/Indian%20Archaeology%20Review/Indian%20Archaeology%201964-65%20A%20Review.pdf.

Indian Archaeology 1965–66: A Review. Available at http://nmma.nic.in/nmma/nmma_doc/Indian%20Archaeology%20Review/Indian%20Archaeology%201965-66%20A%20Review.pdf.

Indian Archaeology 1966–67: A Review. Available at http://nmma.nic.in/nmma/nmma_doc/Indian%20Archaeology%20Review/Indian%20Archaeology%201966-67%20A%20Review.pdf.

Indian Archaeology 1971–72: A Review. Available at http://nmma.nic.in/nmma/nmma_doc/Indian%20Archaeology%20Review/Indian%20Archaeology%201971-72%20A%20Review.pdf.

Indian Archaeology 1975–76: A Review. Available at http://nmma.nic.in/nmma/nmma_doc/Indian%20Archaeology%20Review/Indian%20Archaeology%201975-76%20A%20Review.pdf.

Indian Archaeology 1981–82: A Review. Available at http://nmma.nic.in/nmma/nmma_doc/Indian%20Archaeology%20Review/Indian%20Archaeology%201981-82%20A%20Review.pdf.

Iyer, K. Bharata. 1977. *Animals in Indian Sculpture*. Bombay: Taraporevala.

Jacobi, Hermann, trans. 1884. *Jaina Sūtras*. Part I. *Sacred Books of the East*, edited by F. Max Müller, vol. XXII. Oxford: Clarendon Press.

———, trans. 1895. *Jaina Sūtras*. Part II. *Sacred Books of the East*, edited by F. Max Müller, vol. XLV. Oxford: Clarendon Press.

Jamison, Stephanie W. 1998. 'Rhinoceros Toes, Manu V.17–18, and the Development of the Dharma System'. *Journal of the American Oriental Society*, 118 (2): 249–56.

Jarrige, J.F. 1987–8. 'Excavations at Nausharo'. *Pakistan Archaeology*, 23: 149–203.

Jerdon, T.C. 1867. *The Mammals of India—A Natural History of all the Animals Known to Inhabit Continental India*. Roorkee: Printed for the author by the Thomason College Press.

Jettmar, Karl. 1982. *Rockcarvings and Inscriptions in the Northern Areas of Pakistan*. Islamabad: Institute of Folk Heritage.

Joglekar, P.P. 1994–5. 'Domestic Animals in Ancient India in the Light of Literary and Archaeological Evidence'. *Bhāratī*, 21(1): 1–19.

———. 2000. 'Animal Taxonomy from Ancient and Medieval Indian Literature'. In S.C. Bhattacharya, V.D. Misra, J.N. Pandey, and J.N. Pal, eds, *Peeping through the Past: Professor G.R. Sharma Memorial Volume*, pp. 187–93. Allahabad: University of Allahabad.

———. 2007–8. 'A Fresh Appraisal of the Animal-based Subsistence and Domestic Animals in the Ganga Valley'. *Prāgdhārā*, 18: 309–21.

———. 2012–13. 'A Report on the Faunal Remains from Kalpi, Jalaun District, Uttar Pradesh'. *Prāgdhārā*, 23: 165–74.

Joglekar, P.P., and C.V. Sharda. 2016. 'Report on the Faunal Remains from Madina, Rohtak District, Haryana'. In Manmohan Kumar, Akinori Uesugi, and Vivek Dangi, eds, *Excavations at Madina, District Rohtak, Haryana, India*, pp. 209–47. Kansai University: Research Group for South Asian Archaeology, Archaeological Research Institute.

Joglekar, P.P., and Pankaj Goyal. 2011. 'Faunal Remains from Shikarpur, a Harappan Site in Gujarat, India'. *Iranian Journal of Archaeological Studies*, 1(1): 15–25.

———. 2014. 'Animals'. In D.K. Chakrabarti and Makkhan Lal, eds, *History of Ancient India II: Protohistoric Foundations*, pp. 184–201. New Delhi: Vivekananda International Foundation and Aryan Books International.

Joglekar, P.P., Anil Kumar Dubey, Chandrashekhar, and Sachin Kumar Tiwary. 2013. 'Animal Remains from Madar Dih, District Jaunpur, Uttar Pradesh'. *Purātattva*, 43: 263–8.

Joglekar, P.P., G.S. Abhayan, Jayshree Mungur-Medhi, and Jitendra Nath. 2013. 'A Glimpse of Animal Utilization Pattern at Khirsara, Kachchh, Gujarat, India'. *Heritage: Journal of Multidisciplinary Studies in Archaeology*, 1: 1–15.

Joglekar, P.P., V.D. Misra, J.N. Pal, and M.C. Gupta. 2003. *Mesolithic Mahadaha: The Faunal Remains*. Allahabad: University of Allahabad.

Joshi, Jagat Pati. 1990. *Excavation at Surkotada 1971–72 and Exploration in Kutch*. Memoirs of the Archaeological Survey of India No. 87. New Delhi: Archaeological Survey of India.

Joshi, M.C., and C. Margabandhu. 1976–7. 'Some Terracottas from Excavations at Mathura—A Study'. *Journal of the Indian Society of Oriental Art*, 8: 16–32.

Kala, S.C. 1950. *Terracotta Figurines from Kauśāmbī*. Allahabad: The Municipal Museum.

———. 1992–3. 'Terracottas from Kannauj'. *Prāgdhārā*, 3: 117–21.

Kane, Victoria Stack. 1989. 'Animal Remains from Rojdi'. In Gregory L. Possehl and M.H. Raval, eds, *Harappan Civilization and Rojdi*, pp. 182–4. New Delhi; Bombay; Calcutta: Oxford & IBH Publishing Co. Pvt. Ltd.

Kangle, R.P. 1963. *The Kauṭilīya Arthaśāstra*. Part II. *An English Translation with Critical and Explanatory Notes*. Bombay: University of Bombay.

Karanth, K. Ullas. 2002. *The Way of the Tiger: Natural History and Conservation of the Endangered Big Cat*. Bangalore: Centre for Wildlife Studies.

———. 2011. *The Science of Saving Tigers*. Hyderabad: Universities Press.

Keith, Arthur Berriedale, trans. 1914. *The Yajur Veda (Taittiriya Sanhita)*. Available at http://www.sacred-texts.com/hin/yv/index.htm.

Kenoyer, Jonathan Mark. 2006–7. 'Indus Seals: An Overview of Iconography and Style'. *Ancient Sindh*, 9: 7–30.

———. 2010. 'Master of Animals and Animal Masters in the Iconography of the Indus Tradition'. In Derek B. Counts and Bettina Arnold, eds, *The Master of Animals in Old World Iconography*, pp. 37–58. Budapest: Archaeolingua Alapítvány.

Kosmin, Paul J. 2014. *The Land of the Elephant Kings: Space, Territory, and Ideology in the Seleucid Empire*. Cambridge, Massachusetts: Harvard University Press.

Kumar, Giriraj. 2014. 'Rock Art'. In D.K. Chakrabarti and Makkhan Lal, eds, *History of Ancient India I: Prehistoric Foundations*, pp. 301–46. New Delhi: Vivekananda International Foundation and Aryan Books International.

Kumar, Manmohan. 2016. 'Excavations at Madina, District Rohtak, Haryana'. In Manmohan Kumar, Akinori Uesugi, and Vivek Dangi, eds,

Excavations at Madina, District Rohtak, Haryana, India, pp. 1–25. Kansai University: Research Group for South Asian Archaeology, Archaeological Research Institute.

Kumar, Vijay. 2014. *Flora and Fauna in Harappan Civilization*. New Delhi: Research India Press.

Lahiri, Nayanjot, ed. 2000. *The Decline and Fall of the Indus Civilization*. New Delhi: Permanent Black.

Lal, Braj Basi. 1954–5. 'Excavations at Hastināpura and other Explorations in the Upper Gangā and Sutlej Basins 1950–52'. *Ancient India*, 10 and 11: 5–151.

———. 1993. *Excavations at Śṛiṅgaverapura (1977–86)*, vol. I. Memoirs of the Archaeological Survey of India No. 88. New Delhi: Archaeological Survey of India.

———. 2003. 'Chronology of the Early Harappan Settlement'. In B.B. Lal, Jagat Pati Joshi, B.K. Thapar, and Madhu Bala, eds, *Excavations at Kalibangan: The Early Harappans (1961–69)*, pp. 25–6. Memoirs of the Archaeological Survey of India No. 98. New Delhi: Archaeological Survey of India.

———. 2003–4. 'The *Rigvedic* Flora and Fauna: What Light Do These Throw on the "Aryan Invasion" Debate?' *Purātattva*, 34: 15–19.

———. 2005. *The Homeland of the Aryans: Evidence of Rigvedic Flora and Fauna & Archaeology*. New Delhi: Aryan Books International.

Laurie, William Andrew. 1978. 'The Ecology and Behaviour of the Greater One-Horned Rhinoceros'. PhD dissertation. Cambridge: Selwyn College.

Laurie, W.A., E.M. Lang, and C.P. Groves. 1983. 'Rhinoceros unicornis'. *Mammalian Species*, 211: 1–6.

Lefeber, Rosalind, trans. 1994. *The Rāmāyaṇa of Vālmīki: An Epic of Ancient India*. Vol. IV, *Kiṣkindhākāṇḍa*. Princeton, N.J.: Princeton University Press.

Lindsey, Peter A., Guillaume Chapron, Lisanne S. Petracca, Dawn Burnham, Matthew W. Hayward, Philipp Henschel, Amy E. Hinks, Stephen T. Garnett, David W. Macdonald, Ewan A. Macdonald, William J. Ripple, Kerstin Zander, and Amy Dickman. 2017. 'Relative Efforts of Countries to Conserve World's Megafauna'. *Global Ecology and Conservation*, 10: 243–52.

Lukacs, John R. 1990. 'On Hunter-Gatherers and Their Neighbors in Prehistoric India: Contact and Pathology'. *Current Anthropology*, 31(2): 183–6.

Lukacs, John R., J.N. Pal, and V.D. Misra. 1996. 'Chronology and Diet in Mesolithic North India: A Preliminary Report of New AMS C Dates, δ

C Isotope Values, and Their Significance'. In G.E. Afanas'ev, S. Cleuziou, J.R. Lukacs, and M. Tosi, eds, *Bioarchaeology of Mesolithic India: An Integrated Approach, Colloquim XXXII of the International Union of Prehistoric and Protohistoric Sciences*, pp. 301–11. Forli: ABACO Edizioni.

Lydekker, R. 1886. 'The Fauna of the Karnul Caves'. *Palaeontologica Indica*, X, 4: 23–58.

Macdonell, Arthur Anthony, and Arthur Berriedale Keith. 1958. *Vedic Index of Names and Subjects*, vol. II. Delhi; Varanasi; Patna: Motilal Banarsidass.

Mackay, Ernest. 1931a. 'Figurines and Model Animals'. In Sir John Marshall, ed., *Mohenjo-daro and the Indus Civilization: Being an Official Account of Archaeological Excavations at Mohenjo-daro Carried Out by the Government of India between the Years 1922 and 1927*, vol. I, pp. 338–55. London: Arthur Probsthain.

———. 1931b. 'Seals, Seal Impressions, and Copper Tablets, with Tabulation'. In Sir John Marshall, ed., *Mohenjo-daro and the Indus Civilization: Being an Official Account of Archaeological Excavations at Mohenjo-daro Carried Out by the Government of India between the Years 1922 and 1927*, vol. II, pp. 370–405. London: Arthur Probsthain.

———. 1931c. 'Ivory, Shell, Faience, and Other Objects of Technical Interest'. In Sir John Marshall, ed., *Mohenjo-daro and the Indus Civilization: Being an Official Account of Archaeological Excavations at Mohenjo-daro Carried Out by the Government of India between the Years 1922 and 1927*, vol. II, pp. 562–88. London: Arthur Probsthain.

———. 1938. *Further Excavations at Mohenjo-daro: Being an Official Account of Archaeological Excavations at Mohenjo-daro Carried Out by the Government of India between the Years 1927 and 1931*, 2 vols. Delhi: Manager of Publications.

———. 1943. *Chanhu-Daro Excavations 1935–36*. New Haven, Connecticut: American Oriental Society.

Madella, Marco, and Dorian Q. Fuller. 2006. 'Palaeoecology and the Harappan Civilization of South Asia: A Reconsideration'. *Quaternary Science Reviews*, 25: 1283–301.

Mahadevan, Iravatham. 1977. *The Indus Script: Texts, Concordance and Tables*. New Delhi: Archaeological Survey of India.

Manamendra-Arachchi, Kelum, Rohan Pethiyagoda, Rajith Dissanayake, and Madhava Meegaskumbura. 2005. 'A Second Extinct Big Cat from the Late Quaternary of Sri Lanka'. *The Raffles Bulletin of Zoology*, 12: 423–34.

Mani, B.R. 2000–1. 'Rock Art of Ladakh: Glimpses of Economic and Cultural Life'. *Purākalā*, 11–12: 93–108.

Manuel, Joseph. 2004–5. 'Harappan Environment as One Variable in the Preponderance of Rhinoceros and Paucity of Horse'. *Purātattva,* 35: 21–6.

———. 2008a. 'Depiction of Rhinoceros: Transition from Popular Art to State Sponsored Art'. In B.R. Mani and A. Tripathi, eds, *Expressions in Indian Art: Essays in Memory of Shri M.C. Joshi*, vol. I, pp. 33–8. Delhi: Agam Kala Prakashan.

———. 2008b. 'Why Deny Evidence on Horse?' In K.K. Chakravarty and G.L. Badam, eds, *Rock Art & Archaeology of India: Prof. Shankar Tiwari Commemoration Volume*, pp. 169–77. Delhi: Agam Kala Prakashan.

Margabandhu, C. 1990. 'Miscellaneous Objects of Bone and Ivory'. In Jagat Pati Joshi, ed., *Excavation at Surkotada 1971–72 and Exploration in Kutch*, pp. 339–41. Memoirs of the Archaeological Survey of India No. 87. New Delhi: Archaeological Survey of India.

———. 1991. 'Some Early Historic Circular Plaques (or Medallions) of Terracotta—A Study'. In C. Margabandhu, K.S. Ramachandran, A.P. Sagar, and D.K. Sinha, eds, *Indian Archaeological Heritage (Shri K.V. Soundara Rajan Festschrift)*, vol. I, pp. 261–72. Delhi: Agam Kala Prakashan.

Marshall, Sir John. 1915. 'Excavations at Bhīṭā'. In J.H. Marshall, ed., *Archaeological Survey of India Annual Report 1911–12*, pp. 29–94. Calcutta: Superintendent Government Printing.

———. 1931. *Mohenjo-Daro and the Indus Civilization: Being an Official Account of Archaeological Excavations at Mohenjo-daro Carried Out by the Government of India between the Years 1922 and 1927*, 3 vols. London: Arthur Probsthain.

———. 1951. *Taxila: An Illustrated Account of Archaeological Excavations Carried Out at Taxila Under the Orders of the Government of India between the Years 1913 and 1934*. Vol. II: *Minor Antiquities*. Cambridge: University Press.

Martin, Esmond Bradley, and Chrysee Bradley Martin. 1982. *Run Rhino Run*. London: Chatto and Windus.

McCrindle, J.W., trans. 1877. *Ancient India as Described by Megasthenes and Arrian*. London: Trubner & Co.

———, trans. 1879. *The Commerce and Navigation of the Erythrean Sea*. London : Trubner & Co.

———, trans. 1882. *Ancient India as Described by Ktesias the Knidian*. London: Trubner & Co.

———, trans. 1896. *Invasion of India by Alexander the Great*. Westminster: Archibald Constable and Company.

———, trans. 1901. *Ancient India as Described in Classical Literature*. Westminster: Archibald Constable and Co., Ltd.

McDougal, Charles. 1977. *The Face of the Tiger*. London: Rivington Books.

Mcdermott, James P. 1989. 'Animals and Humans in Early Buddhism'. *Indo-Iranian Journal*, 32 (4): 269–80.

Meadow, Richard H. 1984. 'Notes on the Faunal Remains from Mehrgarh, with a Focus on Cattle (*Bos*)'. In B. Allchin, ed., *South Asian Archaeology 1981*, pp. 34–40. Cambridge: Cambridge University Press.

———. 1989. 'Continuity and Change in the Agriculture of the Greater Indus Valley: The Palaeoethnobotanical and Zooarchaeological Evidence'. In J.M. Kenoyer, ed., *Old Problems and New Perspectives in the Archaeology of South Asia*, pp. 61–74. Wisconsin Archaeological Reports 2. Madison: University of Wisconsin.

———. 1993. 'The Past, Present and Future of Bioarchaeological Studies in Pakistan with Specific Reference to Mohenjodaro and the Indus Civilization'. *Pakistan Archaeology*, No. 28: 183–215.

Meadow, Richard H., and J. Mark Kenoyer. 1997. 'Excavations at Harappa 1994–1995: New Perspectives on the Indus Script, Craft Activities and City Organization'. In R. Allchin and B. Allchin, eds, *South Asian Archaeology 1995*, pp. 139–72. New Delhi: Oxford & IBH.

Mehendale, M.A. 1987. 'The Fauna in the Āraṇyakaparvan of the Mahābhārata'. *Annals of the Bhandarkar Oriental Research Institute*, 68(1/4): 327–44.

Mehta, R.N. 1968. *Excavation at Nagara*. Baroda: M.S. University of Baroda.

Mehta, R.N., and A.J. Patel. 1967. *Excavation at Shamalaji*. Baroda: M.S. University of Baroda.

Misra, V.D., J.N. Pal, and M.C. Gupta. 2003. 'Mesolithic Mahadaha: General Features'. In P.P. Joglekar, V.D. Misra, J.N. Pal, and M.C. Gupta, eds, *Mesolithic Mahadaha: The Faunal Remains*, pp. 1–37. Allahabad: University of Allahabad.

Misra, V.N. 1973. 'Problems of Palaeo-ecology, Palaeo-climate and Chronology of the Microlithic Cultures of North-West India'. In D.P. Agrawal and A. Ghosh, eds, *Radiocarbon and Indian Archaeology*, pp. 58–72. Bombay: Tata Institute of Fundamental Research.

———. 2007. *Rajasthan Prehistoric and Early Historic Foundations*. New Delhi: Aryan Books International.

Mitra, Debala. 1972. *Excavations at Tilaura-Kot and Kodan and Explorations in the Nepalese Terai: Report on the Work Undertaken in 1962 Jointly by the Department of Archaeology, His Majesty's Government of Nepal, and the Archaeological Survey of India*. His Majesty's Government of Nepal: Department of Archaeology.

Momin, K.N., D.R. Shah, and G.M. Oza. 1973. 'Great Indian Rhinoceros Inhabited Gujarat'. *Current Science*, 42, No. 22: 801–3.

Monier-Williams, Sir Monier. 1963. *A Sanskrit–English Dictionary*. Delhi: Motilal Banarsidass.

Mookerji, Radha Kumud. 1960. *Chandragupta Maurya and His Times*. Delhi: Motilal Banarsidass.

Morrison, Kathleen D. 2014. 'Conceiving Ecology and Stopping the Clock: Narratives of Balance, Loss, and Degradation'. In Mahesh Rangarajan and K. Sivaramakrishnan, eds, *Shifting Ground: People, Animals, and Mobility in India's Environmental History*, pp. 39–64. New Delhi: Oxford University Press.

Mountfort, Guy. 1981. *Saving the Tiger*. London: Michael Joseph.

Müller, F. Max, trans. 1879. *The Upanishads*. Part I. *Sacred Books of the East*, edited by F. Max Müller, vol. I. Oxford: Clarendon Press.

Müller, F. Max, and Max Fausböll, trans. 1881. *The Dhammapada and Sutta Nipāta*. *Sacred Books of the East*, edited by F. Max Müller, vol. X. Oxford: Clarendon Press.

Narain, A.K. 1991. 'Gaṇeśa: A Protohistory of the Idea and the Icon'. In Robert L. Brown, ed., *Ganesh: Studies of an Asian God*, pp. 19–48. Albany: State University of New York Press.

Narain, A.K., and T.N. Roy. 1968. *The Excavations at Prahladpur (March–April 1963)*. Varanasi: Banaras Hindu University.

Narain, Lala Aditya. 1970. 'The Neolithic Settlement at Chirand'. *The Journal of the Bihar Research Society*, LVI: 16–35.

———. 1972. 'A Study in The Techniques of Neolithic Bone Tool Making at Chirand and Their Probable Uses'. *The Journal of the Bihar Research Society*, LVIII: 1–24.

Naravane, V.S. 1965. *The Elephant and the Lotus: Essays in Philosophy and Culture*. New York: Asia Publishing House.

Nath, Amarendra. 1997–8. 'Rakhigarhi: A Harappan Metropolis in the Sarsaswati-Drishadvati Divide'. *Purātattva*, 28: 39–45.

Nath, Bhola. 1954–5. 'Animal Remains'. *Ancient India*, 10–11: 107–20.

———. 1959. 'Remains of the Horse and Indian Elephant from the Prehistoric Site of Harappa (West Pakistan). *Proceedings of the First All-India Congress of Zoologists*, Pt. 2, Scientific Papers, pp. 1–14.

———. 1962 and 1963. 'Animal Remains from Rangpur'. *Ancient India*, 18–19: 153–60.

———. 1963. 'Advances in the Study of Prehistoric and Ancient Animal Remains in India: A Review'. *Records of the Zoological Survey of India*, 61 (1–2): 1–63.

———. 1968. 'Animal Remains from Rupar and Bara Sites, Ambala District of East Punjab, India'. *Bulletin of the Indian Museum*, 3 (1–2): 69–116.

———. 1969. 'The Role of Animal Remains in the Early Prehistoric Cultures of India'. *Bulletin of the Indian Museum*, 4 (2): 102–10.

Nath, Bhola and M.K. Biswas. 1969. 'Animal Remains from Alamgirpur'. *Indian Museum Bulletin*, 4 (1): 43–52.

———. 1980. 'Animal Remains from Chirand, Saran District, Bihar'. *Records of the Zoological Survey of India*, 76: 115–24.

Nath, Bhola, and G.V. Sreenivasa Rao. 1985. 'Animal Remains from Lothal Excavations'. In S.R. Rao, ed., *Lothal: A Harappan Port Town 1955–62*, vol. II, pp. 636–50. Memoirs of the Archaeological Survey of India No. 78. New Delhi: Archaeological Survey of India.

Neumayer, Erwin. 1983. *Prehistoric Indian Rock Paintings*. Delhi: Oxford University Press.

———. 1993. *Lines on Stone: The Prehistoric Rock Art of India*. New Delhi: Manohar Publishers & Distributors.

———. 2013. *Prehistoric Rock Art of India*. New Delhi: Oxford University Press.

Olivelle, Patrick. 2002a. '*Abhakṣya* and *Abhojya*: An Exploration in Dietary Language'. *Journal of the American Oriental Society*, 122 (2): 345–54.

———. 2002b. 'Food for Thought: Dietary Rules and Social Organization in Ancient India'. 2001 Gonda Lecture, pp. 5–38. Amsterdam: Royal Netherlands Academy of Arts and Sciences.

———. 2003. *Dharmasūtras: The Law Codes of Āpastamba, Gautama, Baudhāyana and Vasiṣṭha*. Delhi: Motilal Banarsidass Publishers Private Limited.

———. 2006. *Manu's Code of Law: A Critical Edition and Translation of the Mānava-Dharmaśāstra*. New Delhi: Oxford University Press.

———. 2013. 'Talking Animals: Explorations in an Indian Literary Genre'. *Religions of South Asia*, 7: 14–26.

Olivier, R.C.D. 1984. 'Asian Elephant'. In I.L. Mason, ed., *Evolution of Domesticated Animals*, pp. 185–92. London and New York: Longman Group.

Pal, J.N. 1994. 'Mesolithic Settlements in the Ganga Plain'. *Man and Environment*, 19 (1–2): 91–101.

———. 2002. 'The Mesolithic Phase in the Ganga Valley'. In K. Padayya, ed., *Recent Advances in Indian Archaeology*, pp. 60–80. New Delhi: Munshiram Manoharlal.

Pal, Pratapaditya. 1981. *Elephants and Ivories in South Asia*. Los Angeles: Los Angeles County Museum of Art.

Paliwal, B.S. 2003. 'Fossilized Elephant Bones in the Quaternary Gypsum Deposits at Bhadawasi, Nagaur District, Rajasthan'. *Current Science*, 84 (9): 1188–91.

Pande, B.M. 1984. 'Harappan Art: An Experiment in Third Dimension'. In B.B. Lal and S.P. Gupta, eds, *Frontiers of the Indus Civilization: Sir Mortimer Wheeler Commmemoration Volume*, pp. 105–7. New Delhi: Books & Books.

Pant, G.N. 1997. *Horse and Elephant Armour*. Delhi: Agam Kala Prakashan.

Parpola, Asko. 1994. *Deciphering the Indus Script*. Cambridge: Cambridge University Press.

Pathak, Madhusudan Madhavlal. 1968. *Similes in the Rāmāyaṇa*. Baroda: The Maharaja Sayajirao University of Baroda.

Pilikian, Vaughan, ed. and trans. 2006. *Mahābhārata. Book Seven: Droṇa*, vol. I. The Clay Sanskrit Library. New York: New York University Press: JJC Foundation.

———, ed. and trans. 2009. *Mahābhārata. Book Seven: Droṇa*, vol. II. The Clay Sanskrit Library. New York: New York University Press: JJC Foundation.

Pocock, R.I. 1939. *The Fauna of British India, including Ceylon and Burma: Mammalia*. Vol. I: *Primates and Carnivora* (in part). London: Taylor and Francis.

———. 1941. *The Fauna of British India, including Ceylon and Burma: Mammalia*. Vol. II: *Carnivora* (*suborders Aeluroidae and Arctoidae*). London: Taylor and Francis.

Pollock, Sheldon I., trans. 1986. *The Rāmāyaṇa of Vālmīki: An Epic of Ancient India*. Vol. II: *Ayodhyākāṇḍa*. Princeton, N.J.: Princeton University Press.

———, trans. 1991. *The Rāmāyaṇa of Vālmīki: An Epic of Ancient India*. Vol. III: *Araṇyakāṇḍa*. Princeton, N.J.: Princeton University Press.

Possehl, Gregory L. 1999. *Indus Age: The Beginnings*. New Delhi; Calcutta: Oxford & IBH Publishing Co. Pvt. Ltd.

Prakash, Pratibha. 1985. *Terracotta Animal Figurines in the Ganga–Yamuna Valley (600 BC to 600 AD)*. Delhi: Agam Kala Prakashan.

Prasad, A.K. 2014. 'Distinct Dominant Traits in the Rock Art of Eastern India with Special Reference to the Rock Art of Southern Bihar and Adjoining Jharkhand'. In Bansi Lal Malla, ed., *Rock Art Studies: Interpretation through Multidisciplinary Approaches*, vol. I, pp. 117–34. New Delhi: Indira Gandhi National Centre for the Arts and Aryan Books International.

Prasad, Prakash Charan. 1995–6. 'Prehistoric Rock-Paintings in Bihar'. *Purātattva*, No. 26: 87–8.

Prashad, B. 1936. *Animal Remains from Harappa*. Memoirs of the Archaeological Survey of India, No. 51. Delhi: Manager of Publications.

Prater, S.H. 1948. *The Book of Indian Mammals*. Bombay: Bombay Natural History Society.

Puri, Baij Nath. 1963. *India in Classical Greek Writings*. Ahmedabad: The New Order Book Company.

Puri, K.N. 1998. *Excavations at Rairh*. Jaipur: Publication Scheme.

Raikes, Robert L., and Robert H. Dyson Jr. 1961. 'The Prehistoric Climate of Baluchistan and the Indus Valley'. *American Anthropologist*, 63 (2), Part I : 265–81.

Ram, Vikramjit. 2007. *Elephant Kingdom: Sculptures from Indian Architecture*. Usmanpura, India: Mapin Pub.

Rangarajan, Mahesh 2001. *India's Wildlife History: An Introduction*. Delhi: Permanent Black.

Rao, S.R. 1962 and 1963. 'Excavation at Rangpur and other Explorations in Gujarat'. *Ancient India*, 18 and 19: 7–207.

———. 1979. *Lothal: A Harappan Port Town 1955–62*, vol. I. Memoirs of the Archaeological Survey of India No. 78. New Delhi: Archaeological Survey of India.

———. 1985. *Lothal: A Harappan Port Town 1955–62*, vol. II. Memoirs of the Archaeological Survey of India No. 78. New Delhi: Archaeological Survey of India.

Rookmaaker, L.C. 1983. *Bibliography of the Rhinoceros: An Analysis of the Literature on the Recent Rhinoceroses in Culture, History and Biology*. Rotterdam: A.A. Balkema.

———. 1984. 'The Former Distribution of the Indian Rhinoceros (*Rhinoceros unicornis*) in India and Pakistan'. *Journal of the Bombay Natural History Society*, 80(3): 555–63.

———. 1999. 'Records of the Rhinoceros in Northern India'. *Saugetierkundliche Mitteilungen*, 44 (2): 51–78.

———. 2000. 'Records of the Rhinoceros in Pakistan and Afghanistan'. *Pakistan Journal of Zoology*, 32 (1): 65–74.

———. 2002. 'Historical Records of the Rhinoceros (*Rhinoceros unicornis*) in Northern India and Pakistan'. *Zoos Print Journal*, 17 (11): 923–9.

Rouse, W.H.D., trans. 1969a. *The Jātaka or Stories of the Buddha's Former Births*, edited by E.B. Cowell, vol. II. London: Luzac & Company, Ltd.

———. 1969b. *The Jātaka or Stories of the Buddha's Former Births*, edited by E.B. Cowell, vol. IV. London: Luzac & Company, Ltd.

Russell, Nerissa. 1995. 'The Bone Tool Industry at Mehrgarh and Sibri'. In Catherine Jarrige, Jean-François Jarrige, Richard H. Meadow, and Gonzague Quivron, eds, *Mehrgarh Field Reports 1974–1985 from Neolithic Times to the Indus Civilization*, pp. 583–613. Department of Culture and Tourism, Government of Sindh, Pakistan, in Collaboration with the French Ministry of Foreign Affairs.

Saha, U., M. Ghosh, and T.K. Pal. 2004. 'Animal Remains Excavated from Lothal Archaeological Site (Gujarat) and Relevance of the Fauna in This

Ancient Civilization'. *Records of the Zoological Survey of India*, Occasional Paper No. 222: 1–162.

Sahni, Rai Bahadur Daya Ram. 1931. 'HR Area—(*Continued*)'. In Sir John Marshall, ed., *Mohenjo-daro and the Indus Civilization: Being an Official Account of Archaeological Excavations at Mohenjo-daro Carried Out by the Government of India between the Years 1922 and 1927*, vol. I, pp. 187–213. London: Arthur Probsthain.

Sahu, B.P. 1988. *From Hunters to Breeders (Faunal Background of Early India)*. Delhi: Anamika Prakashan.

Sankalia, Hasmukh Dhirajlal. 1965. *Excavations at Langhnaj: 1944–63: Part I: Archaeology*. Poona: Deccan College Postgraduate and Research Institute.

———. 1987. *Prehistoric and Historic Archaeology of Gujarat*. Delhi: Munshiram Manoharlal Publishers Pvt. Ltd.

Sankalia, H.D., S.B. Deo, and Z.D. Ansari. 1969. *Excavations at Ahar (Tambavati) 1961–62*. Poona: Deccan College Postgraduate and Research Institute.

Sant, Urmila. 1997. *Terracotta Art of Rajasthan from Pre-Harappan and Harappan Times to the Gupta Period*. New Delhi: Aryan Books International.

Sathe, Vijay. 2010. 'The Archaeology of Great One-Horned Indian Rhinoceros (*Rhinoceros Unicornis* Linnaeus 1758)'. In Anoop Swarup and S.C. Agrawal, eds, *Indian Civilization through the Millennia*, pp. 22–30. New Delhi: Excel India Publishers.

———. 2015. 'Discovery of a Fossil Bone Bed in the Manjra Valley, District Latur, Maharashtra'. *Bulletin of the Deccan College Post-Graduate and Research Institute*, 75: 1–16.

Schaller, G.B. 1967. *The Deer and the Tiger: A Study of Wildlife in India*. Chicago: The University of Chicago Press.

Scullard, H.H. 1974. *The Elephant in the Greek and Roman World*. Cambridge: Thames and Hudson.

Sen, Asis. 1972. *Animal Motifs in Ancient Indian Art*. Calcutta: Firma K.L. Mukhopadhyay.

Sewell, R.B.S., and B.S. Guha. 1931. 'Zoological Remains'. In Sir John Marshall, ed., *Mohenjo-Daro and the Indus Civilization: Being an Official Account of Archaeological Excavations at Mohenjo-Daro Carried Out by the Government of India between the Years 1922 and 1927*, vol. II. pp. 649–73. London: Arthur Probsthain.

Shaffer, Jim G., and Diane A. Lichtenstein. 1989. 'Ethnicity and Change in the Indus Valley Cultural Tradition'. In J.M. Kenoyer, ed., *Old Problems and New Perspectives in the Archaeology of South Asia*, pp. 117–26. Madison: University of Wisconsin.

Shah, D.R. 1980. 'Animal Remains from Kanewal'. In R.N. Mehta, K.N. Momin, and D.R. Shah, eds, *Excavation at Kanewal*, pp. 74–6. Vadodara: Department of Archaeology and Ancient History, M.S. University of Baroda.

———. 1983. 'The Animal Remains'. In R.C. Gaur, ed., *Excavations at Atranjīkherā: Early Civilization of the Upper Gaṅgā Basin*, pp. 461–71. Aligarh: Centre of Advanced Study, Department of History.

Shamasastry, R., trans. 1967. *Kauṭilya's Arthaśāstra*. Mysore: Mysore Printing and Publishing House.

Sharma, A.K. 1990. 'Animal Bone Remains'. In Jagat Pati Joshi, ed., *Excavation at Surkotada 1971–72 and Exploration in Kutch*, pp. 372–83. Memoirs of the Archaeological Survey of India No. 87. New Delhi: Archaeological Survey of India.

Sharma, G.R. 1969. *Excavations at Kauśāmbī 1949–50*. Memoirs of the Archaeological Survey of India No. 74. Delhi: Manager of Publications.

Sharma, Murari Lal. 2013. 'Sahibee Valley: A New Rock Art Region in North-East Rajasthan'. In Bansi Lal Malla and V.H. Sonawane, eds, *Global Rock Art*, pp. 63–8. New Delhi: Indira Gandhi National Centre for the Arts and Aryan Books International.

———. 2014. 'Salient Features of the Rock Art of Rajasthan: A Detailed Study'. In Bansi Lal Malla, ed., *Rock Art Studies: Interpretation Through Multidisciplinary Approaches*, vol. II, pp. 209–39. New Delhi: Indira Gandhi National Centre for the Arts and Aryan Books International.

Sharma, Murari Lal, and Madan Lal Meena. 2011. 'A Preliminary Study of Newly Discovered Rock Art Site in Alwar District, Rajasthan'. *Kosāla*, 4: 87–99.

Sharma, Priyavrat, ed. and trans. 1994. *Caraka-Saṃhitā: Agniveśa's Treatise Refined and Annotated by Caraka and Redacted by Dṛḍhabala (Text with English Translation), vol. I (Sūtrasthāna to Indriyasthāna) and vol. II (Chikitsāsthānam to Siddhisthānam)*. Varanasi: Chaukhamba Orientalia.

Shastri, Hari Prasad. 1957. *The Ramayana of Valmiki*, vol. II. London: Shanti Sadan.

Shebbeare, E.O. 1953. 'Status of the Three Asiatic Rhinoceros'. *Oryx*, 2 (2): 141–9.

Shelton, Jo-Ann. 2006. 'Elephant as Enemies in Ancient Rome'. *Concentric: Literary and Cultural Studies*, 32 (1): 3–25.

Singh, Arvind Kumar. 2006. *Animals in Early Buddhism*. Delhi: Eastern Book Linkers.

Singh, Birendra Pratap. 2004. *Early Farming Communities of the Kaimur (Excavations at Senuwar) [1986–87, 89–90]*, vol. I. Jaipur: Publication Scheme.

Singh, Gurdip. 1970–1. 'The Indus Valley Culture (Seen in the Context of Post-Glacial Climate and Ecological Studies in North-West India)'. *Purātattva*, 4: 68–76.

Singh, Indra Bir. 2004–5. 'Landform Development and Palaeovegetation in Late Quaternary of Ganga Plain: Implications of Anthropogenic Activity'. *Prāgdhārā*, 15: 5–31.

———. 2005. 'Quaternary Palaeoenvironments of the Ganga Plain and Anthropogenic Activity'. *Man and Environment*, 30 (1): 1–35.

Singh, Mohinder. 1989. 'Terracottas from Thanesar'. In Devendra Handa, ed., *Ajaya-Śrī: Recent Studies in Indology (Prof. Ajay Mitra Shastri Felicitation Volume)*, vol. II, pp. 403–9. Delhi: Sundeep Prakashan.

———. 1990. 'Terracottas from Sugh'. In G. Kuppuram and K. Kumudamani, eds, *Researches in History, Archaeology, Art and Religion (Prof. Upendra Thakur Felicitation Volume)*, vol. I, pp. 5–10. Delhi: Sundeep Prakashan.

Singh, Purshottam. 1994. *Excavations at Narhan (1984–89)*. Varanasi: Banaras Hindu University & Delhi: B.R. Publishing Corporation.

Singh, Sarva Daman. 1965. *Ancient Indian Warfare with Special Reference to the Vedic Period*. Leiden: E.J. Brill.

Singh, Upinder. 2009. *A History of Ancient and Early Medieval India: From the Stone Age to the 12th Century*. Delhi: Pearson Longman.

Sinha, B.P. 1955. 'Elephants in Ancient Indian Army'. *Journal of the Bihar Research Society*, 41 (4): 519–28.

Sinha, B.P., and Sita Ram Roy. 1969. *Vaiśālī Excavations 1958–1962*. Patna, Bihar: Directorate of Archaeology and Museums.

Sinha, K.K. 1967. *Excavations at Sravasti—1959*. Varanasi: Banaras Hindu University.

Sivaramamurti, C. 1974. *Birds and Animals in Indian Sculpture*. New Delhi: National Museum.

Smith, Brian K. 1991. 'Classifying Animals and Humans in Ancient India'. *Man*, 26: 527–48.

Sonawane, V.H. 2002. 'Rock Art of Gujarat: A Regional Study'. *Purākalā*, 13 (1–2): 67–85.

———. 2014. 'Rock Art of Gujarat: A Fresh Look'. In Bansi Lal Malla, ed., *Rock Art Studies: Interpretation through Multidisciplinary Approaches*, vol. II, pp. 181–207. New Delhi: Indira Gandhi National Centre for the Arts and Aryan Books International.

Srivastava, K.M. 1996. *Excavations at Piprahwa and Ganwaria*. Memoirs of the Archaeological Survey of India No. 94. New Delhi: Archaeological Survey of India.

Srivastava, Kamal Shankar. 1989. *The Elephant in Early Indian Art: From Indus Valley Civilization to A.D. 650*. Varanasi, U.P.: Sangeeta Prakashan.

Srivastava, Om Prakash Lal. 1991. *Archaeology of Erich: Discovery of New Dynasties*. Varanasi: Sulabh Prakashan.

Sterndale, R.A. 1884. *Natural History of the Mammalia of India and Ceylon*. Calcutta: Thacker, Spink, and Co.

Stronk, Jan P. 2007. 'Ctesias of Cnidus, a Reappraisal'. *Mnemosyne: A Journal of Classical Studies*, 60: 25–58.

Sukumar, Raman. 2003. *The Living Elephants: Evolutionary Ecology, Behavior, and Conservation*. New Delhi: Oxford University Press.

———. 2011. *The Story of Asia's Elephants*. Mumbai: The Marg Foundation.

Sullivan, H.P. 1964. 'A Re-examination of the Religion of the Indus Civilization'. *History of Religion*, 4 (1): 115–25.

Tewari, D.P. 2002. *Excavations at Chardā*. Lucknow: Tarun Prakashan.

Tewari, Rakesh. 1985. 'Mirzapur ke Chitrit Sailasray (Painted Rock-Shelters of Mirzapur)'. Unpublished PhD Thesis (in Hindi), Avadh University, Faizabad.

———. 1987. 'Rhino-hunt Scene of the Ghora Mangara Rock-shelter, Mirzapur—A Reappraisal'. *Bulletin of Museums & Archaeology in U.P.*, No. 39: 25–9.

———. 1990. *Rock Paintings of Mirzapur*. Lucknow: U.P. State Archaeological Organization.

Thapar, B.K. 1977. 'Climate during the Period of the Indus Civilization: Evidence from Kalibangan'. In D.P. Agrawal and B.M. Pande, eds, *Ecology and Archaeology of Western India*, pp. 67–73. Delhi: Concept Publishing Company.

Thapar, Romila. 2004. 'In Times Past'. In Valmik Thapar, ed., *Tiger: The Ultimate Guide*, pp. 153–67. New York: CDS Books in association with Two Brothers Press.

Thapar, Valmik. 1986. *Tiger: Portrait of a Predator*. London: Collins.

———. 1989. *Tigers: The Secret Life*. London: Elm Tree Books.

———. 1997. *Land of the Tiger: A Natural History of the Indian Subcontinent*. Berkeley: University of California Press.

———. 2002. *The Cult of the Tiger*. New Delhi: Oxford University Press.

———, ed. 2004. *Tiger: The Ultimate Guide*. New York: CDS Books in association with Two Brothers Press.

———, ed. 2013. *Tiger Fire: 500 Years of the Tiger in India*. New Delhi: Aleph Book Company.

Thomas, P.K. 1977. 'Archaeozoological Aspects of the Prehistoric Cultures of Western India'. Unpublished PhD dissertation. Poona: Deccan College.

———. 2000–1. 'Contributions of the Deccan College to Archaeozoological Research'. *Bulletin of the Deccan College Post-Graduate and Research Institute*, 60–1: 77–95.

Thomas, P.K., and P.P. Joglekar. 1994. 'Holocene Faunal Studies in India'. *Man and Environment*, 19 (1–2): 179–203.

———. 1995. 'Faunal Studies in Archaeology'. *Memoirs of the Geological Society of India*, 32: 496–514.

———. 1996. 'Faunal Remains from Balathal, Rajasthan: A Preliminary Report'. *Man and Environment*, 21 (1): 91–7.

Thomas, P.K., P.P. Joglekar, V.D. Mishra, J.N. Pandey, and J.N. Pal. 1995. 'A Preliminary Report of the Faunal Remains from Damdama'. *Man and Environment*, 20 (1): 29–36.

———. 1996. 'Faunal Evidence for the Mesolithic Food Economy of the Gangetic Plain with Special Reference to Damdama'. In G.E. Afanas'ev, S. Cleuziou, J.R. Lukacs, and M. Tosi, eds, *Bioarchaeology of Mesolithic India: An Integrated Approach, Colloquim XXXII of the International Union of Prehistoric and Protohistoric Sciences*, pp. 255–66. Forli: ABACO Edizioni.

Thomas, P.K., P.P. Joglekar, Arati Deshpande-Mukherjee, and S.J. Pawankar. 1995. 'Harappan Subsistence Patterns with Special Reference to Shikarpur, a Harappan Site in Gujarat'. *Man and Environment*, 20 (2): 33–41.

Thomas, P.K., Yoshiyuki Matsushima, and Arati Deshpande. 1996. 'Faunal Remains'. In M.K. Dhavalikar, M.R. Raval, and Y.M. Chitalwala, eds, *Kuntasi: A Harappan Emporium on West Coast*, pp. 297–307. Pune: Deccan College Post-Graduate & Research Institute.

Thsangspa, Tashi Ldawa. 2014. 'Spatial Distribution of Petroglyphs of Ladakh'. In Bansi Lal Malla, ed., *Rock Art Studies: Concept, Methodology, Context, Documentation and Conservation*, vol. I, pp. 177–203. New Delhi: Indira Gandhi National Centre for the Arts and Aryan Books International.

Tisdell, Clem. 2005. 'Elephants and Polity in Ancient India as Exemplified by Kautilya's *Arthasastra* (Science of Polity)'. *Economics, Ecology and the Environment,* Working Paper No. 120: 1–10.

Trautmann, Thomas. R. 1982. 'Elephants and the Mauryas'. In S.N. Mukherjee, ed., *India; History and Thought. Essays in Honour of A.L. Basham*, pp. 254–81. Calcutta: Subarnarekha.

———. 2015. *Elephants and Kings: An Environmental History*. Chicago: The University of Chicago Press.

Tripathi, Vibha, and Ajeet K. Srivastava. 1994. *The Indus Terracottas*. Delhi: Sharada Publishing House.

Uesugi, Akinori. 2011. 'Terrracotta Objects'. In Vasant Shinde, Toshiki Osada, and Manmohan Kumar, eds, *Excavations at Farmana District*

Rohtak, Haryana, India 2006–2008, pp. 385–8. Japan, Kyoto: Indus Project, Research Institute for Humanity and Nature.

Varma, Radha Kant. 1984. 'The Rock-Art of Southern Uttar Pradesh with Special Reference to Mirzapur'. In K.K. Charavarty, ed., *Rock-Art of India: Paintings and Engraving*, pp. 206–13. New Delhi: Arnold-Heinemann.

———. 1996. 'Subsistence Economy of the Mesolithic Folk as Reflected in the Rock-Paintings of the Vindhyan Region'. In G.E. Afanas'ev, S. Cleuziou, J.R. Lukacs, and M. Tosi, eds, *Bioarchaeology of Mesolithic India: An Integrated Approach, Colloquim XXXII of the International Union of Prehistoric and Protohistoric Sciences*, pp. 329–39. Forli: ABACO Edizioni.

———. 2012. *Rock Art of Central India: North Vindhyan Region with Special Reference to Mirzapur and the Adjoining Regions in Uttar Pradesh and Baghelkhand in Madhya Pradesh*. New Delhi: Aryan Books International.

Vats, Madho Sarup. 1940. *Excavations at Harappa: Being an Account of Archaeological Excavations at Harappa Carried Out between the Years 1920–21 and 1933–34*, 2 vols. Delhi.

Verma, B.S. 2007. *Chirand Excavations Report 1961–1964 and 1967–1970*. Bihar: Directorate of Archaeology: Department of Youth, Art and Culture, Government of Bihar.

Verma, Nisha. 1986. *The Terracottas of Bihar*. New Delhi: Ramanand Vidya Bhawan.

Wakankar, V.S. 1973. *Painted Rock Shelters of India*. PhD Thesis. Poona: Deccan College.

Wakankar, Vishnu S., and Robert R.R. Brooks. 1976. *Stone Age Painting in India*. Bombay: D.B. Taraporevala Sons & Co.

Whitney, William Dwight, trans. 1962. *Atharva-Veda Saṁhitā*, 2 vols. Delhi: Motilal Banarsidass.

Wilson, Don E., and Russell A. Mittermeier. 2011. *Handbook of the Mammals of the World*. Vol. II: *Hoofed Mammals*. Barcelona: Lynx Edicions.

Winternitz, Maurice. 1977. *A History of Indian Literature*, vols I and II. New Delhi: Oriental Books Reprint Corporation.

Woodward, F.L., trans. 1954. *The Book of the Kindred Sayings (Sanyutta-Nikāya) or Grouped Suttas*, Part III. London: Luzac & Company Ltd.

———, trans. 1962. *The Book of the Gradual Sayings (Aṅguttara-Nikāya) or More-Numbered Suttas*, vol. II. London: Luzac & Company Ltd.

Wright, Rita P. 2010. *The Ancient Indus Urbanism, Economy and Society*. Cambridge: Cambridge University Press.

Zeuner, F.E. 1952. 'The Microlithic Industry of Gujarat'. *Man*, 52: 129–31.

———. 1959. *The Pleistocene Period Its Climate, Chronology and Faunal Successions*. London: Hutchinson Scientific & Technical.

———. 1963a. *Environment of Early Man with Special Reference to the Tropical Regions*. Baroda: Maharaja Sayajirao University of Baroda.

———. 1963b. *A History of Domesticated Animals*. New York: Harper & Row.

Zimmer, Heinrich. 1946. 'The Elephant'. In Joseph Campbell, ed., *Myths and Symbols in Indian Art and Civilization*, pp. 102–9. New York: Pantheon.

Zimmermann, Francis. 1987. *The Jungle and the Aroma of Meats: An Ecological Theme in Hindu Medicine*. Berkeley: University of California Press.

INDEX

Strabo, 112, 180, 287–8, 296–7, 299
subsistence, 14, 16, 18, 41, 55, 74, 190–4
Sudhābhojana Jātaka, 95, 315
śukladant, 240
Sumatran rhinoceros (*Dicerorhinus sumatrensis*), 38
Surkotada, Gujarat, 54, 69, 73–4, 195, 199–200, 306–7
  elephant remains at, 199–200
  presence of the rhinoceros in, 73
Susīma Jātaka, 256
Suśruta, 29, 34, 101, 102–3, 164, 263, 310
Suśruta Saṃhitā, 102, 155, 263
Sūtrakṛtāṅga, 260, 261
Sutta Nipāta, 25, 93
Suttavibhaṅga, 152, 155, 247–50, 252
symbiotic equations/relationship, 173, 314

Taccha Sūkara Jātaka, 159
Taittirīya Brāhmaṇa, 150
Taittirīya Saṃhitā, 146, 150, 243
Taprobane island (Sri Lanka), 181, 288
Tarsang, 49, 63, 311
Tell Asmar, 86–7
terracotta figures
  animal figurines, 19, 137
  absence of animals amongst, 132
  appliqué motif, use of, 238
  Bhir Mound specimens, 236
  of elephants, 201–9, 212–13, 220, 221, 227, 233–6
  in Ganga–Yamuna Valley, 237
  of gods and goddesses, 225
  hand-modelled, 232
  from Kausambi, 230
  during Maurya–Śuṅga period, 233
  religious significance of, 226
Thanesar, 233
Thapar, Valmik, 8, 31, 144
The Periplus of the Erythrean Sea, 113, 180
tiger (*Panthera tigris tigris*), 3
  adaptability of, 117
  as beasts of prey, 148
  on broken seal, Mohenjodaro, 120
  comparison with lion, 144–5, 173, 179
  cultural fascination with, 117
  depictions in rock paintings, 118
  emergence from Indra's body, 147–8
  ferocity of, 176
  flesh-eating propensity of, 145
  geographical distribution of, 180
  as gifts to kings, 180, 182
  habitats of, 117, 181
  hair of
    significance of, 148
    use in magic performances, 147, 151
  horned tiger, representations of, 119, 123, 126, 128, 137, 310
  hunting 284
  man-eating, 145
  meat
    eating of, 155, 161
    in treatises of Caraka and Suśruta, 164–5
    in *Suśruta Saṃhitā*, 155

ABOUT THE AUTHOR

Shibani Bose, an independent researcher, has been a visiting scholar in the Department of History, University of Minnesota, USA. She has also been a visiting scholar at the University of Wisconsin-Milwaukee, USA. She received her MPhil and PhD from the Department of History, University of Delhi, and has taught at Miranda House, University of Delhi, and also at the University of St. Thomas, Minnesota, USA. Her research interests include archaeology, animal studies, human–animal interactions, and environmental history. She has contributed essays in edited volumes and has co-authored *The Story of India's Unicorns* (2018) with Divyabhanusinh and Asok Kumar Das.